# Lecture Notes in Mathematics

Edited by A. Dold and B. Eckmann



1387

Miodrag Petković

# Iterative Methods for Simultaneous Inclusion of Polynomial Zeros

Springer-Verlag
Berlin Heidelberg New York London Paris Tokyo Hong Kong

**Author**

Miodrag Petković
University of Niš
Faculty of Electronic Engineering
P.O. Box 73, 18000 Niš, Yugoslavia

Mathematics Subject Classification (1980): Primary: 65H05
Secondary: 65G05, 65G10, 30C15

ISBN 3-540-51485-6 Springer-Verlag Berlin Heidelberg New York
ISBN 0-387-51485-6 Springer-Verlag New York Berlin Heidelberg


Printed in Germany

Printing and binding: Druckhaus Beltz, Hemsbach/Bergstr.
2146/3140-543210 – Printed on acid-free paper

**To my sons, Ivan and Vladimir**

# PREFACE

Galois' famous theorem states that a general direct method in terms of explicit formulas exists only for the polynomial equations degree of which is less than five. Because of that, to find the zeros of polynomials of higher degree one must apply some of numerical methods; moreover, such methods are already used for polynomials of degree three or four since the corresponding explicit formulas are remarkably complicated. A great importance of the problem of determining polynomial zeros in the theory and practice (e.g., in the theory of control systems, stability of systems, nonlinear circuits, analysis of transfer functions, various mathematical models, differential and difference equations, eigenvalues problems and other disciplines) has led to the development of a great number of numerical methods in this field. The list of contributors contains many of the most famous names of mathematical history. These numerical methods, which generally take the form of an iterative procedure, have become practically applicable together with the rapid development of digital computers some thirty years ago. In connection with any implementation of the numerical methods on a computer, it is important to note that the selection of zero-finding routine may depend heavily on extramathematical considerations such as speed and memory of the computing equipment and trustworthiness of the result. Anyone using a computing has surely inquired about the effect of rounding error and, eventually, propagated error due to uncertain values of polynomial coefficients. The computed solution of a polynomial equation is only an approximation of the true solution, since there are errors originating from discretization or truncation and from rounding. In connection with this effect we quote Henrici's argumentation given in [**48**]:

"Working with finite word lenght, we cannot hope to identify exactly a complex number such as a zero of a polynomial. We can at best exhibit a circle of arbitrary small radius that contains it. Moreover, it will not be reasonable to assume that the coefficients of the polynomial are known exactly. They, too, must be considered as given only by small circles in the complex plane. If this is so, we cannot distinguish between genuine multiple zeros and clusters of close zeros. Thus the best we can expect any algorithm to do is the following: Given any polynomial of degree N (normalized such that the zeros are not arbitrarily large, but lie, say, in the unit circle) and given a positive number $\epsilon$, the algorithm should produce at most N point sets of diameter $\leq 2\epsilon$ which together contain the zeros; also, it should indicate the number of zeros in each set."

With practical computational problems, a standard question should be "what is the error in the result?" As already pointed out by Wilkinson [**138**], the considerable amount of the applied procedure is to improve the approximate result and also to give error bounds for the improved approximations. The demands of the computer age with its arithmetic of finite precision, have dictated the need for a structure which is refered to as *interval arithmetic*. The calculations in ordinary floating-point arithmetic do not normally produce any information about the accuracy of the obtained result; in order to obtain such information it has been proposed (e.g. Moore [**72**]) to employ interval arithmetic. In particular, the problem of finding complex zeros of a polynomial with automatic error bounds requires *complex interval arithmetic*. Two realizations of interval arithmetics over the complex numbers were introduced. These are the so-called *rectangular* and *circular* complex interval arithmetics. It is interesting to note that circular arithmetic has just been introduced (Gargantini and Henrici [**37**]) as a necessary tool for the construction of an iterative procedure for determining all zeros simultaneously in terms of circular region (the same paper [**37**]).

In this book our attention is restricted to the simultaneous iterative methods for determining complex zeros of a polynomial, realized in complex interval aritmetic. These methods start with the initial complex intervals $Z_1^{(0)}, \ldots, Z_n^{(0)}$ (disks or rectangles) which contain the zeros $\xi_1, \ldots, \xi_n$ of a polynomial, and produce at the $m$-th iteration the sequences $\{Z_1^{(m)}\}, \ldots, \{Z_n^{(m)}\}$ of complex intervals converging to the zeros $\xi_1, \ldots, \xi_n$. Basic characteristic of these iterative interval processes is *inclusion property* regarding the zeros, namely, for all computed approximations $Z_1^{(m)}, \ldots, Z_n^{(m)}$ the inclusions $\xi_i \in Z_i^{(m)}$ hold for each $i = 1, \ldots, n$ and $m = 0, 1, 2, \ldots$. Therefore, we have the automatic determination of the upper error bounds given by radii or semidiagonals of the inclusion approximations. The exact solutions of a polynomial equation are contained in the resulting complex intervals. Besides, there exists the ability to incorporate rounding errors without altering the fundamental structure of the interval method. Finally, these interval methods are suitable for parallel computing. The price to be paid in order to achieve the characteristics of interval methods mentioned above, consists of the increase of numerical operations and the requirement of initial inclusion regions. These regions (in the form of disks or rectangles) can be obtained by means of one of the global search procedures.

The basis for constructing interval methods, presented in this book, is a combination of the fixed point relations and the inclusion property of complex interval arithmetic as well as an extension of the concept of a "complex" number. When the radius or semidiagonal of a complex interval is equal to 0, one obtains a "degenerate" interval which can be identified with the "complex" number equal to the center of the complex interval. Thus, complex interval arithmetic is a generalization or an extension of "complex" arithmetic. Consequently, any iterative method for improving the complex polynomial zeros, realized in complex interval arithmetic, can be easily rearranged to the "point" method in ordinary complex arithmetic. This was one of the starting point in forming the concept of the book which implies a survey of simultaneous methods in complex interval arithmetic. One of the aims of this book is to collect most algorithms for simultaneous inclusion of complex polynomial zeros which have appeared over the last fifteen years and to get the reader introduced to a new, growing and usable brunch of applied mathematics. The book is intended as a text and reference work for scientist, engineers and physicists who are interested in new development and applications, but the presented material is also suitable to anyone with a standard graduate level course in mathematical sciences with some knowledge of computer programming.

Part of the research and the final writing of the book was done while I was a Visiting Professor at the University of Oldenburg. During that visit I have benefited from discussions with Professor Herzberger. I am grateful to his helpful suggestions and valuable remarks.

I am thankful to Biljana Mišić-Ilić and Aleksandra Milošević who read the complete manuscript. Many small errors were eliminated in this manner. I wish to acknowledge with sincere appreciation the assistance I have received from my colleagues at the Faculty of Electronic Engineering - above all, L. V. Stefanović.

My main thanks, however, go to my wife Dr. Ljiljana Petković for her never-failing support, encourangement and comments during the preparation of the manuscript, as well as her assistence in programming a number of numerical examples.

Some of the research was supported by the Deutsche Forschungsgemeinschaft and the Serbian Science Foundation.

*Niš and Oldenburg, Spring 1989.* M. S. Petković

# CONTENTS

CHAPTER 1

# INTRODUCTION

One of the first nonlinear problems the mathematicians had to face in their research and practice concerns the polynomials. In particular, "the problem of determining the zeros of a given polynomial with complex coefficients is a genuine nonlinear problem. At the same time, the problem is simple. It is so simple, in fact, that there is some hope that some day we may be able to solve it perfectly" (Henrici [48, p. 1]). Twenty years after the cited Henrici's ingenious remark, a perfect zero-finding algorithm has not been established yet. Actually, numerous algorithms for determining the zeros of polynomials have been developed. It is not easy to choose the "best" algorithm for a given polynomial equation. Each algorithm usually possesses its own advantages and disadvantages and so, the problem mentioned above appears to be actual at the present time. Henrici has expressed his pessimism in [48]: "I do not feel that the polynomial problem has been solved perfectly, not even from a practical point of view".

Most algorithms compute only one zero at a time. If all zeros are wanted, these algorithms usually work serially as follows: if a zero has been found to sufficient accuracy, the corresponding linear factor is removed from the polynomial by the Horner scheme and the process is employed again to determine a zero of the "deflated" polynomial whose degree is now lowered by one. In many practical applications, it is necessary to find the zeros to any great accuracy. But, if the method of successive *deflations* is used, the polynomial obtained after divisions by the early (inaccurate) linear factors may be falsified to an extent which makes the remaining approximate zeros meaningless. This is a disadvantage of the method of successive removal of linear factors. Besides, as pointed out by Wilkinson [139], "if we use deflation, we need to have a device for ensuring that the zeros are found in increasing order of magnitude...", but, "... there seems to be no simple method of ensuring that Newton's process determines the zeros in increasing order". This is a further difficulty of the deflation. The approximate values of the zeros, produced by deflation, can be employed as initial approximations in some iterative process with the undeflated, original polynomial. This is the so-called "*purification*". However, in certain cases, the accuracy concerning the initial approximations may be such that this purification may fail, particularly in the presence of cluster of zeros (i.e., set of several very close zeros) because the iterative process may converge towards a zero that has already been determined.

The above difficulties can be overcome by determining all zeros simultaneously. There are many different approaches to these procedures as the method of search and exclusion for the simultaneous determination of all zeros (Henrici [51, §6.11]), *qd* algorithm (Henrici [51, §7.6]), methods based on fixed point relations, a globally convergent algorithm that is implemented interactively (Farmer and Loizou [25]) and the others (see, for instance, Wilf [137], Pasquini and Trigiante [82], Jankins and Traub [55], Farmer and Loizou [23],[24]). In this book we deal mainly with the methods based on the fixed point relations. Such approach provides the construction of algorithms with very fast convergence in complex "point" arithmetic as well as in complex interval arithmetic. The procedure is as follows:

Let $\xi_1,\ldots,\xi_n$ be the zeros of a given polynomial and let $z_1,\ldots,z_n$ be their approximations. We consider two types of fixed point relations,

$$\xi_i = F_1(z_1,\ldots,z_{i-1},\xi_i,z_{i+1},\ldots,z_n), \quad (i=1,\ldots,n) \tag{1.1}$$

$$\xi_i = F_2(\xi_1,\ldots,\xi_{i-1},z,\xi_{i+1},\ldots,\xi_n). \tag{1.2}$$

Substituting the exact zeros by their approximations and putting $z = z_i$ in (1.2), from (1.1) and (1.2) we obtain iterative schemes

$$\hat{z}_i = F_1(z_1,\ldots,z_n), \quad (i=1,\ldots,n) \tag{1.3}$$

$$\hat{z}_i = F_2(z_1,\ldots,z_n) \tag{1.4}$$

in (ordinary) complex arithmetic, where $\hat{z}_i$ is a new approximation to the zero $\xi_i$.

Assume now that we have found an array of n complex intervals $Z_1,\ldots,Z_n$ such that $\xi_i \in Z_i$ $(i = 1,\ldots,n)$. Substituting the zeros on the right side of (1.1) and (1.2) by their inclusion regions and using the *subset property*, we obtain

$$\xi_i \in F_1(z_1,\ldots,z_{i-1},Z_i,z_{i+1},\ldots,z_n), \qquad (1.5)$$
$$(i = 1,\ldots,n)$$
$$\xi_i \in F_2(Z_1,\ldots,Z_{i-1},z,Z_{i+1},\ldots,Z_n). \qquad (1.6)$$

Taking the set on the right side of (1.5) and (1.6) as a new (circular or rectangular) approximation $\hat{Z}_i$ to $\xi_i$, we may construct iterative methods

$$\hat{Z}_i = F_1(z_1,\ldots,z_{i-1},Z_i,z_{i+1},\ldots,z_n), \qquad (1.7)$$
$$(i = 1,\ldots,n)$$
$$\hat{Z}_i = F_2(Z_1,\ldots,Z_{i-1},z_i,Z_{i+1},\ldots,Z_n) \qquad (1.8)$$

in complex interval arithmetic, supposing that $\hat{Z}_1,\ldots,\hat{Z}_n$ are also complex intervals and taking $z_i$ to be the center of $Z_i$.

The iterative methods (1.3), (1.4), (1.7) and (1.8) are locally convergent and they must be combined with some of the slow convergent procedures to provide (reasonably good) initial approximations to the zeros. Also, as it will be presented in the following chapters, the iterative methods (1.3) and (1.4) may be combined with the interval methods (1.7) and (1.8). In this manner the efficiency of new *combined* methods is increased because the computations in ordinary floating-point arithmetic require less numerical operations compared to interval arithmetic. The following procedure is used for the construction of combined iterative methods:

1° Find initial disks or rectangles $Z_1^{(0)},\ldots,Z_n^{(0)}$ containing the zeros $\xi_1,\ldots,\xi_n$ of a given polynomial;

2° Using either the iterative method (1.3) or (1.4) (in floating-point arithmetic), compute the point approximations $z_i^{(m)}$ appearing in (1.7) and (1.8) to any desired accuracy (after m iterative steps), starting with the centers $z_i^{(0)}$ of the initial regions $Z_i^{(0)}$ $(i = 1,\ldots,n)$;

$3^{\circ}$ Apply some of the interval methods of the form (1.7) or (1.8) only once, in the final iterative step, dealing with the initial complex intervals $Z_1^{(0)},\ldots,Z_n^{(0)}$ and the improved approximations $z_i^{(m)}$.

The combined iterative formula is of the form

$$Z_i^{(m,1)} = F_2(Z_1^{(0)},\ldots,Z_{i-1}^{(0)}, z_i^{(m)}, Z_{i+1}^{(0)},\ldots,Z_n^{(0)}) \quad (i = 1,\ldots,n).$$

The upper "index" (m,1) indicates that the inclusion disk $Z^{(m,1)}$ is obtained by m point iterations and one interval iteration.

Some of the algorithms of the form (1.8) can be usefully applied for finding k ( < n ) interval approximations to the zeros. Such modified algorithms are of the interest when only a certain group of zeros, clustered around some point in the complex plane, has to be refined. In the special case when k = 1, it is possible to establish computational tests for the existence of a complex zero of a polynomial in a given disk ( §4.5).

Various authors developed the techniques for *a posteriori* error estimates for the approximations of polynomial zeros. These devices mostly use Gerschgorin's theorem (cf. Smith [115], Elsner [21], Braess and Hadeler [11]), Rouché's theorem (cf. Börsch-Supan [10], Gutknecht [42]), fixed point principle (e.g. Börsch-Supan [9], Schmidt [111], Schmidt and Dressel [112]). A quite different approach to error estimates for a given set of approximate zeros uses circular arithmetic, as pointed out by Gargantini and Henrici [37]. Simultaneous iterative methods, realized in circular arithmetic, produce resulting disks that contain the complex zeros of a polynomial. In this manner, not only very close zero approximations (given by the centers of disks) but also the upper error bounds for the zeros (given by the radii of disks) are obtained.

The numerous circular arithmetic algorithms are described in detail in this book (Chapters 3, 4 and 5). For these algorithms a complete convergence analysis is given, including the convergence order as well as the initial conditions that guarantee the convergence. Following the technique introduced by Alefeld and Herzberger [4], the lower bounds of the R-order of convergence of the presented single-step methods are determined.

The circular algorithms can be applied in the presence of multiplicities if multiple zeros are isolated in disjoint disks and their multiplicities are known in advance( § 4.4 and § 5.3). Ceratin problems concerning clusters of zeros can also be solved applying some of the parallel disk iterations, especially in cases when only one cluster is isolated from other zeros ( §4.4 and §5.3).

The initial conditions treated in this book have the form used in most papers due to various contributors and they are connected to the distribution of initial disks. Let $Z_1,..,Z_n$ be the initial disks containing the zeros $\xi_1,\dots,\xi_n$ of a given polynomial of degree N ( $\geq n$) and let $\mu_1,\dots,\mu_n$ $(\mu_1+\cdots+\mu_n = N)$ be the respective multiplicities. If we introduce the notation

$$Z_i = \{ z\colon\ |z-z_i| \leq r_i\} = \{z_i\,;\, r_i\} \quad (i=1,\dots,n)$$

and define

$$\rho = \min_{\substack{i,j\\ i\neq j}} \{ |z|\colon\ z \in z_j - Z_j\} = \min_{\substack{i,j\\ i\neq j}} \{ |z_i - z_j| - r_j\},$$

$$r = \max_{1\leq i\leq n} r_i, \qquad \mu = \max_{1\leq i\leq n} \mu_i, \qquad \tilde{\mu} = \min_{1\leq i\leq n} \mu_i\ ,$$

then the initial convergence conditions can be represented in the form

$$\rho > f(N,\mu,\tilde{\mu})r\ . \tag{1.9}$$

This inequality is a measure of the separation of the disks $Z_i$ from each other and a condition under which the initial disks are disjoint. Actually, the condition (1.9) provides good separatness of starting disks. In practice, it can be weakened, in other words, $f(N,\mu,\tilde{\mu})$ may take remarkably smaller value. Moreover, the considered interval methods can converge even in cases when the initial disks are intersecting. In such cases, in order to define the disk inversion, it is necessary that the inequalities

$$|z_i - z_j| > \max\{r_i\ ,\ r_j\} \quad (i\neq j;\ i,j = 1,\dots,n) \tag{1.10}$$

hold. These inequalities give an explanation for the restriction concerning the initial zero distribution in the form (1.9).

At present, the computational cost of most interval methods is still great, in general. We believe that their efficiency can be increased, especially (*i*) by *microprogramming* the computer in advance so that interval arithmetic operations are built at the machine language level and (*ii*) by creating simultaneous interval algorithms "which will make efficient use of Single Instruction / Multiple Data (SIMD) *parallel computers*. Several attributes of the method facilitate this, namely: all zeros are found simultaneously, $n$ versions of the same algorithm can be run simultaneously on an SIMD computer, and little interprocessor communication would be required in such an implementation. That communication which is required is well adapted for mesh-connected processors, a prevalent architecture" (Ellis and Watson [22]).

This book includes a lot of the author's results as well as related work by other contributors. However, it has to be emphasized that a number of the presented algorithms and many results connected with them have been established by the ideas and results of the fundamental papers due to Henrici, Gargantini, Alefeld and Herzberger, who were the true pioneers in this field of interval mathematics. During the last several years, numerous Chinese mathematicians - above all X. Wang and S. Zheng, gave a remarkable contribution to the development of simultaneous methods. Some of their results are given in Chapter 5.

We now give a short description of the contents of the book consisting six chapters divided into sections.

Chapter 2 contains some basic notions and specialized concepts which are necessary for further applications. Short review and some comments on the problem of finding initial inclusion regions and the number of zeros in these regions, as well as the multiplicities of localized zeros, are presented in § 2.1. The properties and some relations of complex interval arithmetic are given in §2.2. There are two reasonable choices which are considered: rectangles and circular regions (disks) as complex intervals. The operations of circular arithmetic are simpler and produce better results, but rectangular arithmetic has certain advantages in constructing iterative formulas as: (*i*) using the intersection of the former including approximation $Z_i^{(m)}$ (in the form of a rectangle) and the latter one $Z_i^{(m+1)}$ $(m = 0,1,...)$, the monotonicity of the sequences of rectangles $(Z_i^{(m)})$ is provided, that is, the chain of inclusions $Z_i^{(0)} \supseteq Z_i^{(1)} \supseteq Z_i^{(2)} \supseteq \cdots$ holds; (*ii*) rounded complex rectangular arithmetic, constructed by round real interval arithmetic (see Rokne and Lancaster [109], Yohe [140]) takes into account rounding errors appeared in evaluation on a computer which uses fixed length floating-point arithemtic. The R-order of convergence of iterative methods, introduced by Ortega and Rheinboldt [78], is treated in §2.3. The lower bound of the R-order for a class of simultaneous iterative methods for finding polynomial zeros (which are considered in this book) is determined.

The two iterative formulas which do not use any derivatives of a polynomial are considered in Chapter 3. The first of them can be regarded as a circular interval extension of a classical result due to Weierstrass [134]. Two versions of Weierstrass' interval method with quadratic convergence are studied including a complete convergence analysis. Applying the Gauss-Seidel approach, two single-step methods are constructed and the lower bound of the R-order of these methods is found. The combined method of Weierstrass' type is inserted in §3.1 as an example of how the computational cost of interval methods can be decreased. In §3.2 a parallel algorithm in terms of disks is investigated in detail. This algorithm converges cubically and does not use any polynomial derivatives.

In Chapter 4 we present the generalized methods of the root-iteration type, based on the logarithmic derivative $\frac{d^k}{dz^k}(\log P(z))$ of a polynomial P and the k-th root of a circular disk $(k=1,2,...)$. The order of convergence is $k+2$ (§4.1). The corresponding single-step method has the R-order of convergence at least $k+1+\sigma_n(k)$, where $\sigma_n(k) > 1$ is the unique positive root of the equation $\sigma^n - \sigma - k - 1 = 0$ and n is the polynomial degree (§4.2). The numerical stability of the generalized root iterations is discussed in §4.3 taking into consideration the presence of rounding errors. The conditions for preserving the convergence rate are given. Multiple zeros and clusters of zeros are considered in §4.4. A special attention is devoted to the circular algorithm for multiple zeros of the third order, introduced and studied by Gargantini [33], for its great efficiency. In paricular, very efficient combined methods are developed applying the mentioned method of order 3. Further, the problem of one zero cluster is treated in § 4.4. We present a multi-stage composite algorithm that combines iterative methods in circular and ordinary complex arithmetic and is based on "splitting" zeros which belong to a cluster. Starting with the iterative formulas for finding only one zero, a computational test for the existence of a single real or complex zero of a polynomial is established. Also, an algorithm for the inclusion of a simple zero of an entire function is described.

Bell's polynomials and parallel disk iterations are the subject of Chapter 5. First, some properties and relations for Bell's polynomials, necessary to construct a family of simultaneous methods, are presented. The fixed point relation which makes the basis for developing the family of parallel methods, is derived ( § 5.1). Combining the extension of Bell's polynomials in circular arithmetic and a family of iterative methods of the order $k + 1$ $(k = 1,2,\ldots)$, Wang and Zheng have expanded in [127] another family of the order $k + 2$. The properties of this family are studied in § 5.2. Several algorithms for multiple zeros in parallel ( total-step) and serial (single-step) fashion are given. Particularly, Newton's method with q steps, which appears to be a generalization of the interval method presented in [33] (and described in § 4.4), is investigated in depth. The section 5.3 deals with the Halley-like algorithm which is obtained as a special case (for $k = 2$) from the family consedered in § 5.2. This algorithm is of special interest for practical application. For this method the convergence order and the initial convergence conditions in the presence of multiplicity are discussed. Two serial versions using the Gauss-Seidel approach, and numerical stability of the basic Halley-like method are given. Bell's algorithms for a single complex zero in circular arithmetic ( § 5.4) and for all zeros in ordinary complex arithmetic (§ 5.5) complete Chapter 5.

The efficiency of simultaneous methods is considered in Chapter 6. Most of the iterative methods, presented in the previous chapters, are estimated and mutually compared regarding their computational cost and convergence speed. The rate of these methods is of interest in designing a package of routines concerning the iterative methods for improving, simultaneously, approximations to the zeros of a given polynomial, in which automatic procedure selection is desired.

Most of the algorithms presented in Chapters 3, 4 and 5 are illustrated numerically by means of polynomial equations. Except for a few examples, numerical examples were programmed on FORTRAN 77 and realized on the Micro VAX II computer. Double precision arithmetic was frequently used. In the realization of the iterative methods having a very fast convergence, even quad precision arithmetic (about 34 significant decimal digits) was employed.

At the end of the book there is a list of references containing the most important papers and books on the considered subject as well as the other papers used for the book. This Bibliography also includes numerous publications that are generally concerned with the numerical methods for solving polynomial equations.

Theorems, lemmas and formulas are quoted by chapter and number, while remarks and examples are numbered sequentially throughout each chapter. Remarks are concluded by ⊛ and proofs of theorems and lemmas by □. For brevity, the term *simultaneous iterative process* has been replaced in this book by the abbreviation SIP.

Finally, Index of notation and Subject index are added for the sake of easier reading of the book.

CHAPTER 2

# BASIC CONCEPTS

## 2.1. INCLUSION REGIONS FOR POLYNOMIAL ZEROS

Before improving, simultaneously, the zeros of a polynomial by applying a local algorithm with rapid convergence, it is necessary to provide initial regions (in the form of disks or rectangles) containing these zeros and to estimate the multiplicity of zeros. Taking into account the facts that the main purpose of this book is to describe the iterative methods with a *local* convergence and that there exist a number of papers and books concerning the mentioned topic (global search procedures including the proximity tests, the inclusion and exclusion tests, the covering schemes and the others), only a short review and a few comments on the problems of finding initial inclusion regions will be presented in this section.

Consider a polynomial P of degree N,

$$P(z) = a_N z^N + a_{N-1} z^{N-1} + \cdots + a_1 z + a_0 = a_N \prod_{i=1}^{n} (z - \xi_i)^{\mu_i},$$

where the coefficients $\{a_i\}$ are, in general, complex numbers, and the zeros $\{\xi_i\}$ have multiplicities $\{\mu_i\}$ with $\mu_1 + \cdots + \mu_n = N$. Practically, a complete procedure for improving the approximations to the zeros $\xi_1, \ldots, \xi_n$ consists of a three-stage *globally convergent* composite algorithm (see [24]):

(*a*) Find an inclusion region of the complex plane which includes all zeros of a polynomial;

(*b*) Apply a slowly convergent search algorithm to obtain separated initial intervals (disks or rectangles), each containing only one zero and calculate the multiplicity of that zero;

(*c*) Improve these intervals with a rapidly convergent iterative method to any required accuracy.

We will describe each of these stages in short.

(*a*) *INITIAL INCLUSION REGION*

Our aim is to determine an *inclusion radius* for the given polynomial P, i.e., a number R such that the zeros of P satisfy

$$|\xi_i| \le R \quad (i = 1,\ldots,n).$$

The following assertion has been given in [51, p.457] (see, also, Dekker [15]):

**THEOREM 2.1.** *Let* $\lambda_1,\ldots,\lambda_N$ *be positive numbers such that*

$$\lambda_1 + \cdots + \lambda_N < 1 \ [\le 1]$$

*and let*

$$R := \max_{1\le k\le N} \lambda_k^{-1/k} |a_{N-k}/a_N|^{1/k} .$$

*Then R is an inclusion radius for P.*

In particular, taking $\lambda_k = 1/2^k$, from Theorem 2.1 it follows that the disk centered at the origin with the radius

$$R = 2 \max_{1\le k\le N} |a_{N-k}/a_N|^{1/k} \tag{2.1}$$

contains all zeros of the polynomial P. This result can be usefully applied in a procedure of finding very close zeros (see §4.4).

It is possible to use other similar formulas (e.g. [8],[11],[71, § 3.3],[114]), but (2.1) has been found to be sufficient and satisfactory in practice (see [24],[25]). Furthermore, it is often convenient to take for initial inclusion region the smallest possible *square* containing the circle given by (2.1) (cf. [25]).

(*b*) *LOCATION OF ZEROS AND ESTIMATION OF MULTIPLICITY*

Let ε be any positive number and S any initial inclusion region (for instance, the disk about the origin with radius given by (2.1) or the smallest square including this disk). An *ε-covering* of S is any system of closed disks (squares), denoted by $S_1,\ldots,S_m$, of radius (semidiagonal) $\le \varepsilon$ whose union contains S, that is

$$S_1 \cup S_2 \cup \cdots \cup S_m \supseteq S.$$

The covering is said to be *centered* in S if the centers of the covering disks (squares) belong to S. As pointed out in [51, Ch. 6] (see, also [25]), the construction of a *minimal* $\varepsilon$-covering of a given bounded set (i.e., a covering that requires the least number of disks or squares) can raise intricate questions of elementary geometry.

Stewart [118] has shown that there are no particular problems involved in transforming an arbitrary disk in the complex plane into the unit disk. For positive integer N, let $\mathcal{P}_N$ denote the class of all monic polynomials of degree N with complex coefficients whose zeros lie in the unit disks, that is, $|\xi_i| \le 1$ $(i = 1,\ldots,n;\ n \le N)$. The optimal covering of the unit disk by eight disks $S_1,\ldots,S_8$ of radius $r = (1 + 2\cos(2\pi/7))^{-1}$ consists of a disk centered at the origin, surrounded by seven disks centered at the points

$$z_k = \rho e^{2\pi i k/7} \qquad (k = 0,1,\ldots,6),$$

where

$$\rho = \frac{2\cos(\pi/7)}{1 + 2\cos(2\pi/7)} \cong 0.80194$$

(see [49]). It can be shown that from a probabilistic point of view it may be more efficient to use disks whose radii are not all equal. For instance, Lehmer [62] covers the unit disks with a disk of radius $\frac{1}{2}$ centered at 0, surrounded by eight disks of radius $\frac{2}{5}$ centered on a circle of radius $\frac{3}{4}\cos\frac{\pi}{8}$. But, the Lehmer covering is probabilistically not optimal. Optimal coverings, from both a deterministic and a probabilistic point of view, are discussed by Friedli [27].

Continuing a search procedure it is necessary to examine whether or not the covering regions $S_1,\ldots,S_m$ contain a zero of a polynomial. For a given region $S_i$ $(i \in \{1,\ldots,m\})$ the following estimations can be obtained:

(A) $S_i$ may contain a zero;

(B) $S_i$ contains a zero;

(C) $S_i$ does not contain a zero.

The tests (B) and (C) are usually called the *inclusion* and the *exclusion test*, respectively. Henrici [49] has considered a „proximity test" with the property (A) which reacts positively if applied at a point close to a zero of a polynomial. As shown in [51, Ch. 6], the proximity test can be used indirectly as exclusion test.

A simple example of a test with the property (A) is based on the (circular or rectangular) interval extension $\mathbb{P}(S_i)$ of the polynomial P over the given region $S_i$. If $0 \in \mathbb{P}(S_i)$, then $S_i$ *may* contain a zero of P. Obviously, $S_i$ safely includes a zero if $\mathbb{P}(S_i) = \{P(z): z \in S_i\}$ but, if $\mathbb{P}(S_i) \supset \{P(z): z \in S_i\}$ (which is the most frequent case) then it may happen that there is no zero in $S_i$. Of course, if $0 \notin \mathbb{P}(S_i)$ then $S_i$ does not contain a zero of P, which is an exclusion test.

Let $S_i$ be a disk with center $z_i$ and radius $r_i$. The interval extension $\mathbb{P}_T(S_i)$ of P over $S_i$, called the Taylor circular centered form, is given by the disk with the center

$$\text{mid}\,(\mathbb{P}_T(S_i)) = P(z_i)$$

and the radius

$$\text{rad}\,(\mathbb{P}_T(S_i)) = \sum_{k=1}^{n} \frac{|P^{(k)}(z_i)|\,r_i^k}{k!} \quad .$$

Here, $P^{(k)}(z_i)$ is the k-th derivative of P(z) evaluated at point $z_i$. Using the above exclusion test $(0 \in \mathbb{P}(S_i))$ and Taylor's extension, Gargantini [28] has established the following concrete exclusion test:

„Let $z_i$ and $r_i$ denote the center and the radius (semidiagonal) of a disk (square) $S_i$. A disk (square) $S_i$ is free of zeros if

$$|P(z_i)| > \sum_{k=1}^{N} \frac{|P^{(k)}(z_i)|r_i^k}{k!} \quad ."$$

The second exclusion test is derived from the Schur-Cohn-Lehmer criterion ([61]):

„A square $S_i$ is free of zeros if the sequence of polynomials

$Q_j(\zeta)$, $|\zeta| \leq 1$ is such that

$Q_o(0) \neq 0$, $Q_j(0) > 0$ $(j = 1,2,\ldots,N)$. "

The polynomials $Q_1,\ldots,Q_N$ are constructed from the original polynomial $P(z)$ with elementary operations.

A good inclusion test for a given square would be one that computed the number of zeros in a square. Wilf [137] and Krishnamurthy and Venkateswaran [58] have proposed such inclusion tests. The remaining inclusion tests that have appeared in the literature are concerned with the unit circle. A simple inclusion test was originally derived by Lehmer [61] and subsequently improved (computationally) by Stewart [118]. However, a disadvantage of this inclusion test is that it only detects the *presence* of zeros in the unit circle, but not *how many* of them are present.

Farmer and Loizou [25] have employed the inclusion test that was originally proposed by Marden [67, p. 157]. This test gives the number of zeros in the unit circle. The inclusion test described by Gargantini and Münzner [38] can also be usefully applied. In his recent paper Neumaier [74] has presented an existence test for zero clusters and multiple zeros, which can be implemented rigorously on any computer that supports (rounded) complex interval arithmetic.

In the next phase of the search procedure the initial inclusion squares $S_i$, containing any zero, are dissectioned by the median lines parallel to the sides of squares. Thus, each square is subdivided into *four* uniform subsquares with the halved semidiagonals. Therefore, this procedure is called "2-subdivision" ([28]). It is obvious that such dissection is more convenient for squares than for circles as inclusion regions.

Let $S^{(0)}$ be the initial inclusion square with the semidiagonal R given by (2.1) and let $S_i^{(k)}$ be a smaller square obtained in "$2^k$ subdivision" procedure after k dissections. The semidiagonal $r^{(k)}$ of $S_i^{(k)}$ is, obviously, $r^{(k)} = R/2^k$. Then, an algorithm for finding inclusion squares is as follows:

$1^\circ$ Cover each region of interest with a number of smaller regions;

$2^\circ$ Using any appropriate inclusion (exclusion) test, determine which of these smaller regions still contain zeros, discarding any that do not;

$3^\circ$ Repeat steps $1^\circ$ and $2^\circ$ until a prescribed tolerance $\delta$ for the semidiagonals $r^{(k)}$ is reached, that is, $r^{(k)} \le \delta$ for some $k = K$.

The value of $\delta$ has to be sufficiently small so that each of final inclusion squares contains only one zero of P. Denote these squares by $S_1^{(K)}$, $\ldots, S_n^{(K)}$ and assume that all inequalities

$$|\operatorname{mid}(S_i^{(K)}) - \operatorname{mid}(S_j^{(K)})| > 2r^{(K)} \quad (i \neq j)$$

are satisfied. Then each of the squares $S_i^{(K)}$ may be encircled by the disk $D_i^{(K)}$ with center $\operatorname{mid}(S_i^{(K)})$ and radius $r^{(K)}$. The inclusion squares $S_1^{(K)}$, $\ldots, S_n^{(K)}$ or the disks $D_1^{(K)}, \ldots, D_n^{(K)}$ are taken as starting regions applying any iterative method for the simultaneous inclusion of polynomial zeros with rapid convergence.

Finally, it is still necessary to estimate the multiplicity of each zero contained in the square $S_i^{(K)}$ or the disk $D_i^{(K)}$. The center

$$z_i = \operatorname{mid}(S_i^{(K)}) = \operatorname{mid}(D_i^{(K)})$$

is taken as an approximation to a zero. Then we use Lagouanelle's limiting formula ([60])

$$\mu_i = \lim_{z_i \to \xi_i} \frac{P'(z_i)^2}{P'(z_i)^2 - P(z_i)P''(z_i)} . \tag{2.2}$$

Let $I(s)$ denote the integer that is the closest to a real positive number $s$. In practice, the actual computation of (2.2) is

$$\mu_i = I\left(\left|\frac{P'(z_i)^2}{P'(z_i)^2 - P(z_i)P''(z_i)}\right|\right) .$$

To obtain a precise estimation of multiplicity, it is suitable to use an improved approximation with respect to $z_i$, obtained by Newton's formula

$$\hat{z}_i = z_i - \frac{P(z_i)}{P'(z_i)} . \tag{2.3}$$

As it is well known, this formula has only a linear convergence if the multiplicity is greater than 1. To provide a more improved approximation, we may also apply the modified Newton's formula

$$\hat{z}_i = z_i - \frac{P(z_i)}{P'(z_i)} \bigg/ \left(1 - \frac{P(z_i)P''(z_i)}{P'(z_i)^2}\right) , \tag{2.4}$$

which has quadratic convergence.

Other formulas for estimation of multiplicity can be found in the literature (e.g. [8],[15],[123],[133]), but (2.2) yields satisfactory results, especially when we use the approximations improved by (2.3) or (2.4).

In practice, a criterion for the termination of dissection procedure can be established by evaluating (*i*) the total number of zeros contained in all regions applying the inclusion test by Marden or the method based on the principle of argument (see [28],[51],[137]); (*ii*) the sum of multiplicities estimated by (2.2). In both cases the total sum must be equal to the polynomial degree N. Otherwise, the dissection procedure is continued, even if $r^{(k)} < \delta$.

Other useful results concerning the stages (*a*) and (*b*) can be found, for instance, in [41],[47],[48],[52],[53],[59],[81],[106],[115],[122],[136].

(*c*) *ITERATIVE IMPROVEMENT*

The object of the final stage is to improve the interval approximations $S_i^{(K)}$ or $D_i^{(K)}$, obtained in the search procedure (stage (*b*)). We can choose some of the iterative methods with rapid convergence, described herein (Chapters 3, 4 and 5).

## 2.2. COMPLEX INTERVAL ARITHMETIC

We begin this section by introducing real intervals, which are necessary in constructing rectangular complex arithmetic.

A subset of $\mathbb{R}$ of the form

$$A = [a_1, a_2] = \{x : a_1 \le x \le a_2,\quad a_1, a_2 \in \mathbb{R}\}$$

is called a closed *real interval*. The set of all closed real intervals is denoted by $I(\mathbb{R})$.

If $*$ is one of the symbols $+, -, \cdot, :$ , we define arithmetic operations on $I(\mathbb{R})$ by

$$A*B = \{x = a*b:\ a \in A,\ b \in B\} \qquad (A, B \in I(\mathbb{R})).$$

The basic operations on intervals $A = [a_1, a_2]$ and $B = [b_1, b_2]$ may be calculated explicitly as

$$A + B = [a_1+b_1 , a_2+b_2],$$

$$A - B = [a_1-b_2 , a_2-b_1],$$

$$A\cdot B = [\min\{a_1b_1,a_1b_2,a_2b_1,a_2b_2\} , \max\{a_1b_1,a_1b_2,a_2b_1,a_2b_2\}],$$

$$A:B = [a_1,a_2]\cdot[1/b_2 , 1/b_1] \quad (0 \notin B).$$

This follows from the fact that $z = f(x,y)$ with $f(x,y) = x*y$, $* \in \{+ , - , \cdot , :\}$, is a continuous function on a compact set.

The introduced interval arithmetic is an extension of real arithmetic. If we assume that the endpoints of real intervals are computed with infinite precision, then this arithmetic (practically unrealizable) is frequently called *exact interval arithmetic* (Moore [72, Ch. 3]). However, the arithmetic hardware of computers is designed to carry out approximate arithmetic in "fixed-precision". Numbers are represented in the computer by strings of bits of fixed, finite length. Most commonly, so-called floating point arithmetic is used. Therefore, one often requires adding ( or subtracting a "low order bit" to (or from) the right (or left) hand endpoint of a machine computed interval result. The interval result computed in such a way contains the exact interval results. The corresponding arithmetic is called *rounded interval arithmetic* (Moore [72],[73]). More about rounded interval arithmetic can be found in the books [6],[72] and [73], and in the references cited therein.

Computer realization of rounded interval arithmetic can be accomplished in various ways (cf. Yohe [140]), either with a subroutine written in ALGOL, FORTRAN, PASCAL, etc., or by *microprogramming* the computer in advance so that interval arithmetic operations are defined at the machine language level.

Many of the properties and results for real interval arithmetic can be carried over to a complex interval arithmetic. There are two reasonable choices for complex intervals which will now be considered here: (*i*) a complex interval in the form of an axis-parallel rectangle in the complex plane (Alefeld [3], Fischer [26] and others) and (*ii*) a circular disk (Henrici [50], Gargantini and Henrici [37], Krier [57], Hauenschild [46]).

RECTANGULAR ARITHMETIC

Let $A_1, A_2 \in I(\mathbb{R})$. Then the set

$$A = \{a = a_1 + i a_2 : a_1 \in A_1,\ a_2 \in A_2\}$$

of complex numbers is called a *complex interval*. The above set constitutes rectangles in the complex plane with sides parallel to the coordinate axes and it is denoted by $R(\mathbb{C})$, where $\mathbb{C}$ is the set of complex numbers.

Let $* \in \{+, -, \cdot, :\}$ be a binary operation on elements from $I(\mathbb{R})$ and let

$$A = A_1 + i A_2, \quad B = B_1 + i B_2 \quad (A, B \in R(\mathbb{C})).$$

The basic operations of *rectangular arithmetic* are defined by

$$\begin{aligned} A \pm B &= A_1 \pm B_1 + i(A_2 \pm B_2), \\ A \cdot B &= A_1 B_1 - A_2 B_2 + i(A_1 B_2 + A_2 B_1), \\ A : B &= \frac{A_1 B_1 + A_2 B_2}{B_1^2 + B_2^2} + i \frac{A_2 B_1 - A_1 B_2}{B_1^2 + B_2^2} \quad (0 \notin B_1^2 + B_2^2) \end{aligned} \tag{2.5}$$

(see, e.g. [6]).

The case A:B requires special provisions since the division defined as above yields a complex interval that is generally far too large in comparison with the exact range $\left\{ z_1/z_2 : z_1 \in A,\ z_2 \in B \right\}$. Therefore, it is preferable sometimes to apply in practice the definition of division introduced by Rokne and Lancaster [109] as follows:

$$A:B = A \cdot \frac{1}{B}, \tag{2.6}$$

where

$$\frac{1}{B} := \inf \left\{ X \in R(\mathbb{C}) : \left\{ \frac{1}{b} : b \in B \right\} \subseteq X \right\}.$$

In this way a smaller region is obtained but the proposed set of formulas requires considerable computational effort.

Finally, we note that the intersection of two rectangles, if the intersection is not empty, is again a rectangle, which is a convenient property for application.

CIRCULAR ARITHMETIC

The iterative methods, described in this book, are realized in terms of circular regions. For this reason, the operations and properties of *circular arithmetic* will be presented more extensively.

Circular disks as complex intervals were first mentioned by Nickel (see [44, Ch. 3]), but without introducing any operations. The circular disks have been for the first time systematically used by Gargantini and Henrici [37].

Let $c \in \mathbb{C}$ be a complex number and $r \geq 0$. We call

$$Z = \{z: |z-c| \leq r\}$$

a circular disk or a disk only. The set of all disks is denoted by $K(\mathbb{C})$. If $Z$ denotes the disk with center $c$ and radius $r$, we write

$$Z = \{c\,;\,r\}, \qquad c = \operatorname{mid} Z, \qquad r = \operatorname{rad} Z,$$

for brevity.

The operations on $K(\mathbb{C})$ are introduced as generalizations of operations on complex numbers in the following manner:

Let $* \in \{+, -, \cdot, :\}$ be a binary operation on the complex numbers. Then, if $A = \{a\,;\,r_1\}$ and $B = \{b\,;\,r_2\}$ we define

$$A \pm B = \{a \pm b\,;\, r_1 + r_2\}, \tag{2.7}$$

$$\frac{1}{B} = \left\{ \frac{\bar{b}}{|b|^2 - r_2^2}\,;\, \frac{r_2}{|b|^2 - r_2^2} \right\} \quad \text{for } 0 \notin B, \tag{2.8}$$

where the bar denotes the complex conjugate.

Furthermore, if $A_k = \{a_k\,;\,r_k\}$ $(k = 1,\ldots,m)$, then

$$\sum_{k=1}^{m} A_k = \left\{ \sum_{k=1}^{m} a_k\,;\, \sum_{k=1}^{m} r_k \right\}. \tag{2.9}$$

Let $Z = \{z: |z-c| \geq r\}$ be a closed exterior of a circle such that $0 \notin Z$, that is $|c| < r$. Applying the theory of comformal mappings to the mapping of $Z$ by the function $w = \frac{1}{z}$, we get a closed interior of a circle given by

$$Z^{-1} = \left\{ \frac{-\bar{c}}{r^2 - |c|^2}\,;\, \frac{r}{r^2 - |c|^2} \right\}. \tag{2.10}$$

For addition, subtraction and inversion of disks it is obvious that

$$A \pm B = \{z_1 \pm z_2 : z_1 \in A,\ z_2 \in B\},$$

$$\frac{1}{B} = \left\{\frac{1}{z} : z \in B\right\},$$

which means that these operations are "the exact operations".

Unfortunately, the set $\{z_1 z_2 : z_1 \in A,\ z_2 \in B\}$ in general is not a disk. In order to remain within the realm of disks, Gargantini and Henrici [37] defined the multiplication by

$$AB := \{ab\ ;\ |a|r_2 + |b|r_1 + r_1 r_2\}. \tag{2.11}$$

It is, in general, only true that

$$AB \supseteq \{z_1 z_2 : z_1 \in A,\ z_2 \in B\}.$$

Other definitions of product of two disks were introduced by Krier [57] and Hauenschild [46], but their formulas are more complicated compared to (2.11) and have not found any further application.

Further, we have

$$\prod_{k=1}^{m}\{a_k\ ;\ r_k\} = \left\{\prod_{k=1}^{m} a_k\ ;\ \prod_{k=1}^{m}(|a_k| + r_k) - \prod_{k=1}^{m}|a_k|\right\}. \tag{2.12}$$

In particular, if $r_1 = r_2 = \cdots = r_m = r$, then

$$\prod_{k=1}^{m}\{a_k\ ;\ r\} = \left\{\prod_{k=1}^{m} a_k\ ;\ \sum_{k=1}^{m} r^k S_{m-k}\right\}, \tag{2.13}$$

where

$$S_j = \sum_{i_1 < i_2 < \cdots < i_j} |a_{i_1}||a_{i_2}|\cdots|a_{i_j}|, \quad S_0 = 1$$

is the j-th symmetric function of $|a_1|, |a_2|, \ldots, |a_m|$.

As special cases of addition, subtraction and multiplication of disks, we have for arbitrary complex number w

$$w \pm \{c\ ;\ r\} = \{w \pm c\ ;\ r\},$$

$$w \cdot \{c\ ;\ r\} = \{wc\ ;\ |w|r\}.$$

The operations of addition and multiplication on $K(\mathbb{C})$ are commutative and associative and the *subdistributivity*

$$A(B + C) \subseteq AB + AC \qquad (A,B,C \in K(\mathbb{C}))$$

holds.

Using the definitions of inversion and multiplication, the division is defined as

$$A : B = A \cdot \frac{1}{B} \quad \text{for } 0 \notin B. \tag{2.14}$$

A fundamental property of interval computation is *inclusion isotonicity* which forms the basis for almost all applications of interval arithmetic. The following theorem shows this property ([6, Ch. 5]).

**THEOREM 2.2.** *Let $A_k, B_k \in K(C)$ $(k=1,2)$ be such that*

$$A_k \subseteq B_k \quad (k = 1,2).$$

*Then*

$$A_1 * A_2 \subseteq B_1 * B_2$$

*holds for the operations* $* \in \{+, -, \cdot, :\}$.

In particular, if $z_1 \in A$, $z_2 \in B$ $(A,B \in K(\mathbb{C}))$, then

$$z_1 * z_2 \in A * B.$$

Let $A_k \subseteq B_k$ $(k = 1,\dots,m)$. According to Theorem 2.2 it follows

$$\sum_{k=1}^{m} A_k \subseteq \sum_{k=1}^{m} B_k$$

and

$$\prod_{k=1}^{m} A_k \subseteq \prod_{k=1}^{m} B_k .$$

Let $Z = \{c\,;\,r\} \in K(\mathbb{C})$. Then

(*a*) $|Z| := |c| + r$ is called the *absolute value* of $Z$ and

(*b*) $d(A) := 2r$ is called the *width* of $Z$.

The disk $Z_1 = \{c_1\,;\,r_1\}$ contains the disk $Z_2 = \{c_2\,;\,r_2\}$, denoted by $Z_1 \supseteq Z_2$, if and only if

$$|c_1 - c_2| \le r_1 - r_2. \tag{2.15}$$

The disks $Z_1$ and $Z_2$ are disjoint ($Z_1 \cap Z_2 = \emptyset$) if and only if

$$|c_1 - c_2| > r_1 + r_2 . \qquad (2.16)$$

If the inequality (2.16) does not hold, then the intersection $Z_1 \cap Z_2$ is not empty. But, this intersection in general is not a disk, which is a disadvantage of arithmetic of disks in application.

To establish the root iteration methods (Chapter 4), we introduce the definition of the k-th root of a disk. It is well known that the complex-valued range $\{z^{1/k} : z \in Z\}$ $(0 \notin Z \in K(\mathbb{C}))$ is the union of k closed and separated uniform regions in the complex plane. To enable the k-th root of a disk to produce disks enclosing the mentioned regions, the following circular approximation to the set $\{z^{1/k} : z \in Z\}$ was introduced in [97] (see, also [85]):

Let $c = |c| \exp(i\theta)$ and $|c| > r$, i.e., the disk $Z = \{c ; r\}$ does not contain the origin. Then

$$Z^{1/k} := \bigcup_{\lambda=0}^{k-1} \left\{ |c|^{1/k} \exp\left(\frac{\theta+2\lambda\pi}{k}\right) ; \ |c|^{1/k} - (|c|-r)^{1/k} \right\}. \qquad (2.17)$$

Thus, the k-th root of a disk is the union of k disks of the same radius and with the centers which are rotating through the angle $\frac{2\pi}{k}$. Sometimes, it is convenient to write

$$\operatorname{rad} Z^{1/k} = |c|^{1/k} - (|c|-r)^{1/k} = \frac{r}{\sum_{j=0}^{k-1} |c|^{(k-1-j)/k} (|c|-r)^{j/k}} . \qquad (2.18)$$

By formula (2.18) we avoid loss of significant digits due to forming difference $|c|^{1/k} - (|c|-r)^{1/k}$ (when r is very small).

It was proved in [88] that

$$\operatorname{rad}\left(\frac{1}{Z^{1/k}}\right) \le \operatorname{rad}\left(\frac{1}{Z}\right)^{1/k} \quad (0 \notin Z). \qquad (2.19)$$

The inequality (2.19) is stronger if the ratio $r/|c|$ is closer to 1 or 0. The second case appears at iterative schemes where r presents the radius of an inclusion disk.

The following inequality was proved in [84]:

$$\mathrm{rad}\left(\prod_{k=1}^{n}\frac{1}{Z_k}\right) \leq \mathrm{rad}\left(\frac{1}{\prod_{k=1}^{n} Z_k}\right). \tag{2.20}$$

If $Z_1 = \cdots = Z_n$, then

$$\mathrm{rad}\left(\frac{1}{Z}\right)^k \leq \mathrm{rad}\left(\frac{1}{Z^k}\right). \tag{2.20'}$$

As it will be seen later, the inequalities (2.19), (2.20) and (2.20') are helpful in constructing some iterative formulas.

Let us consider a complex function f over a given disk (rectangle) Z. The complex-valued set $\{f(z): z \in Z\}$ is not a disk (rectangle) in general. To deal with complex intervals (disks or rectangles), we introduce an extension F of f, defined on a subset $D \subseteq K(\mathbb{C})$ (or $R(\mathbb{C})$), such that

$$F(Z) \supseteq \{f(z): z \in Z\} \quad \text{for all } Z \in D \quad (\textit{inclusion}),$$

$$F(z) = f(z) \quad \text{for all } z \in Z \quad (\textit{complex resctriction}).$$

We shall say that the complex interval extension F is *inclusion isotone* if the implication

$$Z_1 \subseteq Z_2 \Rightarrow F(Z_1) \subseteq F(Z_2) \tag{2.21}$$

is satisfied for all $Z_1, Z_2 \in D$. In particular, we have

$$z \in Z \Rightarrow F(z) \in F(Z). \tag{2.22}$$

This property makes the basis for the construction of interval methods.

The investigation of all simultaneous interval methods, presented in this book, is performed in circular arithmetic because this arithmetic is considerably simpler than rectangular arithmetic. The obtained results (interval iterative schemes, convergence theorems, initial convergence conditions, analysis of numerical stability and others) can be applied without any difficulty in rectangular arithmetic.

Most frequently the considered iterative methods have the form

$$Z_i^{(m+1)} = G(Z_1^{(m+\lambda)}, \ldots, Z_{i-1}^{(m+\lambda)}, Z_i^{(m)}, Z_{i+1}^{(m)}, \ldots, Z_n^{(m)}) \tag{2.23}$$

$$(i = 1, \ldots, n;\ m = 0, 1, \ldots\ ),$$

where n is the number of different zeros of a given polynomial. In (2.23) we take $\lambda = 1$ when the Gauss-Seidel procedure is applied (the single-step method) and $\lambda = 0$ in the case of the total-step method. All described iterative methods produce the including circular approximations to the polynomial zeros, that is, $Z_1^{(m)},\ldots,Z_n^{(m)}$ are disks. These approximations could be improved using the intersection of the former disk $Z_i^{(m)}$ and the latter one $Z_i^{(m+1)}$, that is, a new approximation is calculated by

$$Z_i^{(m+1)} = G(Z_1^{(m+\lambda)},\ldots,Z_{i-1}^{(m+\lambda)},Z_i^{(m)},Z_{i+1}^{(m)},\ldots,Z_n^{(m)}) \cap Z_i^{(m)} \qquad (2.24)$$

instead of formula (2.23). However, if circular arithmetic is applied, then the iterative formula (2.24) appears to be inconvenient because the intersection of two disks is not a disk in general. On the other hand, the intersection of two rectangles is again a rectangle. Consequently, using (2.24) in rectangular arithmetic the monotonicity of the sequences of rectangles $(Z_i^{(m)})$ $(i = 1,\ldots,n)$ is provided, that is,

$$Z_i^{(0)} \supseteq Z_i^{(1)} \supseteq Z_i^{(2)} \supseteq \cdots .$$

We emphasize a further advantage of the use of rectangles. The rounded complex rectangular arithmetic, constructed by the rounded real interval arithmetic (see Rokne and Lancaster [109] and Yohe [140]) takes into account rounding errors that appear in evaluation on a computer with fixed length floating-point arithmetic. But, in applying circular arithmetic, if the values of the radii of produced disks are close to the accuracy limit of the used (single, double or multiple precision) arithmetic, then any iterative method realized in circular arithmetic cannot guarantee inclusion of zeros. The price to be paid in order to achieve the above advantages of rectangular arithmetic is as follows: (*i*) arithmetic operations with rectangles are complicated and (*ii*) any interval method with disks most frequently produces better results than the same method dealing with rectangles (assuming that, in the case of application of disks, the inclusion of zeros can be provided using double or multiple precision arithmetic).

## 2.3. ORDER OF CONVERGENCE OF ITERATIVE METHODS

Let I be an iterative method which produces a sequence $\{Z^{(m)}\}$ †) of complex intervals belonging to $K(\mathbb{C})$ or $R(\mathbb{C})$, and let each such sequence converge to the limit $Z^* \in K(\mathbb{C})$ (or $R(\mathbb{C})$) assuming that

$$Z^* \subseteq Z^{(m)} \quad (m = 0,1,\ldots). \tag{2.25}$$

One of the posibilities for measuring the deviation of an element $Z^{(m)}$ of the sequence from the limit $Z^*$ can be expressed by a nonnegative real number

$$h^{(m)} = d(Z^{(m)}) - d(Z^*) \tag{2.26}$$

(see Alefeld and Herzberger [6, Apendix A]). In particular, if $Z^*$ is a point in the complex plane, denoted by $\xi$ (for instance, the complex solution of an equation), we have

$$h^{(m)} = d(Z^{(m)}),$$

while the inclusion (2.25) is reduced to $\xi \in Z^{(m)}$. It is obvious that the sequence $\{h^{(m)}\}$ is the *null* sequence.

The well known concept of order of convergence due to Ortega and Rheinboldt [78, Ch. 9] may now be applied to the sequence $\{Z^{(m)}\}$ generated by the iterative method I that converges to $Z^*$ and satisfies (2.25). We first define the *root-convergence factor*, or *R factor*, for short, of such sequences as

$$R_p\{Z^{(m)}\} = \begin{cases} \limsup\limits_{m\to\infty} (h^{(m)})^{1/k} & p = 1 \\ \limsup\limits_{m\to\infty} (h^{(m)})^{1/p^k} & p > 1\,. \end{cases}$$

Let $C(I,Z^*)$ be the set of all sequences produced by the iterative method I for which

$$\lim_{m\to\infty} Z^{(m)} = Z^* \quad \text{and} \quad Z^* \subseteq Z^{(m)} \quad (m = 0,1,\ldots)$$

is valid. The *R factor of I at Z** is defined by

---

†) In this section we use braces { } for the denotation of a sequence. In the remaining sections any sequence is denoted by parentheses ( ).

$$R_p(I,Z^*) = \sup\left\{R_p\{Z^{(m)}\} : \{Z^{(m)}\} \in C(I,Z^*)\right\} .$$

Then the quantity

$$O_R(I,Z^*) = \begin{cases} +\infty & R_p(I,Z^*) = 0 \quad \text{for all } p \geq 1 \\ \inf\left\{p : p \in [1,\infty),\ R_p(I,Z^*) = 1\right\} & \text{otherwise} \end{cases}$$

is called the *R-order of I at* $Z^*$ (see [78, Ch. 9]).

The following result was given in [6, Apendix A]:

**THEOREM 2.3.** *Assume that there exists a* $p \geq 1$ *and a constant* $\gamma$ *such that for all* $\{Z^{(m)}\} \in C(I,Z^*)$ *the inequality*

$$h^{(m+1)} \leq \gamma\,(h^{(m)})^p \qquad (k \geq K = k(\{Z^{(m)}\})) \tag{2.27}$$

*is valid. Then*

$$O_R(I,Z^*) \geq p.$$

If $\{r^{(m)}\}$ is the sequence of radii (semidiagonals) of inclusion disks (rectangles) $Z^{(m)}$, produced by an interval iterative method, then the relation (2.27) becomes

$$r^{(m+1)} \leq \gamma\,(r^{(m)})^p . \tag{2.28}$$

Let $I_1$ and $I_2$ be two iterative processes with the same limit $Z^*$. We say that $I_1$ is *R-faster* than $I_2$ at $Z^*$ if there exists a $p \in [1,\infty)$ such that $R_p(I_1,Z^*) < R_p(I_2,Z^*)$. Furthermore, if $O_R(I_1,Z^*) > O_R(I_2,Z^*)$, then $I_1$ is R-faster than $I_2$ at $Z^*$ (see [78, p. 290]).

The comparison of two iterative methods $I_1$ and $I_2$ in terms of the R-factors consists of two stages. First, we compare the corresponding R-orders $O_R(I_1,Z^*)$ and $O_R(I_2,Z^*)$; if they are different, the method with a larger R-order is R-faster than the other one. If $O_R(I_1,Z^*)=O_R(I_2,Z^*) = \hat{p}$, then we compare the R-factors for $\hat{p}$. If, for instance, $R_{\hat{p}}(I_1,Z^*) < R_{\hat{p}}(I_2,Z^*)$, then $I_1$ is R-faster than $I_2$.

The rest of this section will be devoted to some results about R-orders which are valuable for a class of simultaneous iterative methods in a serial fashion (single-step methods) for finding polynomial zeros, considered extensively in the next chapters.

The R-order of convergence of an iterative method I with the limit point $\xi = (\xi_1,\ldots,\xi_n)$, where $\xi_1,\ldots,\xi_n$ are distinct zeros of a polynomial of degree n, will be denoted by $O_R(I,\xi)$. Let $\{Z_i^{(m)}\}$ $(i = 1,\ldots,n)$ be the sequences of disks or rectangles generated by I, and let $h_i^{(m)}$ be a multiple of the radius (semidiagonal) of disk (rectangle) $Z_i^{(m)}$. The following relations can be derived for a class of single-step iterative methods for the simultaneous inclusion of polynomial zeros (see [96]):

$$h_i^{(m+1)} \le \frac{1}{n-1}(h_i^{(m)})^p\Big(\sum_{j<i} h_j^{(m+1)} + \sum_{j>i}(h_j^{(m)})^q\Big) \tag{2.29}$$

$$(i = 1,\ldots,n;\ p,q \in N).$$

**THEOREM 2.4.** *Assume that the initial complex intervals* $Z_1^{(0)},\ldots,Z_n^{(0)}$ *are chosen so that*

$$h_i^{(0)} \le h = \max_{1\le i\le n} h_i^{(0)} < 1. \tag{2.30}$$

*Then, the R-order of convergence of the iterative method I, for which the relations (2.29) are valid, is given by*

$$O_R(I,\xi) \ge p + t_n(p,q), \tag{2.31}$$

*where* $t_n(p,q)$ *is the unique positive root of the equation*

$$t^n - tq^{n-1} - pq^{n-1} = 0. \tag{2.32}$$

*Proof*. By (2.30) we conclude from the inequalities (2.29) that the sequences $\{h_i^{(m)}\}$ $(i = 1,\ldots,n)$ converge to zero when $m\to\infty$. Consequently, the iterative method I, characterized by the relations (2.29), is convergent.

Following Alefeld and Herzberger [4,5], it can be derived from the relations (2.29) and (2.30) that

$$h_i^{(m+1)} \le h^{s_i^{(m+1)}} \quad (i = 1,\ldots,n;\ m = 0,1,\ldots). \tag{2.33}$$

The column vectors $s^{(m)} = [s_1^{(m)} \cdots s_n^{(m)}]^T$ are successively computed by

$$s^{(m+1)} = A_n(p,q)\, s^{(m)} \tag{2.34}$$

starting with $s^{(0)} = [1 \cdots 1]^T$. The $n\times n$ matrix $A_n(p,q)$ in equation (2.34) is given by

$$A_n(p,q) = \begin{bmatrix} p & q & & & & & \\ & p & q & & & 0 & \\ & & p & q & & & \\ & & & \cdot & \cdot & & \\ & & & & \cdot & \cdot & \\ & 0 & & & & \cdot & \cdot \\ & & & & & p & q \\ p & q & 0 & 0 & \dots & 0 & p \end{bmatrix} \quad (p,q \in N).$$

The characteristic polynomial of the matrix $A_n(p,q)$ is

$$f_n(\lambda;\, p,q) = (\lambda-p)^n - (\lambda-p)q^{n-1} - pq^{n-1}.$$

Substituting $t = \lambda - p$, we obtain

$$g_n(t;\, p,q) = f_n(t+p;\, p,q) = t^n - tq^{n-1} - pq^{n-1}.$$

It is easy to see the graph of the function $y_1(t) = t^n$ and the straight line $y_2(t) = tq^{n-1} + pq^{n-1}$ intersect only at one point for $t>0$. Since $g_n(q;\, p,q) = -pq^{n-1} < 0$ and $g_n(p+q;\, p,q) = (p+q)^n - 2pq^{n-1} - q^n > 0$, the intersecting point belongs to the interval $(q, p+q)$. Therefore, the equation

$$t^n - tq^{n-1} - pq^{n-1} = 0$$

has the unique positive root $t_n(p,q) > q$. The corresponding (positive) eigenvalue of the matrix $A_n(p,q)$ is $p + t_n(p,q)$. The matrix $A_n(p,q)$ is nonnegative and its directed graph is strongly connected ([124, p. 20]), that is, $A_n(p,q)$ is irreducible. By the Perron-Frobenius theorem (see [124, p. 30]) it follows that $A_n(p,q)$ has a positive eigenvalue equal to its spectral radius $\rho(A_n(p,q))$. According to the analysis presented in [4] it can be shown that the lower bound of the R-order of iterative method I, for which the inequalities (2.29) and (2.30) are valid, is given by the spectral radius $\rho(A_n(p,q))$. Therefore, we have

$$O_R(I,\xi) \geq \rho(A_n(p,q)) = p + t_n(p,q),$$

where $t_n(p,q)$ is the unique positive root of equation (2.32). □

To establish a result that gives a more precise lower bound of the R-order compared to $p+q$, we first present the following assertion:

**THEOREM 2.5.** (Deutsch [16]) *Let $A = (a_{ij})$ be a nonnegative irreducible $n \times n$ matrix and let $x$ and $y$ be positive vectors satisfying $Ax = Dx$, $A^T y = Dy$ for some positive diagonal matrix $D = \mathrm{diag}\,(d_1,\dots,d_n)$. If $x$ is not a Perron vector of $A$, then*

$$\rho(A) > \frac{y^T Dx}{y^T x},$$

*where $\rho(A)$ is the spectral radius of the matrix $A$.*

The vector $x > 0$ can be taken in a completely arbitrary manner. Then D is obtained from $d_i = (Ax)_i / x_i$ $(i = 1,\dots,n)$ and $y > 0$ is obtained by solving the homogeneous system $(A^T - D)y = 0$. In a special case, when $x = e \overset{\text{def}}{=} [1 \cdots 1]^T$, then $d_i = \sum_{j=1}^{n} a_{ij}$ and we have the estimation

$$\rho(A) > \frac{y^T De}{y^T e}. \tag{2.35}$$

**THEOREM 2.6.** *Let $I$ be an iterative method with the limit point $\xi$, characterized by (2.29). Then*

$$O_R(I,\xi) > p + q + \frac{pq}{(n-1)(p+q)}.$$

*Proof*. We will apply the estimation (2.35) to the matrix $A_n(p,q)$. We first find

$$D = \mathrm{diag}\,(p+q, \dots, p+q, 2p+q)$$

and

$$y = \frac{\alpha}{q}[p \quad p+q \quad \cdots \quad p+q \quad q]^T \qquad (\alpha > 0);$$

hence

$$y^T A_n(p,q)e = \frac{\alpha}{q}[pq + (n-1)(p+q)]$$

and

$$y^T e = \frac{\alpha}{q}(n-1)(p+q).$$

With respect to Theorems 2.4 and 2.5 we have

$$O_R(I,\xi) \ge \rho(A_n(p,q)) > \frac{y^T A_n(p,q)e}{y^T e} = p + q + \frac{pq}{(n-1)(p+q)}. \qquad \square$$

The dependence of $t_n(p,q)$ on the parameters n, p and q has been considered in the following theorem ([99]):

**THEOREM 2.7.** *The roots $t_n(p,q)$ are strictly decreasing functions of the number of different zeros n and strictly increasing functions of the parameters p and q, that is*

$$t_{n+1}(p,q) < t_n(p,q), \quad t_n(p,q) < t_n(p+\varepsilon,q), \quad t_n(p,q) < t_n(p,q+\varepsilon) \quad (\varepsilon > 0).$$

*P r o o f.* We prove the first statement. According to Theorem 2.6 we have

$$t_n(p,q) = \rho(A_n(p,q)) - p > q + \frac{pq}{(n-1)(p+q)} \; . \tag{2.36}$$

Since

$$t\, g_n(t;\, p,q) = g_{n+1}(t;\, p,q) - q^{n-1}(t+p)(t-q),$$

we find

$$g_{n+1}(t_n(p,q);\, p,q) = q^{n-1}[t_n(p,q) + p][t_n(p,q) - q] > 0$$

because of $t_n(p,q) > q$ (in view of (2.36)). Therefore

$$t_{n+1}(p,q) < t_n(p,q),$$

which means that $t_n(p,q)$ is a strictly decreasing function of n.

The remaining two statements can be proved in a similar manner. □

Equation (2.32) can be rearranged in the form

$$t = q\left(\frac{t+p}{q}\right)^{1/n}$$

or

$$\log t = \log q + \frac{1}{n}\log\left(\frac{t+p}{q}\right) .$$

From the last relation we conclude that $t_n(p,q) \to q$ when $n \to \infty$ so that

$$t_n(p,q) > \lim_{n\to\infty} t_n(p,q) = q$$

and

$$O_R(I,\xi) \geq \rho(A_n(p,q)) > \rho(A_\infty(p,q)) = p+q.$$

Finally, we note that the computation of the best R-orders of sequences, generated by iteration methods, can lead to the problem of finding the cone radius of certain concave operators (see Burmeister and Schmidt [12]). The cone radius is represented as the infimum of spectral radii of all linear operators majorizing the concave operator. In general, this approach gives a more precise bounds although it requires sometimes a lot of computational amount of work.

CHAPTER 3

# ITERATIVE METHODS WITHOUT DERIVATIVES

The iterative formulas without derivatives, based on Weierstrass' classical result [134], appear to be very efficient in the case of real zeros of polynomials (cf. [70]). The basic total-step method of the second order for complex zeros, realized in circular complex arithmetic, will be considered in this chapter. Applying the Gauss-Seidel approach, two accelerated interval methods are obtained and the lower bound for the R-order of convergence of these methods is found. In addition, we describe an interval method of Weierstrass' type with cubic convergence which does not use any polynomial derivatives, too.

## 3.1. WEIERSTRASS' METHOD IN CIRCULAR ARITHMETIC

Consider a monic polynomial of degree $n \geq 3$

$$P(z) = \prod_{j=1}^{n} (z - \xi_j)$$

with simple real or complex zeros $\xi_1, \ldots, \xi_n$. Since

$$P(z) = (z - \xi_i) \prod_{\substack{j=1 \\ j \neq i}}^{n} (z - \xi_j),$$

we obtain the fixed point relation

$$\xi_i = z - \frac{P(z)}{\prod_{\substack{j=1 \\ j \neq i}}^{n} (z - \xi_j)} . \tag{3.1}$$

Let $z_1, \ldots, z_n$ be reasonably good approximations to the zeros $\xi_1, \ldots, \xi_n$. Putting $z = z_i$ and substituting the zeros $\xi_j$ by their approximations $z_j$ $(j \neq i)$ in (3.1), we obtain

$$\hat{z}_i = z_i - \frac{P(z_i)}{\prod_{\substack{j=1\\j\neq i}}^{n}(z_i - z_j)} \quad (i = 1,\ldots,n), \tag{3.2}$$

which is a classical result connected with a constructive proof of the fundamental theorem of algebra, introduced by Weierstrass [134]. In the above formula $\hat{z}_i$ is an improved approximation to the zero $\xi_i$ $(i = 1,\ldots, n)$. According to (3.2) the following iterative procedure can be formulated for approximating, simultaneously, all zeros of the polynomial P:

$$z_i^{(m+1)} = z_i^{(m)} - \frac{P(z_i^{(m)})}{\prod_{\substack{j=1\\j\neq i}}^{n}(z_i^{(m)} - z_j^{(m)})} \quad (i = 1,\ldots,n;\ m = 0,1,\ldots). \tag{3.3}$$

If the starting values $z_1^{(0)},\ldots,z_n^{(0)}$ are sufficiently close to the corresponding zeros, the iterative method (3.3) converges *quadratically* ([17]). Algorithm (3.3) has been rediscovered several times (see, e. g., Durand [18], Dočev [17], Börsch-Supan [9], Kerner [56], Prešić [107], etc.) and it has been derived in various ways. It is well known that (3.2) is a modification of the classical Newton method

$$z^{(m+1)} = z^{(m)} - \frac{P(z^{(m)})}{P'(z^{(m)})} \quad (m = 0,1,\ldots), \tag{3.4}$$

where the derivative P' is replaced by the product of differences of the zero approximations (see [121]).

Assume now that, instead of the initial approximations $z_1^{(0)},\ldots,z_n^{(0)}$, the disjoint disks $Z_j^{(0)} = \{z: |z - z_j^{(0)}| \le r_j^{(0)}\} = \{z_j^{(0)}; r_j^{(0)}\} = \{\text{mid } Z_j^{(0)}; \text{rad } Z_j^{(0)}\}$, containing the zeros $\xi_j$ $(j = 1,\ldots,n)$, have been found. Let us take $z = z_i^{(0)}$ in (3.1). Since $\xi_j \in Z_j^{(0)}$, on the basis of the inclusion isotonicity we obtain from (3.1)

$$\xi_i \in z_i^{(0)} - P(z_i^{(0)}) \prod_{\substack{j=1\\j\neq i}}^{n} \frac{1}{z_i^{(0)} - Z_j^{(0)}} \quad (i = 1,\ldots,n) \tag{3.5}$$

and

$$\xi_i \in z_i^{(0)} - \frac{P(z_i^{(0)})}{\prod_{\substack{j=1\\j\neq i}}^{n}(z_i^{(0)} - Z_j^{(0)})} \quad (i = 1,\ldots,n). \tag{3.6}$$

The relations (3.5) and (3.6) suggest the following iterative interval methods for the simultaneous inclusion of polynomial complex zeros in terms of circular regions:

$$Z_i^{(m+1)} = z_i^{(m)} - P(z_i^{(m)}) \prod_{\substack{j=1 \\ j\neq i}}^{n} \frac{1}{z_i^{(m)} - Z_j^{(m)}}, \qquad (3.7)$$

$$(i = 1,\dots,n; \quad m = 0,1,\dots)$$

$$Z_i^{(m+1)} = z_i^{(m)} - \frac{P(z_i^{(m)})}{\prod_{\substack{j=1 \\ j\neq i}}^{n} (z_i^{(m)} - Z_j^{(m)})}. \qquad (3.8)$$

In connection with the application of formulas (3.7) and (3.8) the following problem arises : if the centers $z_1^{(0)},\dots,z_n^{(0)}$ (or some of them) are not sufficiently close to the zeros $\xi_1,\dots,\xi_n$, or the radii $r_1^{(0)},\dots,r_n^{(0)}$ are too large, then the iterative formulas (3.7) and (3.8) can generate disks with larger radii compared to the initial ones. On the other hand, the application of formulas (3.7) and (3.8) with intersection of circular intervals, $Z_i^{(m+1)} \cap Z_i^{(m)} := Z_i^{(m+1)}$, requires additional calculations. In this section we will investigate the initial conditions under which the sequences $(r_i^{(m)})$ of the radii of disks $Z_i^{(m)}$ $(i = 1,\dots,n)$, generated by (3.7) and (3.8), monotonically converge to 0.

Let $Z_1,\dots,Z_n$ be disjoint disks containing the polynomial zeros. The following two circular extensions are deduced from (3.1) (the iterative formulas (3.7) and (3.8)):

$$Z_i^{\dagger} = z_i - P(z_i) \prod_{\substack{j=1 \\ j\neq i}}^{n} \frac{1}{z_i - Z_j}, \qquad (\dagger)$$

$$Z_i^{*} = z_i - \frac{P(z_i)}{\prod_{\substack{j=1 \\ j\neq i}}^{n} (z_i - Z_j)}, \qquad (*)$$

where $Z_i^{\dagger}$ and $Z_i^{*}$ are the corresponding inclusion disks of the zeros $\xi_i$ $(i = 1,\dots,n)$. Since the disks $Z_1,\dots,Z_n$ are nonoverlapping, it follows that $0 \notin z_i - Z_j$, so that the calculation by $(\dagger)$ is always defined. On the other hand, product of two disks $Z_1$ and $Z_2$ is defined by an enlarged circular region which includes the exact range $\{z_1 z_2 : z_1 \in Z_1, z_2 \in Z_2\}$ (see definition (2.11)). Therefore, product of disks can contain the origin although each of disks does not include the number 0 so that the set defined by (*) is not necessarily a closed disk. Moreover, the inequality (2.20)

$$\operatorname{rad}\left(\prod_i \frac{1}{Z_i}\right) \leq \operatorname{rad}\left(\frac{1}{\prod_i Z_i}\right)$$

points out another advantage of the formula (†). This advantage of Algorithm (†) is dominant in the initial iterations, which provides considerably smaller inclusion disks in the later iterations compared to Algorithm (*)(starting with the same initial disks). Numerical examples veryfy this statement. Nevertheless, the computational cost of Algorithm (†) (that requires the calculation of n-1 inverse disks for one zero) is greater in relation to Algorithm (*) (only one inversion of disk).

The iterative methods (3.7) and (3.8) may be regarded as interval versions of the method (3.2) with error bound; it will be proved that these bounds, given by the radii of inclusion disks, converge quadratically to 0. For easier citation, the iterative methods (3.2), (3.7) and (3.8), as well as their modifications, will be refered to as Weierstrass' methods.

In this section we will carry out a convergence analysis of the iterative methods (3.7) and (3.8). Assume that the zeros $\xi_1,\dots,\xi_n$ of the polynomial P are isolated in the separated disks $Z_i^{(0)} = \{z_i^{(0)}; r_i^{(0)}\}$ $(i = 1,\dots,n)$ and let $Z_i^{(m)} = \{z_i^{(m)}; r_i^{(m)}\}$ be the disk furnished in the m-th iteration. We introduce the notation

$$r^{(m)} = \max_{1 \leq j \leq n} r_j^{(m)},$$

$$\rho^{(m)} = \min_{\substack{i,j \\ i \neq j}} \{|z| : z \in Z_i^{(m)} - Z_j^{(m)}\} = \min_{\substack{i,j \\ i \neq j}} \{|z_i^{(m)} - z_j^{(m)}| - r_j^{(m)}\}.$$

The value $\rho^{(m)}$ is a measure of the separation of the disks $Z_j^{(m)}$ from each other. For simplicity, at the beginning of the analysis we will write $Z_j$, $z_j$, $r_j$, $r$, $\rho$ instead of $Z_j^{(m)}$, $z_j^{(m)}$, $r_j^{(m)}$, $r^{(m)}$, $\rho^{(m)}$, respectively.

CONVERGENCE ANALYSIS OF ALGORITHM (3.7)

Let

$$w_j^{(i)} = \frac{(z_i-\xi_j)(\bar{z}_i-\bar{z}_j)}{|z_i-z_j|^2 - r_j^2}$$

and let

$$S_\mu = \sum_{j_1 < j_2 < \cdots < j_\mu} |w_{j_1}^{(i)}||w_{j_2}^{(i)}|\cdots|w_{j_\mu}^{(i)}| \tag{3.9}$$

be the symmetric function relative to $|w_j^{(i)}|$ $(j = 1,\ldots n;\ j \neq i)$.

**LEMMA 3.1.** *Let the symmetric function $S_\mu$ be defined by (3.9). Then*

$$S_\mu < \binom{n-1}{\mu}\left(1 + \frac{r}{\rho}\right)^\mu \quad (\mu = 1,\ldots,n-1). \tag{3.10}$$

*Proof.* Since $\xi_j \in Z_j$, we have

$$|z_i - \xi_j| < |z_i - z_j| + r_j. \tag{3.11}$$

Further, we have

$$\max_{\substack{j,i\\ j\neq i}} |w_j^{(i)}| = \max \frac{|z_i-\xi_j||\bar{z}_i-\bar{z}_j|}{|z_i-z_j|^2 - r_j^2} < \max \frac{(|z_i-z_j| + r_j)|\bar{z}_i-\bar{z}_j|}{|z_i-z_j|^2 - r_j^2}$$

$$= \max\left(1 + \frac{r_j}{|z_i-z_j| - r_j}\right) < 1 + \frac{r}{\rho},$$

which together with (3.11) proves the lemma. □

Let us introduce the following abbreviations:

$$t_1(n) = \frac{3}{10(n-1)},$$

$$G(n) = \exp(t_1(n)), \tag{3.12}$$

$$Q_1(n) = \frac{1}{2G(n)(G(n)-1)}, \tag{3.13}$$

$$q_1(n) = \sum_{j=0}^{n-2}\left(1 + \frac{r}{\rho}\right)^j\left(1 + \frac{2r}{\rho}\right)^{n-2-j}, \tag{3.14}$$

$$g_1(n) = \frac{1}{t_1(n)Q_1(n)}.$$

**LEMMA 3.2.** *Assume that the following condition*

$$\rho > \frac{r}{t_1(n)} \tag{3.15}$$

*is satisfied. Then*

$$q_1(n) < Q_1(n) \tag{3.16}$$

*and*

$$g_1(n) > 2 \tag{3.17}$$

*hold for each* $n \geq 3$.

*Proof*. Using the property that the sequence

$$v(m) = (1 + \frac{a}{m})^m \quad (a > 0,\ m \in N)$$

is bounded and monotonically increasing, we have

$$v(m) < \lim_{m \to +\infty} v(m) = e^a .$$

According to this, (3.15) and the inequality $1 + x < e^x$ $(x > 0)$, we estimate

$$(1 + \frac{r}{\rho})^m < [1 + t_1(n)]^m < [\exp(t_1(n))]^m = \exp(mt_1(n))$$

and, similarly,

$$(1 + \frac{2r}{\rho})^m < \exp(2mt_1(n)).$$

To estimate $q_1(n)$ we use the last two inequalities and obtain

$$q_1(n) < \exp\left(\frac{3(n-2)}{5(n-1)}\right) \sum_{m=0}^{n-2} \exp(-mt_1(n)) = \frac{e^{3/5} - e^{3/10}}{G(n)(G(n)-1)}$$

$$< \frac{1}{G(n)(G(n)-1)} = Q_1(n).$$

To prove (3.17) we take into account the fact that the sequence (3.14) is monotonically decreasing. Besides, $\lim_{n \to +\infty} g_1(n) = 2$ so that $g_1(n) > 2$ is valid for each $n \geq 3$. □

**LEMMA 3.3.** *Let* $Q_1(n)$ *be defined by (3.13) and let (3.15) hold. Then*

$$\operatorname{rad}\left(P(z_i)\prod_{\substack{j=1\\ j\neq i}}^{n} \frac{1}{z_i - Z_j}\right) < \frac{Q_1(n)r^2}{\rho} \quad (k = 1,\ldots,n). \tag{3.18}$$

*Proof*. Using (2.8) and (3.11) we get

$$\frac{z_i - \xi_j}{z_i - Z_j} = \frac{z_i - \xi_j}{|z_i - z_j|^2 - r_j^2}\{\bar{z}_i - \bar{z}_j ; r_j\}$$

$$\subset \left\{ w_j^{(i)} ; \frac{(|z_i - z_j| + r_j)r_j}{|z_i - z_j|^2 - r_j^2} \right\} = \left\{ w_j^{(i)} ; \frac{r_j}{|z_i - z_j| - r_j} \right\},$$

wherefrom

$$\frac{z_i - \xi_j}{z_i - Z_j} \subset \{ w_j^{(i)} ; \frac{r}{\rho} \}.$$

In view of the last inclusion, we find

$$P(z_i) \prod_{\substack{j=1 \\ j \neq i}}^{n} \frac{1}{z_i - Z_j} = (z_i - \xi_i) \prod_{\substack{j=1 \\ j \neq i}}^{n} \frac{z_i - \xi_j}{z_i - Z_j} \subset (z_i - \xi_i) \prod_{\substack{j=1 \\ j \neq i}}^{n} \{ w_j^{(i)} ; \frac{r}{\rho} \},$$

or, applying (2.13),

$$P(z_i) \prod_{\substack{j=1 \\ j \neq i}}^{n} \frac{1}{z_i - Z_j} \subset (z_i - \xi_i) \left\{ \prod_{\substack{j=1 \\ j \neq i}}^{n} w_j^{(i)} ; \sum_{k=1}^{n-1} \left(\frac{r}{\rho}\right)^k S_{n-1-k} \right\}. \tag{3.19}$$

By virtue of the inequalities (3.10) and (3.16), for the radius of the disk on the right-hand side of (3.19) we can write

$$\sum_{k=1}^{n-1} \left(\frac{r}{\rho}\right)^k S_{n-1-k} < \sum_{k=1}^{n-1} \binom{n-1}{k} \left(\frac{r}{\rho}\right)^k \left(1 + \frac{r}{\rho}\right)^{n-1-k}$$

$$= \left(1 + \frac{2r}{\rho}\right)^{n-1} - \left(1 + \frac{r}{\rho}\right)^{n-1} = \frac{q_1(n) r}{\rho} < \frac{Q_1(n) r}{\rho}.$$

According to this, the inclusion (3.19) becomes

$$P(z_i) \prod_{\substack{j=1 \\ j \neq i}}^{n} \frac{1}{z_i - Z_j} \subset (z_i - \xi_i) \left\{ \prod_{\substack{j=1 \\ j \neq i}}^{n} w_j^{(i)} ; \frac{Q_1(n) r}{\rho} \right\};$$

hence

$$\operatorname{rad}\left( P(z_i) \prod_{\substack{j=1 \\ j \neq i}}^{n} \frac{1}{z_i - Z_j} \right) < |z_i - \xi_i| \frac{Q_1(n) r}{\rho} < \frac{Q_1(n) r^2}{\rho}. \qquad \square$$

The convergence condition and the convergence order of the interval method (3.7) are considered in the following theorem ([103]).

**THEOREM 3.1.** *Let $Z_1^{(0)},\dots,Z_n^{(0)}$ be the initial disks containing the zeros $\xi_1,\dots,\xi_n$, and let the interval sequences $(Z_i^{(m)})$ $(i = 1,\dots,n)$ be defined by (3.7). Then, under the condition*

$$\rho^{(0)} > \frac{r^{(0)}}{t_1(n)}, \tag{3.20}$$

*for each $i = 1,\dots,n$ and $m = 0,1,\dots$ we have*

$1^o$ $\xi_i \in Z_i^{(m)}$;

$2^o$ *the sequence $(r^{(m)})$ monotonically tends to zero and*

$$r^{(m+1)} < \frac{Q_1(n)r^{(m)^2}}{\rho^{(0)} - 5r^{(0)}}.$$

*P r o o f.* Suppose that $\xi_j \in Z_j^{(m)}$ $(j = 1,\dots,n)$ for an arbitrary index m. Using the inclusion isotonicity, from (3.7) it follows

$$\xi_i \equiv z_i^{(m)} - P(z_i^{(m)}) \prod_{\substack{j=1\\ j\neq i}}^{n} \frac{1}{z_i^{(m)} - \xi_j} \in z_i^{(m)} - P(z_i^{(m)}) \prod_{\substack{j=1\\ j\neq i}}^{n} \frac{1}{z_i^{(m)} - Z_j^{(m)}} = Z_i^{(m+1)}.$$

Since $\xi_i \in Z_i^{(0)}$, the assertion $1^o$ of the theorem follows on the basis of mathematical induction.

We will now prove that the interval method (3.7) converges quadratically (assertion $2^o$). Using the estimation

$$\sum_{k=1}^{n-1} \left(\frac{r}{\rho}\right)^k S_{n-1-k} < \frac{q_1(n)r}{\rho},$$

from (3.7) we obtain

$$r_i^{(m+1)} = \operatorname{rad} Z_i^{(m+1)} = \operatorname{rad}\left(P(z_i^{(m)}) \prod_{\substack{j=1\\ j\neq i}}^{n} \frac{1}{z_i^{(m)} - Z_j^{(m)}}\right) < \frac{q_1(n)^{(m)}\, r^{(m)^2}}{\rho^{(m)}},$$

where

$$q_1(n)^{(m)} = \sum_{j=0}^{n-2} \left(1 + \frac{r^{(m)}}{\rho^{(m)}}\right)^j \left(1 + \frac{2r^{(m)}}{\rho^{(m)}}\right)^{n-2-j}.$$

Applying (3.16) we find for m = 0

$$r_i^{(1)} \le r^{(1)} < \frac{q_1(n)^{(0)}\, r^{(0)^2}}{\rho^{(0)}} < \frac{Q_1(n) r^{(0)^2}}{\rho^{(0)}}.$$

Hence, by using (3.17) and (3.20), we get

$$r^{(1)} < \frac{Q_1(n)r^{(0)}}{\dfrac{\rho^{(0)}}{r^{(0)}}} < \frac{Q_1(n)r^{(0)}}{\dfrac{1}{t_1(n)}} = \frac{r^{(0)}}{g_1(n)} < \frac{r^{(0)}}{2} .$$

By a geometric construction it can be seen that the disks $Z_1^{(1)},\ldots,Z_n^{(1)}$ will be disjoint if

$$\rho^{(0)} > |z_i^{(1)} - z_i^{(0)}| + 3r^{(1)} .$$

Since $|z_i^{(1)} - z_i^{(0)}| \le r^{(0)} + r^{(1)}$, using (3.20) and the inequality $r^{(1)} < \frac{1}{2}r^{(0)}$, we obtain

$$\rho^{(0)} > 3r^{(0)} > 4r^{(1)} + r^{(0)} > |z_i^{(1)} - z_i^{(0)}| + 3r^{(1)}.$$

Starting from the inequality (see Gargantini [31],[33])

$$\rho^{(m+1)} \ge \rho^{(m)} - r^{(m)} - 3r^{(m+1)}, \tag{3.21}$$

we find

$$\rho^{(1)} > \frac{r^{(0)}}{t_1(n)} - r^{(0)} - 3r^{(1)} > 2r^{(1)}\left[\frac{10}{3}(n-1) - 1\right] - 3r^{(1)}$$

$$= \left[\frac{20(n-1)}{3} - 5\right]r^{(1)},$$

wherefrom

$$\rho^{(1)} > \frac{10}{3}(n-1)r^{(1)} = \frac{r^{(1)}}{t_1(n)}$$

because $\frac{10}{3}(n-1) > 5$ for $n \ge 3$.

The proof of the statement 2° will be carried out by complete induction. Assume that the following relations are true for some index m:

$$r^{(m)} < \frac{1}{2}r^{(m-1)} , \tag{3.22}$$

$$\rho^{(m)} > \frac{r^{(m)}}{t_1(n)} . \tag{3.23}$$

The above inequalities have already been proved for $m = 1$. We will prove now that these inequalities are valid for index $m + 1$ if (3.22) and (3.23) hold for an arbitrary $m$ ($\ge 1$).

Reasoning as above and by using (3.23) we find

$$r^{(m+1)} < \frac{Q_1(n)\, r^{(m)2}}{\rho^{(m)}} < \frac{1}{2} r^{(m)} .$$

Hence, we conclude that the sequence $(r^{(m)})$ monotonically converges to 0.

Similarly as for $m = 1$, we prove the inequality

$$\rho^{(m)} > |z_i^{(m+1)} - z_i^{(m)}| + 3r^{(m)} ,$$

which means that the disks $Z_1^{(m+1)},\ldots,Z_n^{(m+1)}$ are disjoint at each iteration.

By the same considerations used for $m = 1$, we find

$$\rho^{(m+1)} \geq \rho^{(m)} - r^{(m)} - 3r^{(m+1)} > \frac{r^{(m+1)}}{t_1(n)} .$$

Applying (3.21) and (3.22), we obtain

$$\begin{aligned}
\rho^{(0)} &> \rho^{(m-1)} - r^{(m-1)} - 3r^{(m)} > \rho^{(m-1)} - r^{(m-1)}(1 + \tfrac{3}{2})\\
&> \rho^{(m-2)} - r^{(m-2)} - 3r^{(m-1)} - r^{(m-1)}(1 + \tfrac{3}{2})\\
&> \rho^{(m-2)} - r^{(m-2)}(1 + \tfrac{4}{2} + \tfrac{4}{2^2} - \tfrac{1}{2^2})\\
&> \rho^{(m-3)} - r^{(m-3)} - 3r^{(m-2)} - r^{(m-2)}(1 + \tfrac{4}{2} + \tfrac{4}{2^2} - \tfrac{1}{2^2})\\
&> \rho^{(m-3)} - r^{(m-3)}(1 + \tfrac{4}{2} + \tfrac{4}{2^2} + \tfrac{4}{2^3} - \tfrac{1}{2^3})\\
&\quad \cdots\cdots\\
&> \rho^{(0)} - r^{(0)}\left[1 + 2(1 + \tfrac{1}{2} + \cdots + \tfrac{1}{2^{m-1}}) - \tfrac{1}{2^m}\right].
\end{aligned}$$

Since

$$1 + 2(1 + \frac{1}{2} + \cdots + \frac{1}{2^{m-1}}) - \frac{1}{2^m} < 1 + 2\cdot\frac{1}{1-\frac{1}{2}} = 5,$$

it follows

$$\rho^{(m)} > \rho^{(0)} - 5r^{(0)} ,$$

so that

$$r^{(m+1)} < \frac{Q_1(n)\, r^{(m)2}}{\rho^{(0)} - 5r^{(0)}} .$$

Thus, the convergence order of the iterative process (3.7) is two.

We have fully established Theorem 3.1. □

CONVERGENCE ANALYSIS OF ALGORITHM (3.8)

Now, let us consider the simultaneous method (3.8). First, introduce the following notation:

$$t_2(n) = \frac{2}{7(n-1)} ,$$

$$q_2(n) = \left[2 - (1+\frac{r}{\rho})^{n-1}\right]^{-1} \sum_{j=0}^{n-2} (1+\frac{r}{\rho})^j ,$$

$$Q_2(n) = \frac{1}{2[\exp(t_2(n))-1]},$$

$$g_2(n) = \frac{1}{t_2(n)Q_2(n)} ,$$

where $n = 3,4,\ldots$ .

**LEMMA 3.4.** *Let the following condition*

$$\rho > \frac{r}{t_2(n)} \tag{3.24}$$

*hold. Then the inequalities*

$$q_2(n) \le Q_2(n) \tag{3.25}$$

*and*

$$g_2(n) > 2 \tag{3.26}$$

*are valid for each* $n \ge 3$.

*Proof*. For the properties of the sequence $(v(m))$, defined at the beginning of the proof of Lemma 3.2, under the condition (3.24) we have

$$(1+\frac{r}{\rho})^{n-1} < (1+t_2(n))^{n-1} = (1+\frac{2}{7(n-1)})^{n-1} < \lim_{n\to+\infty}(1+\frac{2}{7(n-1)})^{n-1}$$

$$= e^{2/7} \cong 1.33,$$

hence

$$(1+\frac{r}{\rho})^{n-1} < 2. \tag{3.27}$$

Furthermore, using the elementary inequality $1 + x < e^x$ $(x > 0)$, we find

$$(1+\frac{r}{\rho})^m \le (1+t_2(n))^m < \exp(mt_2(n)).$$

Since $\dfrac{e^{2/7}-1}{2-e^{2/7}} \cong 0.494 < \dfrac{1}{2}$, by means of the previous estimation and

(3.24) we obtain

$$q_2(n) < \frac{\sum_{m=0}^{n-2} \exp(mt_2(n))}{2 - e^{2/7}} < \frac{1}{\exp(t_2(n)) - 1} \cdot \frac{e^{2/7} - 1}{2 - e^{2/7}}$$

$$< \frac{1}{2[\exp(t_2(n)) - 1]} = Q_2(n).$$

In order to prove (3.26) consider the function

$$f(x) = \frac{e^{2x} - 1}{x}$$

on the interval $(0, \frac{1}{2})$. Since

$$(2x-1)e^{2x} + 1 > 0 \quad \text{for } x \in (0, \tfrac{1}{2}),$$

we have

$$f'(x) = \frac{(2x-1)e^{2x} + 1}{x^2} > 0.$$

Besides, $\lim_{x\to 0} f(x) = 2$, so that $f(x) > 2$ for $x \in (0, \frac{1}{2})$. Substituting $x = \frac{1}{7(n-1)} \in (0, \frac{1}{14}]$ (for $n \ge 3$), we find

$$g_2(n) = \frac{1}{t_2(n) Q_2(n)} = 7(n-1)[\exp(\tfrac{2}{7(n-1)}) - 1] > 2. \qquad \square$$

**LEMMA 3.5.** *Assume that (3.24) holds. Then*

$$0 \notin \prod_{\substack{j=1\\ j\neq i}}^{n} (z_i - Z_j) \tag{3.28}$$

*and*

$$\operatorname{rad}\left(\frac{P(z_i)}{\prod_{\substack{j=1\\ j\neq i}}^{n} (z_i - Z_j)}\right) < \frac{Q_2(n) r^2}{\rho} \qquad (i = 1,\dots,n). \tag{3.29}$$

*Proof*. Let $Z_j = \{z_j ; r_j\}$ and $y_{ij} = \frac{r_j}{|z_i - z_j|}$. Using (2.12) we obtain

$$\prod_{j\neq i} (z_i - Z_j) = \prod_{j\neq i} \{z_i - z_j ; r_j\} = \prod_{j\neq i} (z_i - z_j)\left\{1 ; \prod_{j\neq i} (1 + y_{ij}) - 1\right\}. \tag{3.30}$$

To prove (3.28) it is sufficient to show that

$$0 \notin \left\{1 ; \prod_{j\neq i} (1 + y_{ij}) - 1\right\},$$

which is equivalent to the inequality

$$1 > \prod_{j\neq i}(1 + y_{ij}) - 1,$$

that is,

$$\prod_{j\neq i}(1 + y_{ij}) < 2.$$

The last inequality is valid in accordance with (3.27) because $y_{ij} < \frac{r}{\rho}$.

Introduce the notation $b_i = \prod_{j\neq i}(1 + y_{ij})$ for brevity. Using (2.8) and (2.12), we obtain

$$\frac{P(z_i)}{\prod_{j\neq i}(z_i - z_j)} = \frac{P(z_i)}{\prod_{j\neq i}(z_i-z_j)\{1 ; b_i - 1\}} = \frac{\{s_i ; |s_i|(b_i - 1)\}}{b_i(2 - b_i)},$$

where

$$s_i = (z_i - \xi_i)\prod_{j\neq i}\frac{z_i - \xi_j}{z_i - z_j}.$$

Hence, we have

$$\operatorname{rad}\left(\frac{P(z_i)}{\prod_{j\neq i}(z_i - z_j)}\right) = \frac{|s_i|(b_i - 1)}{b_i(2 - b_i)}. \qquad (3.31)$$

We have the following upper bound for $|s_i|$:

$$|s_i| = |z_i - \xi_i|\prod_{j\neq i}\frac{|z_i - \xi_j|}{|z_i - z_j|} < r_i\prod_{j\neq i}\frac{|z_i - z_j| + r_j}{|z_i - z_j|}$$

$$< r\prod_{j\neq i}(1 + \frac{r_j}{|z_i - z_j|}) = rb_i.$$

Taking into consideration the facts that the inequalities

$$1 < b_i < (1 + \frac{r}{\rho})^{n-1} < 2$$

are valid and the function

$$h(x) = \frac{x - 1}{2 - x}$$

is monotonically increasing on the interval (1,2), we estimate

$$\frac{b_i - 1}{2 - b_i} < \frac{(1+\frac{r}{\rho})^{n-1} - 1}{2-(1+\frac{r}{\rho})^{n-1}} = \frac{\frac{r}{\rho}\sum_{m=0}^{n-2}(1+\frac{r}{\rho})^m}{2-(1+\frac{r}{\rho})^{n-1}} = \frac{r}{\rho}q_2(n) < \frac{r}{\rho}Q_2(n),$$

wherefrom, using (3.31) and the bound for $|s_i|$, we obtain (3.29). □

The convergence properties of Algorithm (3.8) are studied in the following theorem :

**THEOREM 3.2.** *Let $Z_1^{(0)},\ldots,Z_n^{(0)}$ be the initial disks containing the zeros $\xi_1,\ldots,\xi_n$, and let the interval sequences $(Z_i^{(m)})$ be produced by (3.8). Then, under the condition*

$$\rho^{(0)} > \frac{r^{(0)}}{t_2(n)}, \tag{3.32}$$

*for each $i = 1,\ldots,n$ and $m = 0,1,\ldots$ we have*

$1^o$ $\xi_i \in Z_i^{(m)}$;

$2^o$ *the sequence $(r^{(m)})$ monotonically tends to zero and*

$$r^{(m+1)} < \frac{Q_2(n)r^{(m)^2}}{\rho^{(0)} - 5r^{(0)}}.$$

*Proof*. The proof of this theorem is similar to the proof of Theorem 3.1 and, for that reason, some parts of it will be omitted.

The proof of assertion $1^o$ is derived by induction and it is same as in Theorem 3.1. The quadratic convergence of Algorithm (3.8) (assertion $2^o$) will be proved in the following manner. Using the estimations from Lemma 3.5 we find from (3.8)

$$r_i^{(m+1)} = \text{rad } Z_i^{(m+1)} = \text{rad}\left(\frac{P(z_i^{(m)})}{\prod\limits_{\substack{j=1\\ j\neq i}}^{n}(z_i^{(m)} - z_j^{(m)})}\right) < \frac{q_2(n)^{(m)}\, r^{(m)^2}}{\rho^{(m)}},$$

where

$$q_2(n)^{(m)} = \left[2 - \left(1+\frac{r^{(m)}}{\rho^{(m)}}\right)^{n-1}\right]^{-1}\sum_{j=0}^{n-2}\left(1+\frac{r^{(m)}}{\rho^{(m)}}\right)^j.$$

Applying (3.25), for m = 0 it follows

$$r_i^{(1)} \leq r^{(1)} < \frac{q_2(n)^{(0)} r^{(0)^2}}{\rho^{(0)}} < \frac{Q_2(n) r^{(0)^2}}{\rho^{(0)}},$$

so that, by (3.32) and (3.26), we obtain

$$r^{(1)} < \frac{Q_2(n)r^{(0)}}{\frac{\rho^{(0)}}{r^{(0)}}} < \frac{Q_2(n)r^{(0)}}{\frac{1}{t_2(n)}} = \frac{r^{(0)}}{g_2(n)} < \frac{1}{2}r^{(0)}. \tag{3.33}$$

Starting from the inequality (3.21), using (3.32) and (3.33) we derive

$$\rho^{(1)} < \frac{7}{2}(n-1)r^{(1)} = \frac{r^{(1)}}{t_2(n)}.$$

By the above consideration and induction, similarly as in Theorem 3.1 we prove for each $m = 1,2,\ldots$

$$\rho^{(m)} > \frac{r^{(m)}}{t_2(n)}, \tag{3.34}$$

$$r^{(m+1)} < \frac{Q_2(n)r^{(m)2}}{\rho^{(m)}} < \frac{1}{2}r^{(m)}. \tag{3.35}$$

Hence, we conclude that the sequence $(r^{(m)})$ monotonically converges to 0.

By successive application of (3.21) and (3.35), we find the same estimate for $\rho^{(m)}$ as in the previous theorem; thus

$$\rho^{(m)} > \rho^{(0)} - 5r^{(0)}. \tag{3.36}$$

Now, by means of (3.35) and (3.36), we obtain

$$r^{(m+1)} < \frac{Q_2(n)r^{(m)2}}{\rho^{(0)} - 5r^{(0)}}.$$

Finally, we observe that the interval process (3.8) is always defined under the condition (3.32). Indeed, since (3.32) implies the inequality $\rho^{(m+1)} > r^{(m+1)}/t_2(n)$ $(m = 0,1,\ldots)$, Lemma 3.5 is applicable for each $m = 0,1,\ldots$, wherefrom there follows

$$0 \notin \prod_{j\neq i}(z_i^{(m)} - z_j^{(m)}). \qquad \square$$

**REMARK 1.** Wang and Zheng have proved in [128] the quadratic convergence of the interval method (3.7) under the initial condition $\rho^{(0)} > 3(n-1)r^{(0)}$, which is slightly weaker than (3.20). ®

SINGLE STEP METHODS

More rapid convergence of the iterative methods (3.7) and (3.8) can be achieved by calculating the new inclusion disk $Z_i^{(m+1)}$ serially (Einzelschrittverfahren), using the already calculated disks $Z_1^{(m+1)}, \ldots, Z_{i-1}^{(m+1)}$ as soon as they are available (the so-called Gauss-Seidel approach). In this manner we obtain the accelerated (single-step) methods:

$$Z_i^{(m+1)} = z_i^{(m)} - P(z_i^{(m)}) \prod_{j<i} \frac{1}{z_i^{(m)} - Z_j^{(m+1)}} \prod_{j>i} \frac{1}{z_i^{(m)} - Z_j^{(m)}}, \tag{$A_1$}$$

$$Z_i^{(m+1)} = z_i^{(m)} - \frac{P(z_i^{(m)})}{\prod_{j<i} (z_i^{(m)} - Z_j^{(m+1)}) \prod_{j>i} (z_i^{(m)} - Z_j^{(m)})} \tag{$A_2$}$$

$$(i = 1,\ldots,n;\ m = 0,1,\ldots).$$

The single-step method of the form $(A_2)$ for real zeros has been analysed by Alefeld and Herzberger [5]. In order to indicate the method which is considered, we will add in writing the belonging index $k \in \{1,2\}$. This index refers also to the corresponding conditions, constants and so on.

Similarly to the analysis presented in [4](see, also,[5],[100]), under the condition

$$\rho^{(0)} > \frac{r^{(0)}}{t_k(n)} \quad (k\ \ \{1,2\}) \tag{$C_k$}$$

we can derive the following relations for the single-step method $(A_k)$:

$$r_i^{(m+1)} \le B_k r_i^{(m)} \left( \sum_{j<i} r_j^{(m+1)} + \sum_{j>i} r_j^{(m)} \right) \quad (i = 1,\ldots,n), \tag{3.37}$$

where $B_k = B_k(\rho^{(0)}, r^{(0)}, n)$ $(k \in \{1,2\})$ is a real constant depending only on the distribution of initial disks, their radii and the polynomial degree n.

Under the condition $(C_k)$ (that is, either (3.20) or (3.32)), it can be shown that the upper bound of $B_k$ is given by

$$B_k < \frac{1}{(n-1)\,t_k(n)\,\rho^{(0)}} \quad (k = 1,2) \tag{3.38}$$

for each $n \geq 3$.

Substituting

$$r_i^{(m)} = \frac{h_i^{(m)}}{(n-1)B_k} \quad (i = 1,\ldots,n;\ k = 1,2)$$

in (3.7), we obtain

$$h_i^{(m+1)} \leq \frac{1}{n-1} h_i^{(m)} \left( \sum_{j<i} h_j^{(m+1)} + \sum_{j>i} h_j^{(m)} \right) \quad (i = 1,\ldots,n). \tag{3.39}$$

Let $h = \max_i h_i^{(0)}$. By means of (3.38) we find

$$h_i^{(0)} \leq h = (n-1)B_k r^{(0)} < 1 \quad (k = 1,2). \tag{3.40}$$

According to (3.37), (3.39) and (3.40) it follows that the sequence $(r^{(m)})$ converges to 0, that is, the accelerated methods $(A_1)$ and $(A_2)$ are convergent under the conditions $(C_1)$ and $(C_2)$, respectively.

Denote the R-order of convergence of the iterative process IP with the limit point $\xi = (\xi_1,\ldots,\xi_n)$ (the vector of the polynomial zeros) by $O_R(IP,\xi)$ (see § 2.3). The relations (3.39) are the same ones as in [6, p. 106] derived for real zeros, so that we can immediately use the result from [6, Ch. 8] concerning the lower bound of the R-order of the iterative methods $(A_1)$ and $(A_2)$ (see, also, Theorem 2.4 for $p = q = 1$). Furthermore, taking into consideration the previous remark relative to the convergence conditions, we have

**THEOREM 3.3.** *Under the conditions $(C_k)$ the iterative single-step method $(A_k)$ $(k \in \{1,2\})$ is convergent with the R-order of convergence*

$$O_R((A_k),\xi) \geq \tau_n + 1,$$

*where $\tau_n \in (1,2)$ is the unique positive root of the equation $\tau^n - \tau - 1 = 0$.*

The convergence of the single-step methods $(A_1)$ and $(A_2)$ is obviously faster if the polynomial degree n is lower. For example, in the extreme case, for $n = 3$, we have $\tau_3 \cong 1.325$, so that the maximal lower bound of the R-order is $\geq \tau_3 + 1 \cong 2.325$. Apart from the faster convergence, the single-step methods $(A_1)$ and $(A_2)$ are more suitable for programming and, also, occupy less digital computer storage space (the new approximations take positions of the previous ones) compared to the basic total-step methods (3.7) and (3.8).

Starting from the total-step methods (3.7) and (3.8) we may construct new algorithms by repeating the Gauss-Seidel procedure. In this manner, the number of numerical operations is reduced.

Let $Z_1^{(0)},\ldots,Z_n^{(0)}$ be initial disjoint disks containing the zeros of the given polynomial P. The accelerated method in terms of disks, based on formula (3.8) and obtained by double application of the Gauss-Seidel approach, has the form

$$U_i^{(m)} = z_i^{(m)} - \frac{P(z_i^{(m)})}{\prod_{j=1}^{i-1}(z_i^{(m)} - U_j^{(m)}) \prod_{j=i+1}^{n}(z_i^{(m)} - Z_j^{(m)})},$$

$$Z_i^{(m+1)} = z_i^{(m)} - \frac{P(z_i^{(m)})}{\prod_{j=1}^{i-1}(z_i^{(m)} - Z_j^{(m+1)}) \prod_{j=i+1}^{n}(z_i^{(m)} - U_j^{(m)})} \tag{3.41}$$

$$(i = 1,\ldots,n;\ m = 0,1,\ldots).$$

The similar algorithm may also be constructed for the method (3.7). We will consider only the iterative scheme (3.41) because the same conclusions concerning the convergence order are valid for the corresponding version of the method (3.7).

Let $\varepsilon_i^{(m)} = \text{rad}\ U_i^{(m)}$. Using similar analysis as in [104], after an extensive but elementary analysis we obtain the following relations

$$\varepsilon_i^{(m)} < b\, r_i^{(m)} \left( \sum_{j=1}^{i-1} \varepsilon_j^{(m)} + \sum_{j=i+1}^{n} r_j^{(m)} \right),$$

$$r_i^{(m+1)} < b\, r_i^{(m)} \left( \sum_{j=1}^{i-1} r_j^{(m+1)} + \sum_{j=i+1}^{n} \varepsilon_j^{(m)} \right) \tag{3.42}$$

$$(i = 1,\ldots,n;\ m = 0,1,\ldots),$$

where $b = b(\rho^{(0)}, r^{(0)}, n)$ is a real constant which depends only on the distribution of initial disks, their radii and the polynomial degree n. Under the condition (3.32) one proves

$$b < \frac{7}{2\rho^{(0)}} \quad \text{for each } n \geq 3. \tag{3.43}$$

Substituting

$$r_i^{(m)} = \frac{v_i^{(m)}}{(n-1)b} \quad \text{and} \quad \varepsilon_i^{(m)} = \frac{\hat{v}_i^{(m)}}{(n-1)b} \quad (i = 1,\ldots,n)$$

in (3.42), we obtain

$$\begin{aligned} \hat{v}_i^{(m)} &< \frac{v_i^{(m)}}{n-1}\left(\sum_{j=1}^{i-1} \hat{v}_j^{(m)} + \sum_{j=i+1}^{n} v_j^{(m)}\right), \\ v_i^{(m+1)} &< \frac{v_i^{(m)}}{n-1}\left(\sum_{j=1}^{i-1} v_j^{(m+1)} + \sum_{j=i+1}^{n} \hat{v}_j^{(m)}\right) \end{aligned} \tag{3.44}$$

$$(i = 1,\ldots,n;\ m = 0,1,\ldots).$$

Let $v = \max_{1 \le i \le n} v_i^{(0)}$. According to (3.43) and (3.24) we find

$$v_i^{(0)} \le v = (n-1)br^{(0)} < 1.$$

By this bound and the relations (3.43) and (3.44) we conclude that the sequences $(r_i^{(m)})$ $(i = 1,\ldots,n)$ converge to 0. In addition, we can write

$$v_i^{(m+1)} \le v^{s_i^{(m+1)}} \qquad (i = 1,\ldots,n;\ m = 0,1,\ldots).$$

The components $s_1^{(m)},\ldots,s_n^{(m)}$ of the vector $s^{(m)} = [s_1^{(m)} \ \ldots \ s_n^{(m)}]^T$ can be successively calculated by

$$s^{(m+1)} = Y_n s^{(m)} \qquad (m = 0,1,\ldots),$$

where $s^{(0)} = [1 \ \cdots \ 1]^T$ is the starting vector and $Y_n$ is the $n \times n$ matrix defined by

$$Y_3 = \begin{bmatrix} 1 & 1 & 1 \\ 1 & 2 & 1 \\ 1 & 1 & 1 \end{bmatrix}, \qquad Y_n = \begin{bmatrix} 1 & 1 & 1 & & & & & & \\ & 1 & 1 & 1 & & & & O & \\ & & 1 & 1 & 1 & & & & \\ & & & \cdot & \cdot & \cdot & & & \\ & & & & \cdot & \cdot & \cdot & & \\ & & O & & & \cdot & \cdot & \cdot & \\ 0 & 0 & & & & & 1 & 1 & 1 \\ 1 & 1 & 0 & & & & 0 & 1 & 1 \\ 1 & 1 & 1 & 0 & & & 0 & 0 & 1 \end{bmatrix} \quad (n \ge 4).$$

Using the same argumentation as in [6, Ch. 8], we prove that the lower bound of $O_R((3.41),\xi)$ is given by

$$O_R((3.41),\xi) \geq \rho(Y_n),$$

where $\rho(Y_n)$ is the spectral radius of the matrix $Y_n$.

It is not easy to determine the characteristic polynomial of the matrix $Y_n$ for a general n. Therefore, in order to calculate $\rho(Y_n)$ we applied a very simple method for finding the *dominant* eigenvalue of $Y_n$ (that is, the spectral radius $\rho(Y_n)$), the so-called *power method*. The values of $\rho(Y_n)$ are displayed in Table 3.1.

| n | 3 | 4 | 5 | 6 | 7 | 8 | 9 | 10 | 15 | 20 |
|---|---|---|---|---|---|---|---|---|---|---|
| $\rho(Y_n)$ | 3.732 | 3.512 | 3.377 | 3.305 | 3.253 | 3.218 | 3.191 | 3.170 | 3.109 | 3.080 |

Table 3.1

Algorithm (3.41), where two successive iterations are applied, can be easily generalized using $q$ repeated steps as follows:

$$z_i^{(m+\frac{\lambda+1}{q})} = z_i^{(m)} - \frac{P(z_i^{(m)})}{\prod_{j=1}^{i-1}(z_i^{(m)} - z_j^{(m+\frac{\lambda+1}{q})}) \prod_{j=i+1}^{n}(z_i^{(m)} - z_j^{(m+\frac{\lambda}{q})})} \tag{3.45}$$

$$(i = 1,\ldots,n;\ \lambda = 0,1,\ldots,q-1;\ m = 0,1,\ldots).$$

**REMARK 2.** In particular, from (3.45) we obtain the single-step method $(A_2)$ for q = 1, and the method (3.41) for q = 2. ●

**REMARK 3.** It is not difficult to show that $O_R((3.45),\xi) \geq q+1$ (the equality occurs only in the limit case when $n \to \infty$). Note that this is merely a rough bound. A more precise lower bound for $O_R((3.45),\xi)$ can be obtained using the spectral radius of the corresponding matrix. ●

The initial conditions (3.20) and (3.32) guarantee the convergence of the iterative methods (3.7), $(A_1)$ and (3.8), $(A_2)$, respectively. But, in practice, initial disks can be chosen under weaker conditions in relation to (3.20) and (3.32) (strong initial conditions (3.20) and (3.2) are required because of the use of rough estimates and weak inequalities in convergence analysis). A great number of numerical examples confirmed this fact; the quotient $\rho^{(0)}/r^{(0)}$ was less than $1/t_k(n)$ $(k = 1,2)$. Moreover, in some cases, initial disks were intersecting but, in spite of that, the implemented methods converged and produced good results.

**EXAMPLE 1.** The interval methods (3.7) and (3.8) were applied for the determination of the eigenvalues of Hessenberg's matrix H (see Stoer and Bulirsch [119]). Gerschgorin's disks were taken as initial regions containing these eigenvalues. It is well known that these disks are of the form $\{a_{ii}\,;\,R_i\}$ $(i = 1,\dots,n)$, where $a_{ii}$ are the diagonal elements of matrix H and $R_i$ are real positive numbers depending only on the elements of H (cf. [39]). The above mentioned methods were tested in the example of matrix

$$H = \begin{bmatrix} 8+12i & 1 & 0 & 0 \\ 0 & 6+9i & 1 & 0 \\ 0 & 0 & 4+6i & 1 \\ 1 & 0 & 0 & 2+3i \end{bmatrix},$$

starting with Gerschgorin's disks

$$Z_1^{(0)} = \{8+12i\,;\,R^{(0)}\}, \qquad Z_2^{(0)} = \{6+9i\,;\,R^{(0)}\},$$

$$Z_3^{(0)} = \{4+6i\,;\,R^{(0)}\}, \qquad Z_4^{(0)} = \{2+3i\,;\,R^{(0)}\}.$$

The characteristic polynomial of the above matrix is

$$f(\lambda) = \lambda^4 - (20+30i)\lambda^3 + (-175+420i)\lambda^2 + (2300-450i)\lambda - 2857 - 2880i\,.$$

The dependence of the largest radius $r^{(m)} = \max_i r_i^{(m)}$ $(m = 1,2,3)$ of the inclusion disks on the initial radii $R^{(0)}$ is given in Table 3.2 for both Algorithms (3.7) and (3.8).

| Algorithm (3.7) | | | | | |
|---|---|---|---|---|---|
| $R^{(0)}$ | 0.4 | 0.6 | 0.8 | 1.0 | 1.2 |
| $r^{(1)}$ | $3.3\times10^{-3}$ | $5.4\times10^{-3}$ | $7.9\times10^{-3}$ | $1.1\times10^{-2}$ | $1.5\times10^{-2}$ |
| $r^{(2)}$ | $3.9\times10^{-7}$ | $1.4\times10^{-6}$ | $3.7\times10^{-6}$ | $8.4\times10^{-6}$ | $1.7\times10^{-5}$ |
| $r^{(3)}$ | $2.8\times10^{-15}$ | $5.1\times10^{-14}$ | $4.5\times10^{-13}$ | $2.7\times10^{-12}$ | $1.3\times10^{-11}$ |

Table 3.2

| Algorithm (3.8) | | | | | |
|---|---|---|---|---|---|
| $R^{(0)}$ | 0.4 | 0.6 | 0.8 | 1.0 | 1.2 |
| $r^{(1)}$ | $3.6 \times 10^{-3}$ | $6.5 \times 10^{-3}$ | $1.2 \times 10^{-2}$ | $3.5 \times 10^{-2}$ | $7.7 \times 10^{-2}$ |
| $r^{(2)}$ | $1.4 \times 10^{-6}$ | $7.3 \times 10^{-6}$ | $3.5 \times 10^{-5}$ | $3.2 \times 10^{-4}$ | $1.6 \times 10^{-3}$ |
| $r^{(3)}$ | $1.3 \times 10^{-13}$ | $5.2 \times 10^{-12}$ | $1.8 \times 10^{-10}$ | $2.2 \times 10^{-8}$ | $1.1 \times 10^{-6}$ |

Table 3.3

Comparing the values of $r^{(m)}$, given in Tables 3.2 and 3.3, we observe that Algorithm (3.7) gave better results than (3.8). Particularly, we display the inclusion disks obtained by Algorithm (3.7) after the third iterative step for $R^{(0)} = 1$.

$$Z_1^{(3)} = \{7.9965050702196818 + 11.9993208810633680\,i\ ;\ 5.03 \times 10^{-13}\},$$

$$Z_2^{(3)} = \{6.0104557911821981 + 9.0020569732912266\,i\ ;\ 2.69 \times 10^{-12}\},$$

$$Z_3^{(3)} = \{3.9895442088178008 + 5.9979430267088056\,i\ ;\ 2.69 \times 10^{-12}\},$$

$$Z_4^{(3)} = \{2.0034949297802799 + 3.0006791189365913\,i\ ;\ 5.03 \times 10^{-13}\}.$$

**EXAMPLE 2.** The zeros $\xi_{1,2} = 1 \pm 2i$, $\xi_3 = -1$, $\xi_4 = 3$ and $\xi_5 = 5i$ of the polynomial

$$P(z) = z^5 - (4+5i)z^4 + (6+20i)z^3 - (4+30i)z^2 - (15-20i)z + 75i$$

are isolated in the disks $Z_i^{(0)} = \{z_i^{(0)}\ ;\ r_i^{(0)}\}$, where $r_i^{(0)} = 0.35$ $(i = 1,\ldots,5)$ and

$$z_1^{(0)} = 1.2 + 2.2i,\quad z_2^{(0)} = 0.8 - 2.2i,\quad z_3^{(0)} = -1.2 - 0.1i,$$

$$z_4^{(0)} = 2.8 + 0.1i,\quad z_5^{(0)} = 0.2 + 4.9i.$$

Starting with these disks, the total-step methods (3.7) and (3.8) as well as their accelerated versions $(A_1)$ and $(A_2)$ were applied for the simultaneous inclusion of the zeros $\xi_1,\ldots,\xi_5$. The largest radii of inclusion disks, produced by (3.7), were

$$r^{(1)} = 0.155,\quad r^{(2)} = 6.21 \times 10^{-3},\quad r^{(3)} = 2.51 \times 10^{-6},$$

while the iterative formila (3.8) furnished

$$r^{(1)} = 0.201,\quad r^{(2)} = 1.91 \times 10^{-2},\quad r^{(3)} = 4.98 \times 10^{-5}.$$

Applying the single-step methods $(A_1)$ and $(A_2)$ to the same initial disks, we obtained considerably better inclusion circular approximations. Algorithm $(A_1)$ produced the following disks after the third iteration:

$$Z_1^{(3)} = \{1.000000\underline{0}364150727 + 2.000000\underline{0}073888955\,i\ ;\ 2.35 \times 10^{-7}\},$$

$$Z_2^{(3)} = \{0.9999999\underline{9}05173809 - 1.9999999\underline{9}16472826\,i\ ;\ 5.59 \times 10^{-8}\},$$

$$Z_3^{(3)} = \{-1.00000000\underline{0}0511335 - 2.06 \times 10^{-10} i\ ;\ 1.66 \times 10^{-9}\},$$

$$Z_4^{(3)} = \{2.9999999999\underline{9}88587 - 1.06 \times 10^{-13} i\ ;\ 1.2 \times 10^{-11}\},$$

$$Z_5^{(3)} = \{-6.46 \times 10^{-15} + 5.0000000000000\underline{0}11\,i\ ;\ 4.56 \times 10^{-14}\}.$$

Algorithm $(A_2)$ gave slightly worse results:

$$Z_1^{(3)} = \{1.00000\underline{1}1175740277 + 2.00000\underline{0}7006123423\,i\ ;\ 4.63 \times 10^{-6}\},$$

$$Z_2^{(3)} = \{0.999999\underline{8}613938640 - 1.999999\underline{7}333866253\,i\ ;\ 9.66 \times 10^{-7}\},$$

$$Z_3^{(3)} = \{-1.0000000\underline{0}36093044 + 1.55 \times 10^{-10} i\ ;\ 3.65 \times 10^{-8}\},$$

$$Z_4^{(3)} = \{2.999999999\underline{9}935545 + 5.2 \times 10^{-11} i\ ;\ 2.11 \times 10^{-10}\},$$

$$Z_5^{(3)} = \{-1.73 \times 10^{-12} + 5.00000000000\underline{0}8316\,i\ ;\ 7.45 \times 10^{-12}\}.$$

The underlined digit corresponds to the power order of radius.

## WEIERSTRASS' METHOD IN RECTANGULAR ARITHMETIC

As it was pointed out before in § 2.2, rectangular complex arithmetic possesses the following useful advantages: *(i)* the intersection of two rectangles $Z_1$ and $Z_2$ is again a rectangle, and the inclusions

$$Z_1 \cap Z_2 \subseteq Z_1, \quad Z_1 \cap Z_2 \subseteq Z_2$$

hold; *(ii)* interval formulas, employing the rounded rectangular arithmetic, take into account rounding errors that appear in any evaluation on a computer.

Let $Z_1^{(0)},\ldots,Z_n^{(0)}$ be the initial rectangles containing the zeros $\xi_1, \ldots, \xi_n$ respectively, and let $z_i^{(0)}$ be the center of the rectangle $Z_i^{(0)}$ $(i = 1,\ldots,n)$. Putting $z = z_i^{(0)}$ in the fixed point relation (3.1), in view of the inclusion isotonicity we obtain

$$\xi_i \in Z_i^{(0)} - \frac{P(z_i^{(0)})}{\prod_{\substack{j=1\\ j\neq i}}^{n} (z_i^{(0)} - Z_j^{(0)})} = W_i^{(0)} \qquad (i = 1,\dots,n).$$

The last relations suggest the following iterative method for the simultaneous inclusion of polynomial zeros in terms of rectangles:

$$W_i^{(m)} = z_i^{(m)} - \frac{P(z_i^{(m)})}{\prod_{\substack{j=1\\ j\neq i}}^{n} (z_i^{(m)} - Z_j^{(m)})},$$

$$Z_i^{(m+1)} = W_i^{(m)} \cap Z_i^{(m)} \qquad (i = 1,\dots,n;\ m = 0,1,\dots). \tag{3.46}$$

This method was considered in detail in [104].

Evidently, $Z_i^{(m+1)} \subseteq Z_i^{(m)}$ $(m = 0,1,\dots)$, and hence

$$Z_i^{(0)} \supseteq Z_i^{(1)} \supseteq Z_i^{(2)} \supseteq \cdots$$

for each $i = 1,\dots,n$. Thus, the monotonicity of the sequences of rectangles $(Z_i^{(m)})$ $(i = 1,\dots,n)$ is provided. Under certain suitable initial conditions the sequences of diagonals of the rectangles $(d(Z_i^{(m)}))$ $(i = 1,\dots,n)$ tend to zero *quadratically* (see [104]).

The efficiency of the iterative method (3.46) can be increased if, calculating the inclusion rectangle $Z_i^{(m+1)}$ for the zero $\xi_i$, we use already calculated rectangles $Z_j^{(m+1)}$ $(j < i)$. In this manner, we obtain the single-step method

$$Z_i^{(m+1)} = \left\{ z_i^{(m)} - \frac{P(z_i^{(m)})}{\prod_{j=1}^{i-1} (z_i^{(m)} - Z_j^{(m+1)}) \prod_{j=i+1}^{n} (z_i^{(m)} - Z_j^{(m)})} \right\} \cap Z_i^{(m)} \tag{3.47}$$

$$(i = 1,\dots,n;\ m = 0,1,\dots).$$

The lower bound of the R-order of convergence of the single-step method (3.47) is the same one as that given in Theorem 3.3.

**EXAMPLE 3.** The total-step method (3.46) and the corresponding single-step method (3.47) were tested in the example of the polynomial from Example 2 (§ 3.1). Starting with the initial rectangles

$$Z_1^{(0)} = [0.9,1.5] + i[1.9,2.5], \quad Z_2^{(0)} = [0.5,1.1] + i[-2.5,1.9],$$
$$Z_3^{(0)} = [-1.5,-0.9] + i[-0.4,0.2], \quad Z_4^{(0)} = [2.5,3.1] + i[-0.2,0.4],$$
$$Z_5^{(0)} = [-0.1,0.5] + i[4.6,5.2],$$

containing the polynomial zeros, the interval method (3.46) produced after the fourth iteration the following inclusion rectangles:

$$Z_1^{(4)} = [0.999943\,,1.000051] + i\,[1.999947\,,2.000054], \qquad \mathrm{sd}\,(Z_1^{(4)}) = 7.63\times10^{-5}$$
$$Z_2^{(4)} = [0.999890\,,1.000111] + i\,[-2.000104\,,-1.999883], \qquad \mathrm{sd}\,(Z_2^{(4)}) = 1.56\times10^{-4}$$
$$Z_3^{(4)} = [-1.000017\,,-0.999982] + i\,[-0.000015\,,0.000019], \qquad \mathrm{sd}\,(Z_3^{(4)}) = 2.48\times10^{-5}$$
$$Z_4^{(4)} = [2.999868\,,3.000136] + i\,[-0.000144\,,0.000123], \qquad \mathrm{sd}\,(Z_4^{(4)}) = 1.89\times10^{-4}$$
$$Z_5^{(4)} = [-0.000001\,,0.000001] + i\,[4.999998\,,5.000001], \qquad \mathrm{sd}\,(Z_5^{(4)}) = 2.03\times10^{-6}.$$

Using the same initial regions, the improved method (3.47) gave considerably smaller rectangles:

$$Z_1^{(4)} = [0.999999141913474\,,1.000000849307834] + i[1.999999157475893\,,2.000000864873012]\,, \qquad \mathrm{sd}\,(Z_1^{(4)}) = 1.21\times10^{-6}$$
$$Z_2^{(4)} = [0.999999967905234\,,1.000000030011304] + i[-2.000000030172363\,,-1.999999968066287]\,, \qquad \mathrm{sd}\,(Z_2^{(4)}) = 4.39\times10^{-8}$$
$$Z_3^{(4)} = [-1.000000001472848\,,-0.999999998548411] + i[-0.000000001461710\,,0.000000001462778]\,, \qquad \mathrm{sd}\,(Z_3^{(4)}) = 2.07\times10^{-9}$$
$$Z_4^{(4)} = [2.999999999998891\,,3.000000000001117] + i[-0.000000000001108\,,0.000000000001118]\,, \qquad \mathrm{sd}\,(Z_4^{(4)}) = 1.57\times10^{-12}$$
$$Z_5^{(4)} = [-0.000000000000006\,,0.000000000000006] + i[4.999999999999994\,,5.000000000000006]\,, \qquad \mathrm{sd}\,(Z_5^{(4)}) = 8.95\times10^{-15}.$$

The values of semidiagonals „sd" in the above lists are presented to give an information about the upper error bound concerning the inclusion rectangles of the zeros.

## WEIERSTRASS' COMBINED METHOD

The main disadvantage of interval algorithms is their great computational cost because interval computations require too much extra operations. Following the idea by Caprani and Madsen [14], we have established in this book a few effective iterative processes for the simultaneous inclusion of polynomial zeros, which combine the efficiency of ordinary floating-point iterations with the accuracy control that may be obtained by the iterations in interval arithmetic. Since interval arithmetic is very costly, it is desirable to apply the interval method as late as possible, at the end of combined procedure, providing in this manner the enclosure of zeros. In fact, interval arithmetic should take the role of an "a posteriori weapon".

Construction of combined iterative methods has been explained in Chapter 1. In particular, we can combine the iterative method (3.3) in ordinary complex arithmetic and either the interval method (3.8) or (3.46). Let $Z_1^{(0)},\ldots,Z_n^{(0)}$ be the initial disks or rectangles containing the zeros $\xi_1,\ldots,\xi_n$, and let $z_i^{(0)} = \text{mid}\, Z_i^{(0)}$ $(i = 1,\ldots,n)$. Weierstrass' combined method is defined by

$$z_i^{(m+1)} = z_i^{(m)} - \frac{P(z_i^{(m)})}{\prod\limits_{\substack{j=1\\j\neq i}}^{n}(z_i^{(m)} - z_j^{(m)})} \qquad (i = 1,\ldots,n;\ m = 0,1,\ldots,M-1),$$

$$Z_i^{(M,1)} = z_i^{(M)} - \frac{P(z_i^{(M)})}{\prod\limits_{\substack{j=1\\j\neq i}}^{n}(z_i^{(M)} - Z_j^{(0)})} \qquad (i = 1,\ldots,n).$$

By virtue of the convergence analysis of the interval method (3.8) we can give the following estimation

$$r^{(M,1)} = O\left((r^{(0)})^{2^M+1}\right),\ ^{*)}$$

where $r^{(0)} = \max\limits_{1\le i\le n} r_i^{(0)}$, $r^{(M,1)} = \max\limits_{1\le i\le n} \text{rad}\, Z_i^{(M,1)}$ and M is the number of the point iterations. The above estimation is more theoretical than practical, but it may be useful in the determination of the necessary number of iterations to satisfy the desired accuracy.

---

*) Sometimes, we will write $\alpha \sim \beta$ instead of $\alpha = O(\beta)$, indicating that the quantities $\alpha$ and $\beta$ are of the same order of magnitude.

Weierstrass' combined method is less efficient compared to the similar combined methods (see § 6.4) and cannot be applied for finding multiple zeros. For these reasons, we will not discuss this method in detail any further. Instead, we present a numerical illustration.

**EXAMPLE 4.** We consider again the polynomial from Example 2 with the same initial disks. The largest radii $r^{(M)}$ and $r^{(M,1)}$ $(M = 1,2,3,4)$ of inclusion disks, obtained by Weierstrass' interval method (3.8) (applying M interval iterations) and by Weierstrass' combined method (applying M point iterations and one (final) interval iteration ) are given in Table 3.4.

| | $M = 1$ | $M = 2$ | $M = 3$ | $M = 4$ |
|---|---|---|---|---|
| Interval method (3.8); $r^{(M)}$ | 0.201 | $1.91 \times 10^{-2}$ | $4.98 \times 10^{-5}$ | $1.54 \times 10^{-9}$ |
| Combined method; $r^{(M,1)}$ | $3.17 \times 10^{-2}$ | $1.73 \times 10^{-4}$ | $9.83 \times 10^{-9}$ | $1.08 \times 10^{-16}$ |

Table 3.4 The largest radii $r^{(M)}$ and $r^{(M,1)}$

We observe from Table 3.4 that the results of the combined method, obtained by M point iterations and one interval iteration, are only slightly worse compared to those produced by the interval method by applying M+1 iterations in interval arithmetic. Accordingly, the efficiency of the combined method is considerably greater (cf. § 6.4).

For the sake of illustration, we give the inclusion disks $Z_1^{(4,1)},\ldots,Z_5^{(4,1)}$ produced by Weierstrass' combined method:

$$Z_1^{(4,1)} = \{0.9999999999999999482 + 1.9999999999999999638\,i\ ;\ 1.08 \times 10^{-16}\},$$

$$Z_2^{(4,1)} = \{1.0000000000000000305 - 2.0000000000000000164\,i\ ;\ 8.68 \times 10^{-17}\},$$

$$Z_3^{(4,1)} = \{-0.9999999999999999963 - 3.54 \times 10^{-17}\,i\ ;\ 7.86 \times 10^{-17}\},$$

$$Z_4^{(4,1)} = \{2.9999999999999999885 + 1.82 \times 10^{-17}\,i\ ;\ 7.25 \times 10^{-17}\},$$

$$Z_5^{(4,1)} = \{2.63 \times 10^{-19} + 5.0000000000000000003\,i\ ;\ 1.03 \times 10^{-18}\}.$$

At the end of this section we want to point to the computational cost of Weierstrass' circular methods. As it will be shown in § 6.3, these methods are not efficient enough compared to the other circular methods because, apart from a slower convergence, they require numerous multiplications of disks (where the absolute values of complex numbers are needed). The efficiency can be slightly increased using rectangular arithmetic (the iterative method (3.46)), where multiplication of rectangles is executed by means of real interval operations. However, in that case, the obtained interval results are worsened. A proper improvement may be attained by applying the presented combined method. Finally, Weierstrass' method as well as its modifications are very efficient if they are used for finding the real zeros only, where real interval arithmetic is employed.

Contrary to the complex interval versions, Weierstrass' iterative method (3.3) in ordinary complex arithmetic is more favourable in relation to the other methods (see § 6.2). Moreover, in the case of real zeros, Algorithm (3.3) belongs to the most efficient ones (cf. [70]).

## 3.2. ON A CUBIC CONVERGENCE METHOD

Braess and Hadeler [11] applied Lagrangean interolation formula for obtaining a method for the simultaneous inclusion of all polynomial zeros by disks in the complex plane. The optimization of these disks leads to a special type of matrix eigenvalue problem. In this section we derive a new simultaneous method which also applies Lagrangean interpolation, but the contraction of inclusion disks is performed by an iterative procedure in circular complex arithmetic.

Suppose that we have found the separated disks $Z_i = \{z_i ; r_i\}$ which contain simple real or complex zeros $\xi_i$ of a monic polynomial $P$ of degree $n$. This polynomial is identical to its Lagrangean interpolation polynomial for the points $z_1,\dots,z_n$ and $\infty$ ($z_i = \text{mid } Z_i$), that is

$$P(z) = \sum_{j=1}^{n} \frac{Q(z)}{Q'(z_j)(z - z_j)} P(z_j) + Q(z),$$

where

$$Q(z) = (z - z_1)(z - z_2)\cdots(z - z_n).$$

Suppose that $P(z_i) \neq 0$ for each $i = 1,\dots,n$. For any zero $\xi_i$ ($i \in \{1,\dots,n\}$) of $P$ we have (taking $z = \xi_i$)

$$\sum_{j=1}^{n} \frac{P(z_j)}{Q'(z_j)(\xi_i - z_j)} = -1$$

or

$$\frac{1}{\xi_i - z_i} \cdot \frac{P(z_i)}{Q'(z_i)} + \sum_{\substack{j=1 \\ j \neq i}}^{n} \frac{P(z_j)}{Q'(z_j)(\xi_i - z_j)} = -1.$$

With the abbreviation

$$W_j = \frac{P(z_j)}{Q'(z_j)} = \frac{P(z_j)}{\prod\limits_{\substack{k=1 \\ k \neq j}}^{n} (z_j - z_k)},$$

we obtain

$$\xi_i \equiv z_i - \frac{W_i}{1 - \sum\limits_{\substack{j=1 \\ j \neq i}}^{n} \frac{W_j}{z_j - \xi_i}} \quad (i = 1, \ldots, n). \tag{3.48}$$

Since $\xi_i \in Z_i$ $(i = 1, \ldots, n)$, on the basis on the inclusion isotonicity from (3.48) it follows

$$\xi_i \in z_i - \frac{W_i}{1 - \sum\limits_{\substack{j=1 \\ j \neq i}}^{n} \frac{W_j}{z_j - Z_i}} \quad (i = 1, \ldots, n). \tag{3.49}$$

Accordingly, if the interval $Z_i$ contains the zero $\xi_i$ of the polynomial P, then the circular region

$$z_i - \frac{W_i}{1 - \sum\limits_{\substack{j=1 \\ j \neq i}}^{n} \frac{W_j}{z_j - Z_i}}$$

also contains $\xi_i$ $(i = 1, \ldots, n)$.

Introduce the following notation :

$$v = \max_{1 \leq j \leq n} |W_j|, \qquad r = \max_{1 \leq j \leq n} r_j,$$

$$\rho = \min_{\substack{i,j \\ i \neq j}} \{ |z_i - z_j| - r_j \}, \qquad \eta = \frac{(n-1) r v}{\rho^2},$$

$$g_i = \sum_{\substack{j=1 \\ j \neq i}}^{n} \frac{W_j}{z_j - z_i}, \qquad G_i = \sum_{\substack{j=1 \\ j \neq i}}^{n} \frac{W_j}{z_j - Z_i}.$$

By the properties of complex circular arithmetic we prove the inequality

$$\left| \frac{1}{z_j - z_i} - \operatorname{mid}\left(\frac{1}{z_j - Z_i}\right) \right| < \frac{r}{\rho^2} - \operatorname{rad}\left(\frac{1}{z_j - Z_i}\right).$$

Hence, according to (2.15), there follows

$$\frac{1}{z_j - Z_i} \subset \left\{ \frac{1}{z_j - z_i} ; \frac{r}{\rho^2} \right\},$$

so that

$$\sum_{\substack{j=1 \\ j \neq i}}^{n} \frac{W_j}{z_j - Z_i} \subset \left\{ \sum_{\substack{j=1 \\ j \neq i}}^{n} \frac{W_j}{z_j - z_i} ; \frac{(n-1) rv}{\rho^2} \right\} = \{ g_i ; \eta \} . \tag{3.50}$$

Before we establish the convergence theorem, we will prove some necessary statements.

**LEMMA 3.6.** *Under the condition*

$$\rho > 3(n-1)r, \tag{3.51}$$

*the inequality*

$$v < \alpha r \tag{3.52}$$

*holds, where* $\alpha = e^{1/3} \cong 1.396$.

*Proof*. The sequence $(a(k))$, defined by $a(k) = (1 + \frac{1}{3k})^k$, is bounded and monotonically increasing so that

$$a(k) < \lim_{k \to +\infty} a(k) = e^{1/3}$$

for each $k \in N$. According to this, for each $j \in \{1,\dots,n\}$ we have the following estimation

$$|W_j| = \frac{|P(z_j)|}{|Q'(z_j)|} = |z_j - \xi_j| \prod_{\substack{k=1 \\ k \neq j}}^{n} \left| \frac{z_j - \xi_k}{z_j - z_k} \right| < r_j \prod_{\substack{k=1 \\ k \neq j}}^{n} \frac{|z_j - z_k| + r_k}{|z_j - z_k|}$$

$$< r\left(1 + \frac{r}{\rho}\right)^{n-1} < r\left(1 + \frac{1}{3(n-1)}\right)^{n-1} < e^{1/3} r,$$

that is

$$v < \alpha r.$$ □

**LEMMA 3.7.** *If (3.51) holds, then the disks* $1 - G_i$ *(i = 1,...,n) do not contain the origin; more precisely, none of these disks intersects the disk centered at the origin with the radius*

$$d(n) = 1 - \frac{1}{3}\left(1 + \frac{1}{3(n-1)}\right)^n.$$

*Proof.* Taking into account the inclusion (3.50), it is sufficient to prove that for any $i \in \{1,\ldots,n\}$ the disk $\{1 - g_i ; \eta\}$ does not intersect the disk $\{0 ; d(n)\}$. This requirement is equivalent to the inequality

$$|1 - g_i| > \eta + d(n). \tag{3.53}$$

Since

$$|1 - g_i| = \left|1 - \sum_{\substack{j=1\\ j\neq i}}^{n} \frac{W_j}{z_j - z_i}\right| > 1 - \sum_{\substack{j=1\\ j\neq i}}^{n} \frac{|W_j|}{|z_j - z_i|} > 1 - \frac{(n-1)v}{\rho},$$

by virtue of (3.51) we find

$$d(n) = 1 - \frac{1}{3}\left(1 + \frac{1}{3(n-1)}\right)^n < 1 - (n-1)\frac{r}{\rho}\left(1 + \frac{r}{\rho}\right)^n < 1 - \frac{(n-1)v(\rho + r)}{\rho^2}$$

$$= 1 - \frac{(n-1)v}{\rho} - \frac{(n-1)vr}{\rho^2} < |1 - g_i| - \eta.$$

Therefore, the inequality (3.53) is valid. □

**REMARK 4.** It can be shown that the sequence $(d(n))$ is bounded and monotonically increasing so that for each $n \geq 3$ we have

$$d(3) \leq d(n) < \lim_{n \to +\infty} d(n) = 1 - \frac{e^{1/3}}{3} \cong 0.535.$$

Since $d(3) \cong 0.47$, it follows that none of the disks $1 - G_i$ $(i = 1,\ldots,n)$ intersects the disk $\{0 ; 0.47\}$ for all $n \geq 3$. ®

Assume that the disjoint disks $Z_i^{(0)} = \{z_i^{(0)} ; r_i^{(0)}\}$, containing the zeros $\xi_i$ $(i = 1,\ldots,n)$, have been found. Relation (3.49) suggests a new interval method for finding, simultaneously, simple complex zeros of a polynomial (Petković [89]).

Let $\rho^{(m)}$, $r^{(m)}$, $W_j^{(m)}$, $v^{(m)}$, $g_i^{(m)}$, $G_i^{(m)}$, $\eta^{(m)}$ be the notations (introduced above) concerning the m-th iterative step, and let

$$\lambda^{(m)} = \frac{r^{(m)}}{\rho^{(m)}}, \qquad \theta(n) = \frac{n + 4/3}{n - 16/9}.$$

**THEOREM 3.4.** *Let the interval sequences* $(Z_i^{(m)})$ $(i = 1,\dots,n)$ *be defined by the iterative formula*

$$Z_i^{(m+1)} = z_i^{(m)} - \frac{W_i^{(m)}}{1 - G_i^{(m)}} \quad (i = 1,\dots,n;\ m = 0,1,\dots). \tag{3.54}$$

*Then, under the condition*

$$\rho^{(0)} > 3(n-1)r^{(0)}, \tag{3.55}$$

*for each* $i = 1,\dots,n$ *and* $m = 0,1,\dots$ *we have*

$1^o$ $\xi_i \in Z_i^{(m)}$;

$2^o$ $r^{(m+1)} < \dfrac{7(n-1)r^{(m)^3}}{(\rho^{(0)} - \theta(n)r^{(0)})^2}$ .

*Proof*. We will prove the assertion $1^o$ by induction. Suppose that $\xi_i \in Z_i^{(m)}$ for $i \in \{1,\dots,n\}$ and $m \geq 1$. On the basis of (3.49) and (3.54), it follows

$$\xi_i \in z_i^{(m)} - \frac{W_i^{(m)}}{1 - \sum\limits_{\substack{j=1\\ j\neq i}}^{n} \dfrac{W_j^{(m)}}{Z_j^{(m)} - Z_i^{(m)}}} = Z_i^{(m+1)}.$$

Since $\xi_i \in Z_i^{(0)}$, we obtain that $\xi_i \in Z_i^{(m)}$ for each $m = 1,2,\dots$ .

Let us prove now that the interval process (3.54) has a cubic convergence (assertion $2^o$).

Using the properties of circular arithmetic and the inclusion (3.50), from (3.54) we obtain

$$r_i^{(m+1)} = \text{rad } Z_i^{(m+1)} < \text{rad}\left(\frac{|W_i^{(m)}|}{\{1 - g_i^{(m)};\ \eta^{(m)}\}}\right) = \frac{|W_i^{(m)}|\ \eta^{(m)}}{|1 - g_i^{(m)}|^2 - \eta^{(m)2}} .$$

Applying (3.52) and (3.55), we find the following bounds

$$v^{(0)} < \alpha r^{(0)},$$

$$\lambda^{(0)} < \frac{1}{3(n-1)},$$

$$\eta^{(0)} < \frac{\alpha(n-1)r^{(0)^2}}{\rho^{(0)2}} < \frac{\alpha}{9(n-1)} .$$

In view of these bounds, for each $i = 1,\dots,n$ we estimate

$$r_i^{(1)} \le r^{(1)} < \frac{v^{(0)}\eta^{(0)}}{\left[1 - \frac{(n-1)v^{(0)}}{\rho^{(0)}}\right]^2 - \eta^{(0)^2}} < \frac{\alpha^2(n-1)r^{(0)^3}}{\rho^{(0)^2}\left([1-\alpha(n-1)\lambda^{(0)}]^2 - \eta^{(0)^2}\right)}$$

$$< \frac{\alpha^2(n-1)r^{(0)^3}}{\rho^{(0)^2}\left[(1-\frac{\alpha}{3})^2 - \frac{\alpha^2}{81(n-1)^2}\right]} < \frac{7(n-1)r^{(0)^3}}{\rho^{(0)^2}} .$$

Hence

$$r^{(1)} < \frac{7(n-1)r^{(0)^3}}{(\rho^{(0)} - \theta(n)r^{(0)})^2}$$

and

$$r^{(1)} < \frac{7(n-1)r^{(0)}}{\left[\frac{\rho^{(0)}}{r^{(0)}}\right]^2} < \frac{r^{(0)}}{\frac{9}{7}(n-1)} .$$

By virtue of (3.55), similarly as in [31],[33] or [88], it can be proved that the disks $Z_1^{(1)},\dots,Z_n^{(1)}$ are disjoint and the following inequality

$$\rho^{(1)} > 3(n-1)r^{(1)}$$

holds.

Using the above consideration and mathematical induction, in the similar manner as in [88] it is proved that the following relations are valid for each $m \ge 1$:

$$r^{(m+1)} < \frac{7(n-1)r^{(m)^3}}{\rho^{(m)^2}} , \tag{3.56}$$

$$\rho^{(m+1)} > 3(n-1)r^{(m+1)} , \tag{3.57}$$

$$\rho^{(m)} > \rho^{(0)} - \theta(n)r^{(0)} . \tag{3.58}$$

By virtue of (3.56) and (3.58) it follows

$$r^{(m+1)} < \frac{7(n-1)r^{(m)^3}}{(\rho^{(0)} - \theta(n)r^{(0)})^2}$$

proving that the sequence $(r^{(m)})$ converges to zero at least cubically.

For example, for an arbitrary n ( > 2) we have

$$r^{(m+1)} < \frac{7(n-1)r^{(m)^3}}{(\rho^{(0)} - \frac{39}{11}r^{(0)})^2} .$$

We proved before that (3.55) implies $\rho^{(m+1)} > 3(n-1)r^{(m+1)}$ $(m = 0,1,\ldots)$. Therefore, Lemma 3.7 is applicable for each $m = 0,1,\ldots,$ so that $0 \notin 1 - G_i^{(m)}$. Thus, under the condition (3.55), the interval method (3.54) is defined in each iterative step.

This completes the proof of Theorem 3.4. □

**REMARK 5.** The interval method (3.54) can be also derived from a family of simultaneous methods presented in [90]. ®

For sufficiently small $r_i^{(m)}$ from (3.54) we obtain the following approximate expression for the center $z_i^{(m+1)}$ of the disk $Z_i^{(m+1)}$:

$$z_i^{(m+1)} = z_i^{(m)} - \frac{W_i^{(m)}}{1 - \sum_{\substack{j=1 \\ j \neq i}}^{n} \frac{W_j^{(m)}}{z_j^{(m)} - z_i^{(m)}}} \quad (i = 1,\ldots,n;\ m = 0,1,\ldots). \tag{3.59}$$

The condition "sufficiently small $r_i^{(m)}$" corresponds to the choice of the initial approximations $z_1^{(0)},\ldots,z_n^{(0)}$ which are sufficiently close to the zeros $\xi_1,\ldots,\xi_n$. Then the sequences $(z_i^{(m)})$, defined by the iterative formula (3.59), converge cubically to $\xi_i$ $(i = 1,\ldots,n)$ ([75]).

Considering the fixed point relation (3.48) we observe that the convergence of the iterative formula (3.59) can be accelerated if we use the Weierstrass correction $z_{W,i}^{(m)} = z_i^{(m)} - W_i^{(m)}$ instead of $z_i^{(m)}$ in the denominator of the sum (because $z_{W,i}^{(m)}$ is closer to the fixed point $\xi_i$ than $z_i^{(m)}$, see (3.48)). The improved method is of the form

$$z_i^{(m+1)} = z_i^{(m)} - \frac{W_i^{(m)}}{1 - \sum_{\substack{j=1 \\ j \neq i}}^{n} \frac{W_j^{(m)}}{z_j^{(m)} - z_i^{(m)} + W_i^{(m)}}} \quad (i = 1,\ldots,n;\ m = 0,1,\ldots), \tag{3.60}$$

and it has the convergence order equal to *four* (Nourein [76]). The increased convergence is obtained by means of only 2n additions as the approximations $z_{W,i}^{(m)}$ use the previously calculated values $W_i^{(m)}$ $(i=1,\dots,n)$. Therefore, the iterative method (3.60) is considerably more efficient than (3.59), as it is shown in § 6.2.

**REMARK 6.** The iterative method (3.54) can be regarded as an interval version of the point iterative process (3.59). In literature the iterative method (3.59) is often ascribed to Nourein (cf . [75]). But, as far as we know, the above formula (3.59) appeared previously in the paper [10] by Börsch-Supan. For that reason, the iterative methods (3.54) , (3 . 59) and (3.60) are refered to as Börsch-Supan's methods. ®

From (3.49) we conclude that for the determination of a new inclusion disk of the zero $\xi_i$, a disk including $\xi_i$ and some approximations $z_j$ $(j \neq i)$ of the remaining zeros are sufficient. This fact enables the construction of an algorithm for the simultaneous inclusion of k zeros of a polynomial P $(1 \leq k \leq n)$ using the disks containing the desired zeros and the initial approximations of the remaining n-k zeros.

Assume that k disjoint disks $Z_1^{(0)},\dots,Z_k^{(0)}$, containing the zeros $\xi_1, \dots,\xi_k$, are known as well as the approximations $z_{k+1}^{(0)},\dots,z_n^{(0)}$ of the remaining zeros. Then, according to (3.49), we can establish the following iterative method for the inclusion of k zeros $\xi_1,\dots,\xi_k$ of P in complex circular arithmetic:

$$Z_i^{(m+1)} = z_i^{(m)} - \frac{W_i^{(m)}}{1 - \sum_{\substack{j=1 \\ j\neq i}}^{k} \dfrac{W_j^{(m)}}{Z_j^{(m)} - z_i^{(m)}} - \sum_{j=k+1}^{n} \dfrac{\hat{W}_j^{(m)}}{z_j^{(0)} - z_i^{(m)}}} \tag{3.61}$$

$$(i = 1,\dots,k;\ m = 0,1,\dots\ ),$$

where

$$W_j^{(m)} = \frac{P(z_j^{(m)})}{\prod_{\substack{\lambda=1 \\ \lambda\neq j}}^{k} (z_j^{(m)} - z_\lambda^{(m)}) \prod_{\lambda=k+1}^{n} (z_j^{(m)} - z_\lambda^{(0)})} \qquad (j = 1,\dots,k),$$

$$\hat{W}_j^{(m)} = \frac{P(z_j^{(0)})}{\prod_{\lambda=1}^{k}(z_j^{(0)} - z_\lambda^{(m)}) \prod_{\substack{\lambda=k+1 \\ \lambda\neq j}}^{n} (z_j^{(0)} - z_\lambda^{(0)})} \qquad (j = k+1,\dots,n).$$

If $k = n$ then the second sum in (3.60) does not exist and (3.61) becomes (3.54).

The following theorem was proved in [116]:

**THEOREM 3.5.** *Let $(Z_i^{(m)})$ be the interval sequences generated by (3.61). Then, for each $i = 1,\dots,k$ $(1 \le k \le n)$ and $m = 0,1,\dots$ we have*

*$1^o$ $\xi_i \in Z_i^{(m)}$;*

*$2^o$ the convergence order of the iterative method (3.61) is two if $k < n$ and three if $k = n$.*

**EXAMPLE 5.** The polynomial

$$P(z) = z^7 + z^5 - 10z^4 - z^3 - z + 10$$

was chosen to illustrate the interval method (3.52) numerically. The zeros of $P$ are $\xi_1 = 2$, $\xi_{2,3} = \pm 1$, $\xi_{4,5} = \pm i$, $\xi_{6,7} = -1 \pm 2i$. The initial disks, containing these zeros, were taken to be $Z_i^{(0)} = \{z_i^{(0)}; 0.3\}$, where

$$z_1^{(0)} = 2.2,\quad z_2^{(0)} = 1.2 + 0.1i,\quad z_3^{(0)} = -0.8 - 0.1i,\quad z_4^{(0)} = 0.1 + 1.2i,$$
$$z_5^{(0)} = -0.1 - 0.8i,\quad z_6^{(0)} = -1.1 + 2.2i,\quad z_7^{(0)} = -1.1 - 1.8i.$$

The largest radii of disks, produced in the first and second iteration, were $r^{(1)} \cong 5.03 \times 10^{-2}$ and $r^{(2)} \cong 2.77 \times 10^{-5}$. The third iterative step furnished the following disks:

$$Z_1^{(3)} = \{2.00000000000000001\underline{6} - 1.11 \times 10^{-17} i\ ;\ 5.09 \times 10^{-17}\},$$
$$Z_2^{(3)} = \{1.00000000000000000\underline{5}3 + 1.29 \times 10^{-17} i\ ;\ 7.15 \times 10^{-16}\},$$
$$Z_3^{(3)} = \{-1.00000000000000000\underline{0}3 - 2.06 \times 10^{-18} i\ ;\ 3.12 \times 10^{-17}\},$$
$$Z_4^{(3)} = \{3.06 \times 10^{-18} + 0.99999999999999999\underline{9}9 i\ ;\ 2.12 \times 10^{-17}\},$$
$$Z_5^{(3)} = \{1.63 \times 10^{-18} - 0.99999999999999999\underline{8}1 i\ ;\ 1.09 \times 10^{-16}\},$$
$$Z_6^{(3)} = \{-1.00000000000000000\underline{0} + 2.00000000000000000\underline{0} i\ ;\ 3.61 \times 10^{-18}\},$$
$$Z_7^{(3)} = \{-1.00000000000000000\underline{0} - 2.00000000000000000\underline{0} i\ ;\ 7.92 \times 10^{-18}\}.$$

| | m = 2 | m = 3 |
|---|---|---|
| | **the case k < n** | |
| $z_1^{(m)}$ | $\{-2.99999987 - 1.04\times10^{-6}i \; ; \; 1.76\times10^{-5}\}$ | $\{-2.99999999999997 + 3.97\times10^{-14}i \; ; \; 1.24\times10^{-12}\}$ |
| $z_2^{(m)}$ | $\{-0.99999743 - 2.25\times10^{-7}i \; ; \; 7.78\times10^{-5}\}$ | $\{-1.00000000000051 - 6.05\times10^{-13}i \; ; \; 4.16\times10^{-11}\}$ |
| $z_3^{(m)}$ | $\{-4.33\times10^{-5} + 1.99995755\,i \; ; \; 2.28\times10^{-4}\}$ | $\{2.38\times10^{-10} + 2.00000000023532\,i \; ; \; 2.51\times10^{-9}\}$ |
| $z_4^{(m)}$ | $\{-2.00000187 + 1.00001197\,i \; ; \; 7.47\times10^{-5}\}$ | $\{-2.00000000000029 + 1.00000000000861\,i \; ; \; 8.32\times10^{-11}\}$ |
| $z_5^{(m)}$ | $\{-1.99996633 - 1.00000883\,i \; ; \; 1.90\times10^{-4}\}$ | $\{-2.00000000010139 - 0.00000000007488\,i \; ; \; 8.20\times10^{-10}\}$ |
| | **the case k = n** | |
| $z_1^{(m)}$ | $\{-2.99999949 + 2.78\times10^{-7}i \; ; \; 5.14\times10^{-6}\}$ | $\{-3.00000000000000000 + 1.32\times10^{-18}i \; ; \; 1.86\times10^{-17}\}$ |
| $z_2^{(m)}$ | $\{-0.99999979 + 3.55\times10^{-7}i \; ; \; 1.30\times10^{-5}\}$ | $\{-1.00000000000000000 + 5.33\times10^{-19}i \; ; \; 6.87\times10^{-17}\}$ |
| $z_3^{(m)}$ | $\{1.0\times10^{-5} + 2.00000155\,i \; ; \; 2.63\times10^{-5}\}$ | $\{-3.98\times10^{-16} + 2.00000000000000066\,i \; ; \; 2.15\times10^{-15}\}$ |
| $z_4^{(m)}$ | $\{-2.00000073 + 0.99999811\,i \; ; \; 1.55\times10^{-5}\}$ | $\{-2.00000000000000002 + 0.99999999999999999\,i \; ; \; 2.18\times10^{-16}\}$ |
| $z_5^{(m)}$ | $\{-1.99999976 - 0.99999607\,i \; ; \; 2.61\times10^{-5}\}$ | $\{-2.00000000000000004 - 0.99999999999999997\,i \; ; \; 5.70\times10^{-16}\}$ |
| $z_6^{(m)}$ | $\{1.99999487 + 0.99999081\,i \; ; \; 2.57\times10^{-5}\}$ | $\{2.00000000000000235 + 0.99999999999999944\,i \; ; \; 4.46\times10^{-15}\}$ |
| $z_7^{(m)}$ | $\{1.99997621 - 0.99999509\,i \; ; \; 6.61\times10^{-5}\}$ | $\{2.00000000000000276 - 0.99999999999999856\,i \; ; \; 1.91\times10^{-14}\}$ |
| $z_8^{(m)}$ | $\{0.99998576 + 7.13\times10^{-6}i \; ; \; 9.40\times10^{-5}\}$ | $\{1.00000000000000176 - 4.75\times10^{-16}i \; ; \; 3.04\times10^{-14}\}$ |
| $z_9^{(m)}$ | $\{-6.70\times10^{-7} - 1.99999991\,i \; ; \; 5.84\times10^{-6}\}$ | $\{-3.64\times10^{-19} - 2.00000000000000000\,i \; ; \; 4.16\times10^{-17}\}$ |

Table 3.5

**EXAMPLE 6.** The polynomial of degre $n = 9$

$$P(z) = z^9 + 3z^8 - 3z^7 - 9z^6 + 3z^5 + 9z^4 + 99z^3 + 297z^2 - 100z - 300$$

$$= (z+3)(z^2-1)(z^2+4)(z^2+4z+5)(z^2-4z+5),$$

with the zeros $-3$, $\pm 1$, $\pm 2i$, $\pm 2 \pm i$, was solved. First, the iterative method (3.61) was applied for improving 5 inclusion disks for the zeros $\xi_1 = -3$, $\xi_2 = -1$, $\xi_3 = 2i$, $\xi_4 = -2+i$ and $\xi_5 = -2-i$ ($k = 5 < n$, quadratic convergence) and then, for the simultaneous inclusion of all zeros of P ($k = n = 9$, cubic convergence). The following complex numbers were taken as the centers of starting disks as well as the approximations of the remaining zeros:

$$z_1^{(0)} = -3.2 + 0.2i\,, \quad z_2^{(0)} = -1.2 - 0.2i\,, \quad z_3^{(0)} = 0.1 + 0.7i\,,$$

$$z_4^{(0)} = -1.9 + 1.3i\,, \quad z_5^{(0)} = -1.8 - 0.8i\,, \quad z_6^{(0)} = 2.3 + 1.1i\,,$$

$$z_7^{(0)} = 1.9 - 0.7i\,, \quad z_8^{(0)} = 1.2 + 0.2i\,, \quad z_9^{(0)} = 0.2 - 2.2i\,.$$

In the procedure of finding the inclusion disks, the initial circular regions $Z_i^{(0)} = \{z_i^{(0)}; 0.35\}$ $(i = 1,\dots,k)$ were chosen, first $k = 5$ and then $k = n = 9$. The disks containing the zeros $\xi_1,\dots,\xi_5$, produced after the first iteration applying Algorithm (3.61) for $k = 5$ and $k = 9$ are, naturally, the same ones. The largest radius was $r^{(1)} = 8.12 \times 10^{-2}$. The improved disks obtained after the second and third iteration are given in Table 3.5. Comparing the radii $r_1^{(m)},\dots,r_5^{(m)}$ in both cases, we observe that considerably better results were generated in the case when $k = n$ $(= 9)$.

CHAPTER 4

# GENERALIZED ROOT ITERATIONS

Using the fixed point relation based on the logarithmic derivative of the k-th order of a polynomial P and the definition of the k-th root of a disk, a family of interval methods, the so-called *root iterations*, for the simultaneous inclusion of polynomial zeros is established. The convergence order of the basic method in parallel fashion is $k+2$ $(k \geq 1)$. The single-step method and the methods for multiple zeros and zero clusters are also considered. The root iterations are used for the construction of some tests for existence of a polynomial zeros, which are computationally verifiable.

## 4.1. GENERALIZED ROOT ITERATIONS FOR SIMPLE ZEROS

Consider a monic polynomial of degree $n \geq 3$

$$P(z) = z^n + a_{n-1}z^{n-1} + \cdots + a_1 z + a_0 = \prod_{j=1}^{n} (z - \xi_j) \qquad (a_i \in \mathbb{C})$$

with simple real or complex zeros $\xi_1, \ldots, \xi_n$. Let $Z_j = \{z_j ; r_j\}$ $(j = 1, \ldots, n)$ be disjoint disks that contain the zeros $\xi_j$ of the polynomial P. Let us introduce the following notation:

$$r = \max_{1 \leq j \leq n} r_j, \qquad \beta(k,n) = \begin{cases} 2n & k = 1 \\ k(n-1) & k > 1, \end{cases}$$

$$\rho = \min_{\substack{i,j \\ i \neq j}} \{|z| : z \in Z_i - Z_j\} = \min_{\substack{i,j \\ i \neq j}} \{|z_i - z_j| - r_j\}.$$

Assume that the condition

$$\rho > \beta(k,n) r \tag{4.1}$$

holds.

**LEMMA 4.1.** *If (4.1) holds, then $P'(z) \neq 0$ for all $z \in \bigcup_{j=1}^{n} Z_j$ .*

*Proof.* Consider an arbitrary disk $Z_i$ ($i \in \{1,\ldots,n\}$). Since all zeros of P are simple, $P'(\xi_i) \neq 0$ holds. For other points $z\,(\neq \xi_i)$ from $Z_i$ we have

$$|P'(z)/P(z)| = \left|\sum_{j=1}^{n} \frac{1}{z-\xi_j}\right| > \left|\frac{1}{z-\xi_i}\right| - \sum_{\substack{j=1\\ j\neq i}}^{n} \left|\frac{1}{z-\xi_j}\right|$$

$$> \frac{1}{r} - \frac{n-1}{\rho - r} = \frac{\rho - nr}{r(\rho - r)} > 0$$

because (4.1) implies $\rho > nr$.

Applying the above consideration for each $i = 1,2,\ldots,n$, we prove the assertion of Lemma 4.1. □

SELECTION OF THE K-TH ROOT

Let us introduce the function $z \mapsto h_k(z)$ by

$$h_k(z) = \frac{(-1)^{k-1}}{(k-1)!} \cdot \frac{d^k}{dz^k}[\log P(z)] \qquad (k = 1,2,\ldots).$$

It is easy to show that

$$h_k(z) = \sum_{j=1}^{n} \frac{1}{(z-\xi_j)^k} .$$

Assume that all zeros of the polynomial P, except $\xi_\nu$ , are known. If z is any complex number, the unknown zero $\xi_\nu$ is equal to one of the k values of

$$z - \frac{1}{\left[h_k(z) - \sum_{\substack{j=1\\ j\neq\nu}}^{n} \frac{1}{(z-\xi_j)^k}\right]^{1/k}} . \tag{4.2}$$

Therefore, if $k > 1$ then the problem of the choice of appropriate k-th root appears. A similar problem was solved for $k = 2$ in [31]. We will now consider this problem for $k \geq 2$. In the following, we will use the abbreviation

$$q_\nu = h_k(z) - \sum_{\substack{j=1\\ j\neq\nu}}^{n} \frac{1}{(z-\xi_j)^k} \qquad (k \geq 2) .$$

**LEMMA 4.2.** *If (4.1) holds, then the value of the k-th root to be chosen in (4.2), for $z = z_\nu$, is that which satisfies*

$$| P'(z_\nu)/P(z_\nu) - q_\nu^{1/k} | \le \frac{n-1}{\rho} . \tag{4.3}$$

*Proof.* Let $(z_\nu - \xi_\nu)^{-1} = |z_\nu - \xi_\nu|^{-1} \exp(i\alpha_\nu)$ and let $\{q_\nu\}_*^{1/k}$ be the value of $q_\nu^{1/k}$ equal to $(z_\nu - \xi_\nu)^{-1}$. Then, we have

$$|P'(z_\nu)/P(z_\nu) - q_\nu^{1/k}| = \left| \sum_{j=1}^{n} \frac{1}{z_\nu - \xi_j} - \left( \sum_{j=1}^{n} \frac{1}{(z_\nu - \xi_j)^k} - \sum_{\substack{j=1 \\ j\ne\nu}}^{n} \frac{1}{(z_\nu - \xi_j)^k} \right)^{\frac{1}{k}} \right|$$

$$= \left| \sum_{j=1}^{n} \frac{1}{z_\nu - \xi_j} - \left( \frac{1}{(z_\nu - \xi_\nu)^k} \right)^{\frac{1}{k}} \right| = \left| \sum_{j=1}^{n} \frac{1}{z_\nu - \xi_j} - \frac{1}{|z_\nu - \xi_\nu|} \exp\left(i\left(\alpha_\nu + \frac{2\lambda\pi}{k}\right)\right) \right|$$

$$(\lambda = 0, 1, \ldots, k-1).$$

For $\lambda = 0$ we have $q_\nu^{1/k} = \{q_\nu\}_*^{1/k}$ so that

$$|P'(z_\nu)/P(z_\nu) - \{q_\nu\}_*^{1/k}| = \left| \sum_{j=1}^{n} \frac{1}{z_\nu - \xi_j} - \frac{1}{z_\nu - \xi_\nu} \right|$$

$$= \left| \sum_{\substack{j=1 \\ j\ne\nu}}^{n} \frac{1}{z_\nu - \xi_j} \right| < \sum_{\substack{j=1 \\ j\ne\nu}}^{n} \left| \frac{1}{z_\nu - \xi_j} \right| < \frac{n-1}{\rho} .$$

For $\lambda = 1, \ldots, k-1$ we obtain the remaining $k-1$ values of $q_\nu^{1/k}$, for which we have

$$|P'(z_\nu)/P(z_\nu) - q_\nu^{1/k}| = \left| \sum_{j=1}^{n} \frac{1}{z_\nu - \xi_j} - \frac{1}{z_\nu - \xi_\nu} \exp\left(i \frac{2\lambda\pi}{k}\right) \right|$$

$$= \left| \frac{1}{z_\nu - \xi_\nu} - \frac{1}{z_\nu - \xi_\nu} \exp\left(i \frac{2\lambda\pi}{k}\right) + \sum_{\substack{j=1 \\ j\ne\nu}}^{n} \frac{1}{z_\nu - \xi_j} \right|$$

$$> \left| \frac{1}{z_\nu - \xi_\nu} \right| \left| 1 - \exp\left(i \frac{2\lambda\pi}{k}\right) \right| - \sum_{\substack{j=1 \\ j\ne\nu}}^{n} \left| \frac{1}{z_\nu - \xi_j} \right| > \frac{\left|1 - \exp\left(i\frac{2\lambda\pi}{k}\right)\right|}{r} - \frac{n-1}{\rho}$$

$$= \frac{2 \sin \frac{\lambda\pi}{k}}{r} - \frac{n-1}{\rho} \ge \frac{2 \sin \frac{\pi}{k}}{r} - \frac{n-1}{\rho} .$$

If $x \in (0, \frac{\pi}{2})$ then $\sin x > \frac{x}{\pi}$ holds. For $x = \frac{\pi}{k}$ $(k \ge 2)$ we obtain $k \sin \frac{\pi}{k} > 1$. Applying the last inequality we find

$$|P'(z_\nu)/P(z_\nu) - q_\nu^{1/k}| > \frac{2\sin\frac{\pi}{k}}{r} - \frac{n-1}{\rho} > \frac{2}{kr} - \frac{n-1}{\rho} = \frac{2\rho - k(n-1)r}{kr\rho}$$

$$> \frac{2k(n-1)r - k(n-1)r}{kr\rho} > \frac{n-1}{\rho} .$$

Thus, when n-1 zeros are known exactly, then there is no ambiguity in the choice of the value of the k-th root since only one satisfies (4.3). □

Let us put $z = z_\nu$ in (4.2). Since $\xi_j \in Z_j = \{z_j ; r_j\}$ $(j = 1,\dots,n)$, according to the inclusion isotonicity property we obtain the inclusion relation

$$\xi_\nu \in z_\nu - \frac{1}{\left[h_k(z_\nu) - \sum_{\substack{j=1\\ j\neq\nu}}^{n} \left(\frac{1}{z_\nu - Z_j}\right)^k\right]^{1/k}} \quad (\nu = 1,\dots,n), \tag{4.4}$$

supposing that the denominator does not contain zero. We will establish a criterion for the choice of the k-th root of a disk in (4.4), which is similar with the criterion defined by Lemma 4.2.

Since $\xi_\nu \in Z_\nu$ and

$$h_k(z_\nu) - \sum_{\substack{j=1\\ j\neq\nu}}^{n} \frac{1}{(z_\nu - \xi_j)^k} \equiv \frac{1}{(z_\nu - \xi_\nu)^k} ,$$

on the basis of the inclusion isotonicity it follows

$$\frac{1}{(z_\nu - \xi_\nu)^k} \in h_k(z_\nu) - \sum_{\substack{j=1\\ j\neq\nu}}^{n} \left(\frac{1}{z_\nu - Z_j}\right)^k = Q_\nu .$$

Assume that the disk $Q_\nu$ does not contain the origin. Then $Q_\nu^{1/k}$ is the union of k circular disks (see § 2.2), one of which contains $(z_\nu - \xi_\nu)^{-1}$. Let this disk be denoted by $\{Q_\nu\}_*^{1/k} = \{y_\nu^* ; d_\nu^*\}$. Denote the remaining k-1 disks $Q_\nu^{1/k}$, which are not containing $(z_\nu - \xi_\nu)^{-1}$, with $\{y_\nu ; d_\nu\}$. The criterion for the choice of the "proper" disk $\{Q_\nu\}_*^{1/k}$ is given by means of the following

**LEMMA 4.3.** *If* $\rho > k(n-1)r$ *and*

$$d_\nu^* < \frac{\rho - k(n-1)r}{2kr\rho} , \tag{4.5}$$

*then*

$$|P'(z_\nu)/P(z_\nu) - y_\nu^*| \le \frac{n-1}{\rho} + d_\nu^*, \tag{4.6}$$

$$|P'(z_\nu)/P(z_\nu) - y_\nu| \ge \frac{n-1}{\rho} + 3d_\nu^*. \tag{4.7}$$

*Proof.* Since $P'(z_\nu)/P(z_\nu) - (z_\nu - \xi_\nu)^{-1} \in \{P'(z_\nu)/P(z_\nu) - y_\nu^*; d_\nu^*\}$, we have

$$|P'(z_\nu)/P(z_\nu) - y_\nu^*| \le |P'(z_\nu)/P(z_\nu) - \frac{1}{z_\nu - \xi_\nu}| + d_\nu^*$$

$$= \left| \sum_{j=1}^{n} \frac{1}{z_\nu - \xi_j} - \frac{1}{z_\nu - \xi_\nu} \right| + d_\nu^* < \frac{n-1}{\rho} + d_\nu^* .$$

Using (4.5), similarly as in the proof of Lemma 4.2, we obtain

$$|P'(z_\nu)/P(z_\nu) - y_\nu| \ge |P'(z_\nu)/P(z_\nu) - \frac{1}{z_\nu - \xi_\nu}\exp(i\frac{2\lambda\pi}{k})| - d_\nu^*$$

$$> \frac{|1 - \exp(i\frac{2\lambda\pi}{k})|}{|z_\nu - \xi_\nu|} - \sum_{\substack{j=1 \\ j\ne\nu}}^{n} \left|\frac{1}{z_\nu - \xi_j}\right| - d_\nu^* > \frac{2\sin\frac{\lambda\pi}{k}}{r} - \frac{n-1}{\rho} - d_\nu^*$$

$$> \frac{2}{kr} - \frac{n-1}{\rho} - d_\nu^* = \frac{2\rho - k(n-1)r}{kr\rho} - 4d_\nu^* + 3d_\nu^*$$

$$> \frac{2\rho - k(n-1)r}{kr\rho} - \frac{2(\rho - k(n-1)r)}{kr\rho} + 3d_\nu^* = \frac{n-1}{\rho} + 3d_\nu^* .$$

According to (2.15) and the relation

$$|P'(z_\nu)/P(z_\nu) - y_\nu^*| \le (\frac{n-1}{\rho} + 2d_\nu^*) - d_\nu^*,$$

we conclude that the disk $\{Q_\nu\}_*^{1/k} = \{y_\nu^*; d_\nu^*\}$, containing $(z_\nu - \xi_\nu)^{-1}$, is interior of the disk

$$\Gamma_\nu = \left\{ \frac{P'(z_\nu)}{P(z_\nu)} ; \frac{n-1}{\rho} + 2d_\nu^* \right\} .$$

Further, on the basis of (2.16) and the relation

$$|P'(z_\nu)/P(z_\nu) - y_\nu| > (\frac{n-1}{\rho} + 2d_\nu^*) + d_\nu^*$$

it follows that all the remaining k-1 disks $Q_\nu^{1/k} = \{y_\nu; d_\nu\}$, which do not contain $(z_\nu - \xi_\nu)^{-1}$, lie out of the disk $\Gamma_\nu$. Therefore, among k disks $Q_\nu^{1/k}$ we will choose that whose center satisfies (4.6). □

In view of (4.6) and (4.7) there follows that we have to choose the disk whose center minimizes $|P'(z_\nu)/P(z_\nu) - y_\nu|$. Computing this expression (for k values of $y_\nu$) we use the values $P(z_\nu)$ and $P'(z_\nu)$, which are evaluated anyway in the computation of $h_k(z_\nu)$. For brevity, the described criterion for the choice of root will be denoted by CCR.

In addition, we note that $d_\nu = d_\nu^*$ .

## FAMILY OF INTERVAL METHODS

Before deriving generalized methods, we will prove some necessary inequalities.

**LEMMA 4.4.** *If k is natural number and $0 < x < 1$, then*

$$1 - (1 - x)^k \le kx. \tag{4.8}$$

*Proof*. The inequality (4.8) can be obtained from Bernoulli's inequality

$$(1 + t)^k \ge 1 + kt \quad (k \in \mathbb{N};\ t > -1)$$

substituting $t = -x$. The equality in (4.8) holds only for $k = 1$. □

At the m-th iteration we denote by $z_j^{(m)}$ the center and by $r_j^{(m)}$ the radius of the disk $Z_j^{(m)}$. We also define

$$r^{(m)} = \max_{1 \le j \le n} r_j^{(m)}, \qquad \theta(k,n) = \begin{cases} 3n & k = 1 \\ k(n-1) & k > 1, \end{cases}$$

$$\rho^{(m)} = \min_{\substack{i,j \\ i \ne j}} \{|z_i^{(m)} - z_j^{(m)}| - r_j^{(m)}\} \quad (m = 0,1,\ldots).$$

In order to approximate the sum

$$\sum_{\substack{j=1 \\ j \ne i}}^{n} \left(\frac{1}{z_i^{(m)} - Z_j^{(m)}}\right)^k \quad (k \ge 1),$$

let us find an enclosing disk for the circular region

$$\left(\frac{1}{z_i^{(m)} - Z_j^{(m)}}\right)^k,$$

assuming that $r^{(m)}$ is sufficiently small. We write for simplicity $r_i, z_i, Z_i$, $r$ and $\rho$ instead of $r_i^{(m)}$, $z_i^{(m)}$, $Z_i^{(m)}$, $r^{(m)}$ and $\rho^{(m)}$, respectively.

**REMARK 1.** The sums

$$\sum_{j \ne i} \left(\frac{1}{z_i - Z_j}\right)^k \quad \text{and} \quad \sum_{j \ne i} \frac{1}{(z_i - Z_j)^k}$$

are both circular extensions of $\sum_{j\neq i} \frac{1}{(z_i - \xi_j)^k}$. But, with respect to (2.20'), the first sum yields a smaller disk and so, it will be taken in the coming interval formulas.®

Sometimes, we will write $(z_i - Z_j)^{-k}$ instead of $(1/(z_i - Z_j))^k$ understanding strictly that the inverse operation is performed *before* raising to the k-th power.

Since

$$(z_i - Z_j)^{-1} = \frac{1}{\{z_i - z_j ; r_j\}} = \frac{\{\overline{z_i - z_j} ; r_j\}}{|z_i - z_j|^2 - r_j^2} \quad (i \neq j),$$

using (2.12) we obtain

$$(z_i - Z_j)^{-k} = \frac{\left\{(\overline{z_i - z_j})^k ; (|z_i - z_j| + r_j)^k - |z_i - z_j|^k\right\}}{(|z_i - z_j|^2 - r_j^2)^k} . \tag{4.9}$$

Applying the inequality

$$(|z_i - z_j| + r_j)^k - |z_i - z_j|^k < kr_j(|z_i - z_j| + r_j)^{k-1},$$

we find

$$\text{rad } (z_i - Z_j)^{-k} < \frac{kr_j(|z_i - z_j| + r_j)^{k-1}}{(|z_i - z_j| - r_j)^k (|z_i - z_j| + r_j)^k} < \frac{kr_j}{\rho^{k+1}} < \frac{kr}{\rho^{k+1}} .$$

For sufficiently small $r_j$ one obtains

$$\text{mid } (z_i - Z_j)^{-k} \cong (z_i - z_j)^{-k}.$$

Furthermore, since $\rho < |z_i - z_j|$ and $\frac{r}{\rho} < 1$, using (4.8) we find

$$\left| \frac{(\overline{z_i - z_j})^k}{(|z_i - z_j|^2 - r_j^2)^k} - (z_i - z_j)^{-k} \right| + \frac{(|z_i - z_j| + r_j)^k - |z_i - z_j|^k}{(|z_i - z_j|^2 - r_j^2)^k}$$

$$= \frac{|z_i - z_j|^k - (|z_i - z_j| - r_j)^k}{|z_i - z_j|^k (|z_i - z_j| - r_j)^k} = \frac{1 - (1 - r_j/|z_i - z_j|)^k}{(|z_i - z_j| - r_j)^k} < \frac{1 - (1 - \frac{r}{\rho})^k}{\rho^k} < \frac{kr}{\rho^{k+1}} .$$

Therefore

$$\left| \frac{(\overline{z_i - z_j})^k}{(|z_i - z_j|^2 - r_j^2)^k} - (z_i - z_j)^{-k} \right| < \frac{kr}{\rho^{k+1}} - \frac{(|z_i - z_j| + r_j)^k - |z_i - z_j|^k}{(|z_i - z_j|^2 - r_j^2)^k} .$$

According to (2.15) and (4.9), we have the following inclusion

$$(z_i - z_j)^{-k} \subseteq \left\{(z_i-z_j)^{-k} \,;\, \frac{kr}{\rho^{k+1}}\right\}; \tag{4.10}$$

hence

$$\sum_{\substack{j=1\\ j\neq i}}^{n} (z_i - z_j)^{-k} \subseteq \left\{\sum_{\substack{j=1\\ j\neq i}}^{n} (z_i-z_j)^{-k} \,;\, \frac{k(n-1)r}{\rho^{k+1}}\right\} \subseteq \{c_i \,;\, \eta\}, \tag{4.11}$$

where

$$c_i = \sum_{\substack{j=1\\ j\neq i}}^{n} (z_i-z_j)^{-k}, \qquad \eta = \frac{\beta(k,n)r}{\rho^{k+1}} .$$

Under the condition (4.1) we will now prove two inequalities.

**LEMMA 4.5.** *If (4.1) holds, then for all $k \in \mathbb{N}$ and $\lambda = 1,\dots,n$*

$$\sum_{\substack{j=1\\ j\neq \lambda}}^{n} \left|(z_\lambda-\xi_j)^{-k} - (z_\lambda-z_j)^{-k}\right| < \eta \tag{4.12}$$

*and*

$$|h_k(z_\lambda) - c_\lambda| > \eta . \tag{4.13}$$

*Proof*. Let us prove first the inequality (4.12). We set

$$m_{\lambda,j} = \min\{|z_\lambda-\xi_j|, |z_\lambda-z_j|\}, \qquad M_{\lambda,j} = \max\{|z_\lambda-\xi_j|, |z_\lambda-z_j|\}$$

and find

$$\sum_{\substack{j=1\\ j\neq \lambda}}^{n} \left|(z_\lambda-\xi_j)^{-k} - (z_\lambda-z_j)^{-k}\right| = \sum_{\substack{j=1\\ j\neq \lambda}}^{n} \left|\frac{(z_\lambda-z_j)^k - (z_\lambda-\xi_j)^k}{(z_\lambda-\xi_j)^k (z_\lambda-z_j)^k}\right|$$

$$= \sum_{\substack{j=1\\ j\neq \lambda}}^{n} \frac{|\xi_j-z_j| \left|\sum_{q=0}^{k-1} (z_\lambda-z_j)^{k-1-q}(z_\lambda-\xi_j)^q\right|}{|z_\lambda-\xi_j|^k |z_\lambda-z_j|^k}$$

$$< r\sum_{\substack{j=1\\ j\neq \lambda}}^{n} \frac{\sum_{q=0}^{k-1} |z_\lambda-z_j|^{k-1-q}|z_\lambda-\xi_j|^q}{|z_\lambda-\xi_j|^k |z_\lambda-z_j|^k} < r\sum_{\substack{j=1\\ j\neq \lambda}}^{n} \frac{k M_{\lambda,j}^{k-1}}{(m_{\lambda,j} M_{\lambda,j})^k} < \frac{k(n-1)r}{\rho^{k+1}} \leq \eta .$$

Using (4.12) and the inequality $\rho > 2r$, which follows from (4.1), we obtain

$$|h_k(z_\lambda) - c_\lambda| = \Big|(z_\lambda - \xi_\lambda)^{-k} + \sum_{\substack{j=1\\ j\neq\lambda}}^{n} (z_\lambda - \xi_j)^{-k} - \sum_{\substack{j=1\\ j\neq\lambda}}^{n} (z_\lambda - z_j)^{-k}\Big|$$

$$> |z_\lambda - \xi_\lambda|^{-k} - \sum_{\substack{j=1\\ j\neq\lambda}}^{n} |(z_\lambda - z_j)^{-k} - (z_\lambda - \xi_j)^{-k}| > \frac{1}{r^k} - \frac{\beta(k,n)r}{\rho^{k+1}}$$

$$> \frac{2}{\rho^k} - \frac{\beta(k,n)r}{\rho}\cdot\frac{1}{\rho^k} > \frac{2}{\rho^k} - \frac{1}{\rho^k} = \frac{1}{\rho^k} > \frac{\beta(k,n)r}{\rho^{k+1}} = \eta\ .$$

This proves the inequality (4.13). □

Assume that we have found the initial disjoint disks $Z_1^{(0)},\ldots,Z_n^{(0)}$ containing the zeros $\xi_1,\ldots,\xi_n$ of the polynomial P. The relation (4.4) suggests the following generalized *root iterations* for finding these complex zeros:

$$Z_i^{(m+1)} = z_i^{(m)} - \frac{1}{\left[h_k(z_i^{(m)}) - \sum\limits_{\substack{j=1\\ j\neq i}}^{n} \left(\dfrac{1}{z_i^{(m)} - Z_j^{(m)}}\right)^k\right]_*^{1/k}} \tag{4.14}$$

$$(i = 1,\ldots,n;\ m = 0,1,\ldots;\ k = 1,2,\ldots),$$

where $z_i^{(m)} = \operatorname{mid} Z_i^{(m)}$ and the symbol $*$ denotes the disk (among k disks) satisfying (4.6). The convergence properties of the generalized root iterations are given in the following theorem (Petković [88]):

**THEOREM 4.1.** *Let the sequences of disks* $(Z_i^{(m)})$ $(i = 1,\ldots,n)$ *be produced by (4.14). Then, under the condition*

$$\rho^{(0)} > \beta(k,n)r^{(0)}, \tag{4.15}$$

*for each* $i = 1,\ldots,n$ *and* $m = 0,1,\ldots$ *we have*

$1^o$ $\xi_i \in Z_i^{(m)}$;

$2^o$ $r^{(m+1)} < \dfrac{\theta(k,n)(r^{(m)})^{k+2}}{(\rho^{(0)} - 3r^{(0)})^{k+1}}$ $(k = 1,2,\ldots)$.

*P r o o f*. Assume that $\xi_i \in Z_i^{(m)}$ $(i = 1,\dots,n)$. Evidently,

$$z_i^{(m)} - \frac{1}{\left[h_k(z_i^{(m)}) - \sum\limits_{\substack{j=1\\ j\neq i}}^{n} (z_i^{(m)} - \xi_j)^{-k}\right]_*^{1/k}} \equiv \xi_i,$$

where the symbol $*$ denotes the (complex) value of the k-th root equal to $(z_i^{(m)} - \xi_i)^{-1}$. Since

$$\sum_{\substack{j=1\\ j\neq i}}^{n} (z_i^{(m)} - \xi_j)^{-k} \in \sum_{\substack{j=1\\ j\neq i}}^{n} (z_i^{(m)} - Z_j^{(m)})^{-k},$$

from (4.14) one obtains $\xi_i \in Z_i^{(m+1)}$. Since $\xi_i \in Z_i^{(0)}$, the assertion $1^\circ$ follows by mathematical induction.

Let us prove now that the interval method (4.14) has the order of convergence equal $k+2$ (the assertion $2^\circ$). Let $u_i^{(m)} = h_k(z_i^{(m)}) - c_i^{(m)}$. According to (2.18), (2.19) and the inclusion (4.11), from (4.14) we obtain

$$r_i^{(m+1)} = \operatorname{rad} Z_i^{(m+1)} \leq \operatorname{rad}\left(\frac{1}{\{u_i^{(m)};\ \eta^{(m)}\}^{1/k}}\right)$$

$$\leq \operatorname{rad}\left\{\frac{\bar{u}_i^{(m)}}{|u_i^{(m)}|^2 - \eta^{(m)2}};\ \frac{\eta^{(m)}}{|u_i^{(m)}|^2 - \eta^{(m)2}}\right\}^{1/k}$$

$$= \frac{\eta^{(m)}}{[\,|u_i^{(m)}|^2 - \eta^{(m)2}]^{1/k} \sum\limits_{\lambda=0}^{k-1} |u_i^{(m)}|^{\frac{k-1-\lambda}{k}} (|u_i^{(m)}| - \eta^{(m)})^{\frac{\lambda}{k}}}.$$

By virtue of Lemma 4.5, we find

$$|u_i^{(0)}| > \frac{1}{r^{(0)k}} - \frac{1}{\rho^{(0)k}} > \frac{1}{r^{(0)k}} - \frac{1}{[\beta(k,n)r^{(0)}]^k} = \frac{1}{r^{(0)k}}(1-s)$$

and

$$\eta^{(0)} = \frac{\beta(k,n)r^{(0)}}{\rho^{(0)k+1}} < \frac{1}{\rho^{(0)k}} < \frac{1}{[\beta(k,n)r^{(0)}]^k} = \frac{s}{r^{(0)k}},$$

where $s = [\beta(k,n)]^{-k}$. Then, for each $i = 1,\dots,n$ we have

$$r_i^{(1)} \leq r^{(1)} < \frac{\eta^{(0)}}{\left[\frac{(1-s)^2}{r^{(0)2k}} - \frac{s^2}{r^{(0)2k}}\right]^{\frac{1}{k}} \sum\limits_{\lambda=0}^{k-1} \frac{(1-s)^{\frac{k-1-\lambda}{k}}}{r^{(0)k-1-\lambda}} \cdot \frac{(1-2s)^{\frac{\lambda}{k}}}{r^{(0)\lambda}}}$$

$$= \frac{\beta(k,n)r^{(0)^{k+2}}}{\rho^{(0)^{k+1}}} \cdot \frac{\left(\frac{1-s}{1-2s}\right)^{\frac{1}{k}} - 1}{s} .$$

If we define

$$f(k,n) = \frac{\left(\frac{1-s}{1-2s}\right)^{\frac{1}{k}} - 1}{s} ,$$

then

$$\max_{\substack{k \\ n \geq 3}} f(k,n) = \max_{k} f(k,3) = \begin{cases} f(1,3) & k = 1 \\ f(2,3) & k > 1 . \end{cases}$$

Since $f(1,3) = \frac{3}{2}$ and $f(2,3) = 16\,[\,(\frac{15}{14})^{1/2} - 1\,] < 1$, we have

$$\max_{\substack{k \\ n \geq 3}} f(k,n) \leq \gamma(k) = \begin{cases} \frac{3}{2} & k = 1 \\ 1 & k > 1 , \end{cases}$$

so that

$$r^{(1)} < \frac{\theta(k,n)r^{(0)^{k+2}}}{\rho^{(0)^{k+1}}}$$

because of $\beta(k,n)\gamma(k) = \theta(k,n)$. Hence

$$r^{(1)} < \frac{\theta(k,n)r^{(0)^{k+2}}}{(\rho^{(0)} - 3r^{(0)})^{k+1}} ,$$

and

$$r^{(1)} < \frac{\theta(k,n)r^{(0)}}{\left[\frac{\rho^{(0)}}{r^{(0)}}\right]^{k+1}} < \frac{\gamma(k)r^{(0)}}{[\beta(k,n)]^{k}} < \frac{\gamma(k)r^{(0)}}{[\beta(k,3)]^{k}} = \frac{r^{(0)}}{\omega(k)} ,$$

where

$$\omega(k) = \frac{[\beta(k,3)]^{k}}{\gamma(k)} = \begin{cases} 4 & k = 1 \\ (2k)^{k} & k > 1 . \end{cases}$$

The disks $Z_i^{(1)}$ $(i = 1,\dots,n)$ are disjoint if

$$\rho^{(0)} > |z_i^{(1)} - z_i^{(0)}| + 3r^{(1)} ,$$

which can be observed from a geometric construction considering "the worst case". Since $|z_i^{(1)} - z_i^{(0)}| \leq r^{(0)} + r^{(1)}$, using (4.15) and the inequality $r^{(1)} < r^{(0)}/\omega(k)$, we obtain

$$\rho^{(0)} > \left(\frac{4}{\omega(k)} + 1\right) r^{(0)} > 4r^{(1)} + r^{(0)} > |z_i^{(1)} - z_i^{(0)}| + 3r^{(1)}.$$

Starting from the inequality (see [31],[33],[131])

$$\rho^{(1)} \geq \rho^{(0)} - r^{(0)} - 3r^{(1)}, \tag{4.16}$$

we find

$$\rho^{(1)} > \beta(k,n) r^{(0)} - r^{(0)} - \frac{3r^{(0)}}{\omega(k)} > \omega(k) r^{(1)} [\beta(k,n) - 1 - \frac{3}{\omega(k)}],$$

wherefrom

$$\rho^{(1)} > \beta(k,n) r^{(1)}. \tag{4.17}$$

We will prove now the assertion 2° by induction. Assume that the following relations are true:

$$r^{(m)} < \frac{\theta(k,n) {r^{(m-1)}}^{k+2}}{{\rho^{(m-1)}}^{k+1}}, \tag{4.18}$$

$$r^{(m)} < \frac{r^{(m-1)}}{\omega(k)}, \tag{4.19}$$

$$\rho^{(m)} > \beta(k,n) r^{(m)}, \tag{4.20}$$

$$\rho^{(m)} \geq \rho^{(m-1)} - r^{(m-1)} - 3r^{(m)}. \tag{4.21}$$

The relations (4.18) - (4.21) have already been proved for $m = 1$. We show that these relations are valid for index $m + 1$.

By the above consideration and by (4.20) we obtain

$$r^{(m+1)} < \frac{\theta(k,n) {r^{(m)}}^{k+2}}{{\rho^{(m)}}^{k+1}} < \frac{r^{(m)}}{\omega(k)}.$$

The same reasoning used for $m = 1$ yields

$$\rho^{(m+1)} \geq \rho^{(m)} - r^{(m)} - 3r^{(m+1)} > \beta(k,n) r^{(m+1)}.$$

Finally, using (4.19) and (4.21), in a similar manner as in the proof of Theorem 3.1 we obtain

$$\rho^{(m)} > \rho^{(0)} - 3r^{(0)},$$

so that

$$r^{(m+1)} < \frac{\theta(k,n) {r^{(m)}}^{k+2}}{(\rho^{(0)} - 3r^{(0)})^{k+1}}.$$

It remains to be shown that, under the condition (4.15), the interval process (4.14) is always defined. We have already proved that (4.15) implies $\rho^{(m+1)} > \beta(k,n)r^{(m+1)}$ $(m = 0,1,\ldots)$. Therefore, Lemma 4.5 is applicable for each $m = 0,1,\ldots$ and the inequalities

$$|h_k(z_i^{(m)}) - c_i^{(m)}| > \eta^{(m)} \quad (m = 0,1,\ldots)$$

hold. The last inequalities imply that $0 \notin h_k(z_i^{(m)}) - \{c_i^{(m)}; \eta^{(m)}\}$, which means that the radicand in (4.14) does not include the origin.

This completes the proof of the theorem. □

**REMARK 2.** For $k = 1$ the inequality $\rho^{(0)} > k(n-1)r^{(0)}$ implies (4.17) when $n > 3$, but (4.17) does not follow if $n = 3$. Because of this, the condition $\rho^{(0)} > 2nr^{(0)}$ for $k = 1$ is introduced, which gets us to operate with the function $(k,n) \mapsto \beta(k,n)$. ®

**REMARK 3.** Especially, for $k = 1$, we obtain from (4.14)

$$z_i^{(m+1)} = z_i^{(m)} - \frac{1}{\dfrac{P'(z_i^{(m)})}{P(z_i^{(m)})} - \displaystyle\sum_{\substack{j=1\\ j\neq i}}^{n} \frac{1}{z_i^{(m)} - z_j^{(m)}}} \quad (i = 1,\ldots,n;\ m = 0,1,\ldots), \tag{4.22}$$

which is the *third* order interval method considered in [28],[37],[50],[51, § 6.13]. For $k = 2$ the square root iteration method of the *fourth* order

$$z_i^{(m+1)} = z_i^{(m)} - \frac{1}{\left[h_2(z_i^{(m)}) - \displaystyle\sum_{\substack{j=1\\ j\neq i}}^{n} \left(\frac{1}{z_i^{(m)} - z_j^{(m)}}\right)^2\right]_*^{1/2}} \tag{4.23}$$

$$(i = 1,\ldots,n;\ m = 0,1,\ldots)$$

follows from (4.14), where $h_2(z) = [P'(z)^2 - P(z)P''(z)]/P(z)^2$ (Gargantini [31]). The symbol $*$ indicates that one of the two disks is chosen.®

**EXAMPLE 1.** The SIP (4.14) for $k = 3$ (with the convergence order five) was applied for solving the polynomial equation

$$z^{12} - (2+5i)z^{11} - (1-10i)z^{10} + (12-25i)z^9 - 30z^8 - z^4 + (2+5i)z^3 + (1-10i)z^2 - (12-25i)z + 30 = 0,$$

whose roots are $\xi_{1,2} = \pm 1$, $\xi_{3,4} = \pm i$, $\xi_{5,6} = \sqrt{2}/2 \pm i\sqrt{2}/2$, $\xi_{7,8} = -\sqrt{2}/2 \pm i\sqrt{2}/2$, $\xi_9 = 2i$, $\xi_{10} = 3$ and $\xi_{11,12} = 1 \pm 2i$.

The initial regions which contain these roots were given by the disks $Z_j^{(0)} = \{z_j^{(0)}; r_j^{(0)}\}$, where $|z_j^{(0)} - \xi_j| \cong 0.2$ and $r_j^{(0)} = 0.3 \quad (j = 1,\ldots,12)$.

As expected, the results obtained after the first iteration depend on the initial distribution of disks so that they cannot be taken as reliable measure of the convergence rate. The largest radius was less than $10^{-2}$.

After the second iteration the following inclusion disks were obtained (table 4.1):

| $i$ | $\mathrm{Re}\,\{z_i^{(2)}\}$ | $\mathrm{Im}\,\{z_i^{(2)}\}$ | $r_i^{(2)}$ |
|---|---|---|---|
| 1 | $-1.0000000000000\underline{0}1643$ | $2. \times 10^{-14}$ | $1.07 \times 10^{-13}$ |
| 2 | $1.00000000000000\underline{0}513$ | $1.05 \times 10^{-14}$ | $6.30 \times 10^{-14}$ |
| 3 | $5.19 \times 10^{-15}$ | $-0.999999999999\underline{9}99052$ | $1.36 \times 10^{-13}$ |
| 4 | $-1.18 \times 10^{-18}$ | $0.9999999999999999\underline{9}3$ | $5.52 \times 10^{-17}$ |
| 5 | $0.7071067811865\underline{4}9505$ | $0.7071067811865\underline{5}0125$ | $3.55 \times 10^{-14}$ |
| 6 | $0.7071067811865\underline{5}8256$ | $-0.7071067811865\underline{3}9654$ | $5.74 \times 10^{-14}$ |
| 7 | $-0.7071067811865\underline{6}0905$ | $0.70710678118\underline{6}608532$ | $1.05 \times 10^{-12}$ |
| 8 | $-0.7071067811865\underline{5}8237$ | $-0.7071067811865\underline{3}9299$ | $8.02 \times 10^{-14}$ |
| 9 | $9.23 \times 10^{-14}$ | $1.99999999999\underline{9}949261$ | $1.08 \times 10^{-12}$ |
| 10 | $-5.27 \times 10^{-15}$ | $3.000000000000\underline{0}72356$ | $2.66 \times 10^{-13}$ |
| 11 | $1.000000000000000000$ | $-2.000000000000000000$ | $7.35 \times 10^{-18}$ |
| 12 | $1.00000000000000\underline{0}931$ | $2.00000000000000\underline{0}173$ | $4.92 \times 10^{-15}$ |

Table 4.1

The underlined digit in Table 4.1 corresponds to the power order of radius.

For comparison, the above equation was also solved using the square root method (4.23) with the same initial approximations. The largest radii, produced after the first and second iteration were $r^{(1)} = 1.55 \times 10^{-2}$ and $r^{(2)} = 9.2 \times 10^{-9}$.

**REMARK 4.** The k-th root method for finding a simple real root of an eqaution $f(x) = 0$, with the order of convergence equal to $k + 1$, has been considered in [83]. This method is defined by the formula

$$x^{(m+1)} = x^{(m)} - f(x^{(m)}) / [h_k(x^{(m)})]^{1/k} \quad (m = 0,1,\ldots).$$

If k is even, the sign in front of the root coincides to the sign of $P(x^{(m)})P'(x^{(m)})$. ®

ROOT ITERATIONS IN COMPLEX ARITHMETIC

For sufficiently small $r_i^{(m)}$ $(i = 1,\ldots,n)$, from (4.14) we obtain the following approximate expression for the center $z_i^{(m+1)}$ of the disk $Z_i^{(m+1)}$:

$$z_i^{(m+1)} = z_i^{(m)} - \frac{1}{\left[h_k(z_i^{(m)}) - \sum_{\substack{j=1\\j\neq i}}^{n} (z_i^{(m)} - z_j^{(m)})^{-k}\right]_*^{1/k}} \tag{4.24}$$

$$(i = 1,\ldots,n;\ m = 0,1,\ldots;\ k = 1,2,\ldots).$$

The symbol $*$ denotes the value of the k-th root which satisfies (4.3).

Formula (4.24) defines the generalized iterative process in ordinary (point) complex arithmetic for improving, simultaneously, approximations to the complex zeros of a polynomial. The convergence order of the root iterations (4.24) is $k + 2$ (Petković [88]).

Particularly, for $k = 1$, the well known iterative method of the *third* order follows from (4.24):

$$z_i^{(m+1)} = z_i^{(m)} - \frac{1}{\dfrac{P'(z_i^{(m)})}{P(z_i^{(m)})} - \sum_{\substack{j=1\\j\neq i}}^{n} (z_i^{(m)} - z_j^{(m)})^{-1}} \tag{4.25}$$

$$(i = 1,\ldots,n;\ m = 0,1,\ldots).$$

To our knowledge the last formula appeared for the first time in the paper [65] by Maehly (see the remarks in [9] and [51, § 6.12]). For that reason, in this book the iterative method (4.25) and its modifications will be refered to as Maehly's methods. The SIP (4.25) has been considered later by Börsch-Supan [9], Ehrlich [19], Weisenhorn [135], Aberth [1], and others.

Let $N(z) = P(z)/P'(z) = 1/h_1(z)$ denote Newton's correction. The convergence rate of the SIP (4.25) can be increased using the Newton approximations $z_{N,j}^{(m)} = z_j^{(m)} - N(z_j^{(m)})$ in (4.25) instead of $z_j^{(m)}$, as follows:

$$z_i^{(m+1)} = z_i^{(m)} - \frac{1}{N(z_i^{(m)})^{-1} - \sum_{\substack{j=1\\j\neq i}}^{n} [z_i^{(m)} - z_j^{(m)} + N(z_j^{(m)})]^{-1}} \tag{4.26}$$

$$(i = 1,\ldots,n;\ m = 0,1,\ldots).$$

The convergence order of the improved method (4.26) is *four* (Nourein [77]). This improvement is obtained by means of few additional operations because the values $P(z_j^{(m)})$ and $P'(z_j^{(m)})$, necessary for the calculation of $z_{N,j}^{(m)}$, have already been calculated. Therefore, the SIP (4.26) is very efficient in view of a computational amount of work (see § 6.2).

The total-step methods (4.25) and (4.26) can be accelerated by applying the Gauss-Seidel approach. In this manner we obtain the single-step methods

$$z_i^{(m+1)} = z_i^{(m)} - \frac{1}{\dfrac{P'(z_i^{(m)})}{P(z_i^{(m)})} - \sum_{j=1}^{i-1}(z_i^{(m)} - z_j^{(m+1)})^{-1} - \sum_{j=i+1}^{n}(z_i^{(m)} - z_j^{(m)})^{-1}}$$

$$(i = 1,\ldots,n;\ m = 0,1,\ldots), \tag{4.27}$$

and

$$z_i^{(m+1)} = z_i^{(m)} - \frac{1}{N(z_i^{(m)})^{-1} - \sum_{j=1}^{i-1}(z_i^{(m)} - z_j^{(m+1)})^{-1} - \sum_{j=i+1}^{n}(z_i^{(m)} - z_{N,j}^{(m)})^{-1}}$$

$$(i = 1,\ldots,n;\ m = 0,1,\ldots). \tag{4.28}$$

Alefeld and Herzberger have proved in [4] that $O_R((4.27),\xi) \geq 2+\sigma_n\ (>3)$, where $\sigma_n > 1$ is the unique positive zero of the polynomial

$$p_n(\sigma) = \sigma^n - \sigma - 2.$$

The single-step method (4.28) has been studied by Milovanović and Petković [69]; it has been proved that $O_R((4.28),\xi) \geq 2(1+\tau_n)$, where $\tau_n > 1$ is the unique positive zero of the polynomial

$$g_n(\tau) = \tau^n - \tau - 1.$$

For k = 2 we obtain from (4.24) the square root method of the *fourth* order,

$$z_i^{(m+1)} = z_i^{(m)} - \frac{1}{\left[h_2(z_i^{(m)}) - \sum_{\substack{j=1 \\ j\neq i}}^{n}(z_i^{(m)} - z_j^{(m)})^{-2}\right]_*^{1/2}} \tag{4.29}$$

$$(i = 1,\ldots,n;\ m = 0,1,\ldots).$$

The correction term in the form of a sum modifies the well known Ostrowski's method [79] (with cubic convergence)

$$z^{(m+1)} = z^{(m)} - \frac{1}{\left[h_2(z^{(m)})\right]_*^{1/2}} \quad (m = 0,1,\ldots)$$

and enables the simultaneous determination of all zeros of a polynomial. The function $h_2(z)$ will be refered to as *Ostrowski's function*.

The square root method (4.29) and its modifications were studied in detail in [102]. Here we give only a review of the modified methods. For simplicity, the iteration index will be omitted; we will write $\hat{z}_i$ and $z_i$ instead of $z_i^{(m+1)}$ and $z_i^{(m)}$. By means of the functions $h_1(z) = P'(z)/P(z)$ and $s(z) = P''(z)/P'(z)$ we define

$$h_2(z) = h_1(z)[h_1(z) - s(z)] \quad \text{(Ostrowski's function)},$$

$$N(z) = 1/h_1(z) \quad \text{(Newton's correction)},$$

$$H(z) = [h_1(z) - \tfrac{1}{2}s(z)]^{-1} \quad \text{(Halley's correction)}.$$

We recall that the above-mentioned corrections appear in the iterative formulas

$$z^{(m+1)} = z^{(m)} - N(z^{(m)}) \quad \text{(Newton's method)},$$

$$z^{(m+1)} = z^{(m)} - H(z^{(m)}) \quad \text{(Halley's method)},$$

which have quadratic and cubic convergence, respectively.

The convergence order of the basic total-step method (4.29) (shortly TS) can be increased substituting the approximation $z_j$ in (4.29) by the improved Newton's and Halley's approximations $z_{N,j} = z_j - N(z_j)$ and $z_{H,j} = z_j - H(z_j)$. The corresponding methods are denoted by (TSN) and (TSH), indicating the use of Newton's and Halley's corrections. Applying the Gauss-Seidel procedure to the total-step methods (TS), (TSN) and (TSH), the single-step methods (SS), (SSN) and (SSH) are established. The improved methods of square root type are presented below:

$$\text{(SS)} \qquad \hat{z}_i = z_i - \frac{1}{\left[h_2(z_i) - \sum_{j=1}^{i-1}(z_i - \hat{z}_j)^{-2} - \sum_{j=i+1}^{n}(z_i - z_j)^{-2}\right]_*^{1/2}},$$

(TSN) $$\hat{z}_i = z_i - \frac{1}{\left[h_2(z_i) - \sum_{\substack{j=1 \\ j\neq i}}^{n} (z_i - z_{N,j})^{-2}\right]_*^{1/2}},$$

(SSN) $$\hat{z}_i = z_i - \frac{1}{\left[h_2(z_i) - \sum_{j=1}^{i-1} (z_i - \hat{z}_j)^{-2} - \sum_{j=i+1}^{n} (z_i - z_{N,j})^{-2}\right]_*^{1/2}},$$

(TSH) $$\hat{z}_i = z_i - \frac{1}{\left[h_2(z_i) - \sum_{\substack{j=1 \\ j\neq i}}^{n} (z_i - z_{H,j})^{-2}\right]_*^{1/2}},$$

(SSH) $$\hat{z}_i = z_i - \frac{1}{\left[h_2(z_i) - \sum_{j=1}^{i-1} (z_i - \hat{z}_j)^{-2} - \sum_{j=i+1}^{n} (z_i - z_{H,j})^{-2}\right]_*^{1/2}}.$$

The convergence order of the total-step square root methods (TSN) and (TSH) with Newton's and Halley's corrections is *five* and *six*, respectively. The lower bound of the R-order of convergence $O_R((SS),\xi)$ was determined in [100], and $O_R((SSN),\xi)$ and $O_R((SSH),\xi)$ in [102]. We summarize these results in the following theorem :

**THEOREM 4.2.** *The R-order of the single-step methods (SS), (SSN) and (SSH) is given by*

$$O_R((SS),\xi) \geq 3 + t_n \in (4,5),$$

$$O_R((SSN),\xi) \geq 3 + x_n \in (5,6),$$

$$O_R((SSH),\xi) \geq 3 + y_n \in (6,7),$$

*where* $t_n$, $x_n$ *and* $y_n$ *are the unique positive roots of the equations*

$$t^n - t - 3 = 0,$$

$$x^n - 2^{n-1}x - 3\cdot 2^{n-1} = 0,$$

$$y^n - 3^{n-1}y - 3^n = 0,$$

*respectively.*

**EXAMPLE 2.** The presented iterative methods of square root type were tested in the example of the polynomial

$$P(z) = z^5 - (4+5i)z^4 + (6+20i)z^3 - (4+30i)z^2 + (-15+20i)z + 75i,$$

| | i | Re $\{z_i^{(2)}\}$ | Im $\{z_i^{(2)}\}$ |
|---|---|---|---|
| (TS) | 1 | 0.999999380197767821 | 2.000001707170553462 |
| | 2 | 1.000000279303052643 | -2.000000176446057521 |
| | 3 | -0.999999790801744628 | $1.47 \times 10^{-7}$ |
| | 4 | 3.000000008454234552 | $-2.06 \times 10^{-7}$ |
| | 5 | $5.12 \times 10^{-7}$ | 5.000000353285864895 |
| (SS) | 1 | 1.000000160088381563 | 1.999999846637151023 |
| | 2 | 1.000000232361937907 | -1.999999875334209145 |
| | 3 | -0.999999999974857274 | $8.58 \times 10^{-11}$ |
| | 4 | 3.000000000074857274 | $-1.53 \times 10^{-11}$ |
| | 5 | $-1.25 \times 10^{-15}$ | 5.000000000000000117 |
| (TSN) | 1 | 0.999999616667618872 | 2.000000554250890694 |
| | 2 | 1.000000113100207197 | -1.999999987734416132 |
| | 3 | -1.000000225669099023 | $4.27 \times 10^{-7}$ |
| | 4 | 3.000000036009140354 | $4.78 \times 10^{-9}$ |
| | 5 | $-4.04 \times 10^{-9}$ | 4.999999989567260054 |
| (SSN) | 1 | 0.999999944040282847 | 1.999999964167704765 |
| | 2 | 0.999999998785935964 | -2.000000000153604734 |
| | 3 | -1.000000000002193334 | $-3.77 \times 10^{-12}$ |
| | 4 | 2.999999999999888187 | $6.01 \times 10^{-14}$ |
| | 5 | $-1.61 \times 10^{-15}$ | 5.000000000000000783 |
| (TSH) | 1 | 0.999999999931345461 | 1.999999999885598444 |
| | 2 | 0.999999999988968412 | -1.999999999991093962 |
| | 3 | -1.000000000053598353 | $-6.91 \times 10^{-11}$ |
| | 4 | 3.000000000031266106 | $5.52 \times 10^{-11}$ |
| | 5 | $-4.84 \times 10^{-11}$ | 5.000000000045326267 |
| (SSH) | 1 | 1.000000000028365003 | 1.999999999977318455 |
| | 2 | 1.000000000000004753 | -2.000000000000354773 |
| | 3 | -1.000000000000001134 | $6.88 \times 10^{-16}$ |
| | 4 | 2.999999999999999874 | $-2.41 \times 10^{-16}$ |
| | 5 | $4.85 \times 10^{-19}$ | 5.000000000000000000 |

Table 4.2

whose zeros are $\xi_{1,2} = 1 \pm 2i$, $\xi_3 = -1$, $\xi_4 = 3$ and $\xi_5 = 5i$. As the initial approximations to these zeros the following complex number were taken:

$$z_1^{(0)} = 1.8 + 1.3i, \quad z_2^{(0)} = 1.8 - 1.3i, \quad z_3^{(0)} = -1.8 - 0.7i,$$

$$z_4^{(0)} = 3.7 + 0.7i, \quad z_5^{(0)} = 0.7 + 4.3i.$$

Before calculating new approximations, the values $h_1(z_i^{(m)})$ and $s(z_i^{(m)})$ $(i = 1, \dots, n)$, necessary for the calculation of Ostrowski's function, were found. The same values were used for the calculation of Newton's and Halley's corrections.

In spite of rough initial approximations, the considered square root methods demonstrated good behavior and very fast convergence. Numerical results, obtained after the second iteration, are displayed in Table 4.2.

Let $\mathbf{z}^{(m)} = (z_1^{(m)}, \dots, z_n^{(m)})$ be the vector of approximations to the zeros at the m-th iteration. Take Euclid's norm,

$$\mathbf{e}^{(m)} := \| \mathbf{z}^{(m)} - \xi \|_E = \left( \sum_{i=1}^{n} |z_i^{(m)} - \xi_i|^2 \right)^{1/2},$$

as a measure of closeness of approximations with regard to the exact zeros.

In the presented example for the initial approximations we have even $e^{(0)} \cong 5.35$! For this reason, the results of the first iteration are not so good: the values $e^{(1)}$ for the considered methods belong to the range [0.069,0.184]. But, the second iteration produces significantly better approximations to the zeros. The values $e^{(2)}$ for every applied method are given in Table 4.3

| | TS | SS | TSN | SSN | TSH | SSH |
|---|---|---|---|---|---|---|
| $e^{(2)}$ | $1.97 \times 10^{-6}$ | $3.48 \times 10^{-7}$ | $8.38 \times 10^{-7}$ | $6.66 \times 10^{-8}$ | $1.82 \times 10^{-10}$ | $3.63 \times 10^{-11}$ |

Table 4.3

From Table 4.2 and Table 4.3 we observe very fast convergence of the square root methods with Halley's correction.

## 4.2. SINGLE STEP ROOT ITERATIONS

Assume again that we have found the initial nonoverlapping disks $Z_1^{(0)}, \ldots, Z_n^{(0)}$ containing the zeros $\xi_1, \ldots, \xi_n$ of a polynomial $P$ of degree $n \geq 3$. The SIP (4.14), realized in a parallel fashion, can be improved if we immediately employ already calculated disks from the current iteration in a serial fashion; we then have

$$Z_i^{(m+1)} = z_i^{(m)} - \frac{1}{\left[h_k(z_i^{(m)}) - \sum_{j=1}^{i-1} (z_i^{(m)} - Z_j^{(m+1)})^{-k} - \sum_{j=i+1}^{n} (z_i^{(m)} - Z_j^{(m)})^{-k}\right]_*^{1/k}}$$

$$(i = 1,\ldots,n;\ m = 0,1,\ldots;\ k = 1,2,\ldots). \qquad (4.30)$$

In this section we are concerned with the convergence properties of the single-step method (4.30). The convergence analysis will be carried out using the R-order of convergence, defined in § 2.3. We continue to use the notations introduced in § 4.1 as well as some relations derived in connection to the total-step method (4.14). For simplicity, at the beginning of our study we will omit the iteration index and write $Z_i, \hat{Z}_i$, $z_i$, $\hat{z}_i$, $r_i$, $\hat{r}_i$, $r$, $\hat{r}$, $\rho$ instead of $Z_i^{(m)}$, $Z_i^{(m+1)}$, $z_i^{(m)}$, $z_i^{(m+1)}$, $r_i^{(m)}$, $r_i^{(m+1)}$, $r^{(m)}$, $r^{(m+1)}$, $\rho^{(m)}$, respectively. In addition, we recall that, under the condition

$$\rho^{(0)} > \beta(k,n) r^{(0)}, \qquad (4.31)$$

the relations (4.19), (4.20) and (4.21) have been proved in § 4.1 for each $m = 1,2,\ldots$ .

Assume again that $\rho > \beta(k,n)r$ holds. According to (4.19) we have

$$\hat{r}_j \leq \hat{r} < \frac{r}{\omega(k)}.$$

Since $\rho \leq |z_i - z_j| - r_j$ and $|z_j - \hat{z}_j| < r_j + \hat{r}_j \leq r_j + \hat{r}$, we find

$$\frac{1}{|z_i-\hat{z}_j|} < \frac{1}{|z_i-z_j| - |z_j-\hat{z}_j|} < \frac{1}{|z_i-z_j| - r_j - \hat{r}_j} \leq \frac{1}{\rho - \hat{r}},$$

$$\frac{1}{|z_i-\hat{z}_j| - \hat{r}_j} < \frac{1}{|z_i-z_j| - |z_j-\hat{z}_j| - \hat{r}_j} < \frac{1}{|z_i-z_j| - r_j - 2\hat{r}_j} \leq \frac{1}{\rho - 2\hat{r}}.$$

We will now determine an inclusion disk for the disk $(z_i - \hat{Z}_j)^{-k}$ applying the above bounds. By (2.8) and (2.12) we obtain

$$\left|\operatorname{mid}(Z_i - \hat{Z}_j)^{-k} - (z_i - \hat{z}_j)^{-k}\right| + \operatorname{rad}(Z_i - \hat{Z}_j)^{-k}$$

$$= \left|\frac{\overline{(z_i-\hat{z}_j)}^k}{(|z_i-\hat{z}_j|^2 - \hat{r}_j^2)^k} - (z_i - \hat{z}_j)^{-k}\right| + \frac{\sum_{\lambda=1}^{k} \binom{k}{\lambda} |z_i - \hat{z}_j|^{k-\lambda} \hat{r}_j^{\lambda}}{(|z_i-\hat{z}_j|^2 - \hat{r}_j^2)^k}$$

$$= \frac{1 - (1 - \hat{r}_j/|z_i-\hat{z}_j|)^k}{(|z_i-\hat{z}_j| - \hat{r}_j)^k} < \frac{1 - (1 - \hat{r}_j/(\rho - \hat{r}))^k}{(\rho - 2\hat{r})^k} < \frac{k\hat{r}_j}{(\rho - \hat{r})(\rho - 2\hat{r})^k} .$$

Hence, by means of (2.15) we find

$$(z_i - \hat{Z}_j)^{-k} \subseteq \left\{ (z_i - \hat{z}_j)^{-k} ; \frac{k\hat{r}_j}{(\rho - \hat{r})(\rho - 2\hat{r})^k} \right\} . \tag{4.32}$$

The inclusion

$$(z_i - Z_j)^{-k} \subseteq \left\{ (z_i - z_j)^{-k} ; \frac{kr_j}{\rho^{k+1}} \right\} , \tag{4.33}$$

similar to (4.32), has already been obtained in § 4.1.

Using (4.32) and (4.33) we obtain the following inclusion for the radicand in (4.30) (omitting the iteration index):

$$h_k(z_i) - \sum_{j=1}^{i-1} (z_i - \hat{Z}_j)^{-k} - \sum_{j=i+1}^{n} (z_i - Z_j)^{-k} \subseteq \{u_i ; \eta_i\},$$

where

$$u_i = h_k(z_i) - \sum_{j=1}^{i-1} (z_i - \hat{z}_j)^{-k} - \sum_{j=i+1}^{n} (z_i - z_j)^{-k},$$

$$\eta_i = \frac{k}{(\rho - \hat{r})(\rho - 2\hat{r})^k} \sum_{j=1}^{i-1} \hat{r}_j + \frac{k}{\rho^{k+1}} \sum_{j=i+1}^{n} r_j .$$

Let $s = [\beta(k,n)]^{-k}$ , as in § 4.1. Using the proof of Lemma 4.5 and the identity

$$h_k(z) = \sum_{j=1}^{n} (z - \xi_j)^{-k} ,$$

we furnish the lower bound for $|u_i|$:

$$|u_i| = \left| (z_i-\xi_i)^{-k} + \sum_{j=1}^{i-1} \left[(z_i-\xi_j)^{-k} - (z_i-\hat{z}_j)^{-k}\right] + \sum_{j=i+1}^{n} \left[(z_i-\xi_j)^{-k} - (z_i-z_j)^{-k}\right]\right.$$

$$> \frac{1}{r_i^k} - \frac{k}{\rho(\rho-\hat{r})^k}\sum_{j=1}^{i-1}\hat{r}_j - \frac{k}{\rho^{k+1}}\sum_{j=i+1}^{n} r_j > \frac{1}{r_i^k} - \frac{k(n-1)r}{\rho^{k+1}},$$

whence

$$|u_i| > \frac{1}{r_i^k} - \frac{1}{[\beta(k,n)r_i]^k} = \frac{1}{r_i^k}(1-s). \tag{4.34}$$

The upper bound for $\eta_i$ is

$$\eta_i < \frac{k(n-1)r}{\rho^{k+1}} < \frac{1}{\rho^k} < \frac{1}{[\beta(k,n)r_i]^k} < \frac{s}{r_i^k}. \tag{4.35}$$

The convergence order of the single-step method (4.30) is treated in the following theorem (Petković and Stefanović [100]):

**THEOREM 4.3.** *Under the condition (4.31), the iterative interval method (4.30) is convergent with the R-order of convergence*

$$O_R((4.30),\xi) \geq k+1+\sigma_n(k),$$

*where $\sigma_n(k) > 1$ is the unique positive root of the equation*

$$\sigma^n - \sigma - k - 1 = 0.$$

*Proof*. Condition (4.31) implies (4.19) and hence, for $\omega(k) \geq 4$ it follows that the sequence $(r^{(m)})$ converges to zero. Thus, the initial condition (4.31) provides the convergence of Algorithm (4.30).

By virtue of (2.18) and (2.19), using at the same time the bounds (4.34) and (4.35), we get from (4.30)

$$r_i^{(m+1)} \leq \text{rad}\left(\frac{1}{\{u_i^{(m)};\eta_i^{(m)}\}^{1/k}}\right) \leq \text{rad}\left\{\frac{\bar{u}_i^{(m)}}{|u_i^{(m)}|^2-\eta_i^{(m)2}};\frac{\eta_i^{(m)}}{|u_i^{(m)}|^2-\eta_i^{(m)2}}\right\}^{\frac{1}{k}}$$

$$= \frac{\eta_i^{(m)}}{[|u_i^{(m)}|^2-\eta_i^{(m)2}]^{1/k}\sum_{\lambda=0}^{k-1}|u_i^{(m)}|^{\frac{k-1-\lambda}{k}}(|u_i^{(m)}|-\eta_i^{(m)})^{\frac{\lambda}{k}}}$$

$$< \frac{\eta_i^{(m)}}{\left[\dfrac{(1-s)^2-s^2}{r_i^{(m)2k}}\right]^{1/k}\sum_{\lambda=0}^{k-1}\dfrac{(1-s)^{\frac{k-1-\lambda}{k}}}{r_i^{(m)k-1-\lambda}}\cdot\dfrac{(1-2s)^{\frac{\lambda}{k}}}{r_i^{(m)}}}$$

$$< \frac{k f(k,n) r_i^{(m)^{k+1}}}{[\rho^{(m)} - r^{(m+1)}][\rho^{(m)} - 2r^{(m+1)}]^k} \left( \sum_{j=1}^{i-1} r_j^{(m+1)} + \sum_{j=i+1}^{n} r_j^{(m)} \right),$$

where $f(k,n)$ is introduced before in § 4.1 and bounded by

$$\max_{\substack{k \\ n\geq 3}} f(k,n) \leq \gamma(k) = \begin{cases} \frac{3}{2} & k = 1 \\ 1 & k > 1 \end{cases}.$$

Hence

$$k \max_{\substack{k \\ n\geq 3}} f(k,n) \leq \theta(k) = \begin{cases} \frac{3}{2} & k = 1 \\ k & k > 1 \end{cases},$$

so that

$$r_i^{(m+1)} < \frac{\theta(k) r_i^{(m)^{k+1}}}{[\rho^{(m)} - r^{(m+1)}][\rho^{(m)} - 2r^{(m+1)}]^k} \left( \sum_{j=1}^{i-1} r_j^{(m+1)} + \sum_{j=i+1}^{n} r_j^{(m)} \right).$$

Starting from (4.21), we find by (4.19) and (4.20)

$$\rho^{(m)} > \rho^{(0)} - \frac{\omega(k) + 3}{\omega(k) - 1} r^{(0)} \geq \rho^{(0)} - \frac{7}{3} r^{(0)},$$

and hence, we can write for each $m = 0,1,\ldots$

$$\frac{\theta(k)}{[\rho^{(m)} - r^{(m+1)}][\rho^{(m)} - 2r^{(m+1)}]^k} < \frac{\theta(k)}{[\rho^{(0)} - 3r^{(0)}]^{k+1}} = A(k).$$

This yields

$$r_i^{(m+1)} < A(k) r_i^{(m)^{k+1}} \left( \sum_{j=1}^{i-1} r_j^{(m+1)} + \sum_{j=i+1}^{n} r_j^{(m)} \right).$$

Substituting

$$r_i^{(m)} = \frac{v_i^{(m)}}{[(n-1)A(k)]^{\frac{1}{k+1}}} \quad (i = 1,\ldots,n) \tag{4.36}$$

in the previous relation, we obtain

$$v_i^{(m+1)} < \frac{1}{n-1} v_i^{(m)^{k+1}} \left( \sum_{j=1}^{i-1} v_j^{(m+1)} + \sum_{j=i+1}^{n} v_j^{(m)} \right). \tag{4.37}$$

From (4.36) it follows

$$v_i^{(m)} = r_i^{(m)} [(n-1)A(k)]^{\frac{1}{k+1}} \leq r^{(0)} [(n-1)A(k)]^{\frac{1}{k+1}} \quad (m = 0,1,\ldots).$$

We will now prove that $v_i^{(m)} < 1$ $(i = 1,\ldots,n;\ m = 0,1,\ldots)$, which is equivalent to the inequality

$$\frac{(n-1)\,\theta(k)\,r^{(0)^{k+1}}}{[\rho^{(0)} - 3r^{(0)}]^{k+1}} < 1,$$

that is,

$$\left[\frac{\rho^{(0)}}{r^{(0)}} - 3\right]^{k+1} > (n-1)\,\theta(k). \tag{4.38}$$

Applying the inequalities

$$[\beta(1,n) - 3]^2 > \tfrac{3}{2}(n-1) = (n-1)\,\theta(1),$$

$$[\beta(k,n) - 3]^{k+1} > k(n-1) = \beta(k,n) = (n-1)\,\theta(k) \quad (k > 1),$$

whose proofs are elementary, we estimate

$$\left[\frac{\rho^{(0)}}{r^{(0)}} - 3\right]^{k+1} > [\beta(k,n) - 3]^{k+1} > (n-1)\,\theta(k).$$

This proves (4.38) and, consequently, $v_i^{(m)} < 1$ holds for each $i = 1,\ldots,n$ and $m = 0,1,\ldots$ . According to these inequalities we conclude from (4.37) that the sequences $(v_i^{(m)})$ $(i = 1,\ldots,n)$ converge to zero and

$$v_i^{(0)} \le v < 1 \quad (i = 1,\ldots,n;\ v = \max_i v_i^{(0)}) \tag{4.39}$$

hold.

The inequlities (4.37) and (4.39) are sufficient for the application of Theorem 2.4. Taking $p = k + 1$ and $q = 1$, we immediately obtain the lower bound for $O_R((4.30),\xi)$, given in Theorem 4.3. □

The increase of convergence of the single-step method (4.30) with respect to the total-step method (4.14) is given by the size $\sigma_n(k) - 1 \sim \frac{\log(k+1)}{n}$,

| k \ n | 3 | 4 | 5 | 6 | 7 | 8 | 9 | 10 |
|---|---|---|---|---|---|---|---|---|
| 1 | 0.521 | 0.353 | 0.267 | 0.215 | 0.180 | 0.154 | 0.135 | 0.121 |
| 2 | 0.672 | 0.453 | 0.341 | 0.274 | 0.229 | 0.196 | 0.172 | 0.153 |
| 3 | 0.796 | 0.534 | 0.401 | 0.321 | 0.268 | 0.230 | 0.201 | 0.179 |
| 4 | 0.904 | 0.603 | 0.452 | 0.361 | 0.301 | 0.258 | 0.225 | 0.200 |

Table 4.3

so that the accelerating effect of the single-step method is slight for large n. The values of the improving size $\sigma_n(n) - 1$ for $k = 1(1)4$ and $n = 3(1)10$ are given in Table 4.3.

For sufficiently small radii of inclusion disks $Z_i^{(m)}$ $(i = 1,\ldots,n)$, we obtain from (4.30) the approximate expression for the centers of $Z_i^{(m+1)}$:

$$z_i^{(m+1)} = z_i^{(m)} - \frac{1}{\left[h_k(z_i^{(m)}) - \sum_{j=1}^{i-1}(z_i^{(m)} - z_j^{(m+1)})^{-k} - \sum_{j=i+1}^{n}(z_i^{(m)} - z_j^{(m)})^{-k}\right]_*^{1/k}}$$

$$(i = 1,\ldots,n;\ m = 0,1,\ldots;\ k = 1,\ldots). \quad (4.40)$$

The symbol $*$ indicates the choice of the appropriate value of the k-th root of a complex number in the denominator of (4.40). A criterion for the choice of this value is given by (4.3).

Let $e_i^{(m)} = |z_i^{(m)} - \xi_i|$ $(i = 1,\ldots,n;\ m = 0,1,\ldots)$. Since $e_i^{(m)} = O(r_i^{(m)})$, the relation of the same form as (4.37) can be easily derived for the point iterative process (4.40) and hence, we have the following assertion:

**THEOREM 4.4.** *The order of convergence of the single-step method (4.40) is at least $k + 1 + \sigma_n(k)$, where $\sigma_n(k)$ is the unique positive root of the equation*

$$\sigma^n - \sigma - k - 1 = 0.$$

Two special cases, which are obtained from (4.40) for $k = 1$ and $k = 2$, have been discussed before in the previous section.

**EXAMPLE 3.** The iterative interval formulas (4.14) and (4.30) for $k = 2$ were applied for solving the equation

$$z^5 + (-4+i)z^4 + (6-4i)z^3 + (-4+6i)z^2 - (15+4i)z - 15i = 0$$

whose roots are $\xi_1 = -1$, $\xi_{2,3} = 1 \mp 2i$, $\xi_4 = 3$, $\xi_5 = -i$. The following initial disks were taken:

$$Z_1^{(0)} = \{-0.7 + 0.5i\ ;\ 0.8\}, \quad Z_2^{(0)} = \{1.3 - 2.3i\ ;\ 0.8\},$$

$$Z_3^{(0)} = \{1.4 + 2.4i\ ;\ 0.8\}, \quad Z_4^{(0)} = \{2.6 + 0.4i\ ;\ 0.8\},$$

$$Z_5^{(0)} = \{0.2 - 1.3i\ ;\ 0.8\}.$$

For these disks the value $\rho^{(0)}/r^{(0)} \cong 0.857$ is much smaller than $k(n-1)=8$, which follows from the condition (4.15) of Theorems 4.1 and 4.3. Moreover, the disks $Z_2^{(0)}$ and $Z_5^{(0)}$ are overlapping. Although the initial disks are badly separated, the square root interval methods demonstrated good behavior. The radii of inclusion disks, obtained after the first iteration, were in the range $[3.05\times 10^{-2}, 6.86\times 10^{-2}]$ for Algorithm (4.14) and $[3.53\times 10^{-3}, 6.6\times 10^{-2}]$ for Algorithm (4.30). After the second iteration Algorithm (4.14) gave the disks with radii in the range $[1.6\times 10^{-9}, 9.\times 10^{-7}]$, while Algorithm (4.30) (with the R-order at least $3+\sigma_5(2) \cong 4.34$) produced the following circular regions:

$$Z_1^{(2)} = \{-1.0000000001\underline{5} + 9.8\times 10^{-11} i \; ; \; 4.12\times 10^{-10}\},$$

$$Z_2^{(2)} = \{1.0000000\underline{0}591 - 1.999999997\underline{6}1\, i \; ; \; 8.48\times 10^{-9}\},$$

$$Z_3^{(2)} = \{0.99999999999\underline{2}5 + 2.0000000000\underline{0}9\, i \; ; \; 5.98\times 10^{-10}\},$$

$$Z_4^{(2)} = \{2.99999999999\underline{8} + 2.07\times 10^{-13} i \; ; \; 4.22\times 10^{-12}\},$$

$$Z_5^{(2)} = \{-1.69\times 10^{-17} - 0.99999999999999999\underline{6}2 \; ; \; 1.46\times 10^{-16}\}.$$

The quotient $d = \rho^{(0)}/r^{(0)}$ can be taken as a measure of separation of the initial disks and of the size of their radii. The total error relative to circular approximations can be expressed (via Euclid's norm) by

$$\varepsilon^{(m)} = \left( \sum_{i=1}^{n} r_i^{(m)^2} \right)^{1/2} .$$

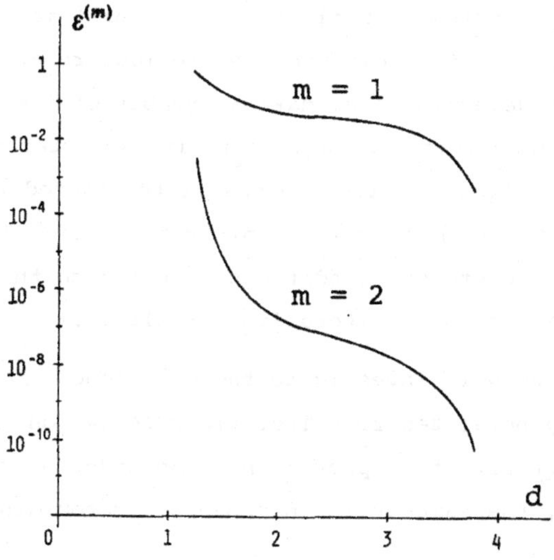


Fig. 1

Changing the size of initial disks and their disposition, the inclusion disks for the roots of the above equation were determined by the single-step square root interval method (4.30). For d in the range (0.857 , 3.025) it was obtained

$$\varepsilon^{(1)} \quad (8.5\times 10^{-3}\ , 9.\times 10^{-2}), \qquad \varepsilon^{(2)} \quad (2.23\times 10^{-12}, 4.2\times 10^{-7}).$$

Smaller value of $\varepsilon^{(m)}$ corresponds to large d with rare exceptions which may appear when the initial disks are small but they are not well separated.

The dependence of $\varepsilon^{(m)}$ with respect to d is qualitatively displayed (by a fitting curve) in Fig. 4.1.

## 4.3. NUMERICAL STABILITY OF THE GENERALIZED ROOT ITERATIONS

As it was pointed out in [33], any iterative procedure for finding the zeros of a function f must terminate in a finite number of iterations in the presence of rounding errors. The termination will happen as soon as the absolute value of the rounding error involved in the evaluation of f in the proximity of a zero is of the same order of magnitude as $|f|$.

Let $P, P', \dots, P^{(k)}$ denote the given polynomial and its derivatives and let $\Delta P, \Delta P', \dots, \Delta P^{(k)}$ be upper bounds for the absolute value of the rounding error appearing in the evaluation of $P, P', \dots, P^{(k)}$, respectively. These errors can be evaluated according to the rounding-error analysis introduced by Wilkinson [139] and depend on the number of significant digits of the mantissa for floating-point arithmetic operations (see also Igarishi [54]). According to the previous remarks, the stopping criterion which determines the maximal number of iterations, is based on the comparison of $\Delta P(z_i)$ with $|P(z_i)|$ in each iterative step and for each $i = 1,\dots,n$. The iterative process is terminated when any of the inequalities $\Delta P(z_i) < |P(z_i)|$ $(i = 1,\dots,n)$ is not valid. The last inequalities are also very useful as conditions under which the inversion of certain disks, appearing in the iterative formula (4.14), is defined.

In this section we will investigate the dependence of the convergence order of the generalized root iterations (4.14) on the absolute value of rounding errors. It is proved that the order of convergence of (4.14) remains $k+2$, the same as in the absence of rounding errors, if:

(I) the absolute value of the rounding errors with which the polynomial is evaluated in the proximity of a zero is small compared to the absolute value of the polynomial itself; and (II) the rounding error generated in the evaluation of $h_k$ is small as compared to the radius of the disk including the exact zero. If the condition (II) is weakened so that the rounding error in the evaluation of $h_k$ is independent of the disk size, although "reasonably small", then the convergence order reduces to $k+1$.

ITERATIVE METHOD IN THE PRESENCE OF ROUNDING ERRORS

The application of the iterative interval formula (4.14) requires single errors concerning the evaluation of $P, P', \ldots, P^{(k)}$ to be included in determination of $h_k$ because of the presence of rounding errors. Accordingly, we have to replace $P$ by the disk $Q_0 = \{P ; \Delta P\}$ and the derivatives $P^{(i)}$ by the disks $Q_i = \{P^{(i)} ; \Delta P^{(i)}\}$ $(i = 1, \ldots, k)$ with the centers $P, P', \ldots, P^{(k)}$ and the radii $\Delta P, \Delta P', \ldots, \Delta P^{(k)}$. Circular extension of the scalar quantity $h_k(z)$, obtained by the above substitutions, will be denoted by $H_k(z)$. According to the properties of circular arithmetic it follows $h_k(z)$ $H_k(z)$. Since in circular arithmetic the inversion is an exact operation, while the multiplication is not, in order to obtain the disk $H_k(z)$ as small as possible, before substituting $P, P', \ldots, P^{(k)}$ by the corresponding disks $Q_0, Q_1, \ldots, Q_k$ it is necessary to perform cancelations and single divisions of the addends in numerator of developed expression of $h_k$ by the denominator $P^k$. For instance, we have

$$h_2 = \frac{P'^2 - P''P}{P^2}, \quad H_2 = \left(\frac{Q_1}{Q_0}\right)^2 - \frac{Q_2}{Q_0} \quad \text{with} \quad \operatorname{rad} H_2 \leqq \operatorname{rad} \frac{Q_1^2 - Q_2 Q_0}{Q_0^2},$$

$$h_3 = \frac{P'''P^2 - 3PP'P'' + 2P'^3}{2P^3}, \quad H_3 = \frac{1}{2}\left[\left(\frac{Q_3}{Q_0}\right) - 3\frac{Q_1}{Q_0}\frac{Q_2}{Q_0} + 2\left(\frac{Q_1}{Q_0}\right)^3\right]$$

with

$$\operatorname{rad} H_3 \leqq \operatorname{rad} \frac{Q_3 Q_0^2 - 3Q_0 Q_1 Q_2 + 2Q_1^3}{2Q_0^3},$$

and so forth.

In the following, we will operate with total rounding error $\delta = \operatorname{rad} H_k$, which incorporates the single errors concerning the evaluation of $P, P', \ldots, P^{(k)}$.

In our analysis of the numerical stability of the iterative scheme (4.14) we will require $\Delta P < |P|$ because of the inversion of the disk $Q_0 = \{P ; \Delta P\}$ (that appears in the evaluation of $H_k$). Also, as it has already been said, the stopping criterion which determines the maximal number of iterations, is based on the comparison of $\Delta P(z_i^{(m)})$ with $|P(z_i^{(m)})|$ for each i.

In this section we will use some of the notations introduced in the previous sections. At the m-th iteration we denote the radius of $H_k(z_i^{(m)})$ by $\delta_i^{(m)}$. Also, we introduce the abbreviations

$$\delta^{(m)} = \max_{1 \le i \le n} \delta_i^{(m)}, \qquad \beta(k,n) = \begin{cases} 4n & k = 1 \\ k(n-1) & k > 1 \end{cases}, \qquad d^{(m)} = \frac{\beta(k,n) r^{(m)}}{\rho^{(m)k+1}},$$

$$Y_i^{(m)} = \sum_{\substack{j=1 \\ j \ne i}}^{n} (z_i^{(m)} - Z_j^{(m)})^{-k}, \qquad c_i^{(m)} = \sum_{\substack{j=1 \\ j \ne i}}^{n} (z_i^{(m)} - z_j^{(m)})^{-k}.$$

In new iterative formula only the scalars $h_k(z_i^{(m)})$ are replaced by the disks $H_k(z_i^{(m)})$ $(i = 1,\dots,n;\ m = 0,1,\dots)$, compared to (4.14). Thus, the generalized root iterations for polynomial complex zeros in the presence of rounding errors have the form

$$z_i^{(m+1)} = z_i^{(m)} - \frac{1}{[H_k(z_i^{(m)}) - Y_i^{(m)}]_*^{1/k}} \qquad (i = 1,\dots,n;\ m = 0,1,\dots). \quad (4.41)$$

According to the mentioned conditions (I) and (II), we will distinguish two cases: 1) $\delta^{(m)} \le r^{(m)} / \rho^{(m)k+1}$ and 2) $\delta^{(m)} \le \min\{1, 1/\rho^{(m)k+1}\}$, assuming that $\Delta P(z_i^{(m)}) < |P(z_i^{(m)})|$ $(i = 1,\dots,n;\ m = 0,1,\dots)$ ( (I) condition) holds in both cases.

For simplicity, we will omit the iteration index always when there is no possibility of any confusion. Also, we will often write only $\beta$ and $\Delta P_i$ instead of $\beta(k,n)$ and $\Delta P(z_i)$, respectively.

CONVERGENCE ANALYSIS: THE CASE $\delta \le r/\rho^{k+1}$

The iterative formula (4.41) will be defined for each $i = 1,\dots,n$ and $m = 0,1,\dots$ if $0 \notin H_k(z_i^{(m)}) - Y_i^{(m)}$. We will now consider the conditions for the fulfilment of the last claim. Let

$$u_i = \operatorname{mid} H_k(z_i) - c_i, \qquad \eta_i = d + \delta_i, \qquad \eta = \max_{1 \le i \le n} \eta_i = d + \delta, \qquad s = 1/\beta^k.$$

**LEMMA 4.6.** *Assume that*

$$\Delta P_i < |P(z_i)|, \quad \delta_i \le r/\rho^{k+1} \quad (i = 1,\ldots,n)$$

*and*

$$\rho > \beta(k,n)r \tag{4.42}$$

*hold. Then, for each* $i = 1,\ldots,n$ *and* $k = 1,2,\ldots,$ *the following inequalities are valid:*

$$|u_i| > \eta_i, \tag{4.43}$$

$$|u_i| > \frac{1-2s}{r^k}, \tag{4.44}$$

$$\eta_i < \frac{2s}{r^k}. \tag{4.45}$$

*Proof.* From (4.42) we obtain $\rho \ge 4r$, $(\beta r)/\rho < 1$ and $\rho > ((\beta+1)r)/2$. Using these bounds, (4.12), the estimate

$$|\operatorname{mid} H_k(z_i) - h_k(z_i)| \le \delta_i,$$

as well as the identity

$$h_k(z) = \sum_{j=1}^{n} (z-\xi_j)^{-k},$$

we find for each i

$$|u_i| = |\operatorname{mid} H_k(z_i) - c_i| \ge |h_k(z_i) - c_i| - \delta_i$$

$$= \left|(z_i-\xi_i)^{-k} + \sum_{\substack{j=1\\ j\ne i}}^{n}(z_i-\xi_j)^{-k} - \sum_{\substack{j=1\\ j\ne i}}^{n}(z_i-z_j)^{-k}\right| - \delta_i$$

$$\ge |z_i-\xi_i|^{-k} - \sum_{\substack{j=1\\ j\ne i}}^{n}|(z_i-\xi_j)^{-k} - (z_i-z_j)^{-k}| - \delta_i$$

$$> \frac{1}{r^k} - d - \frac{r}{\rho^{k+1}} > \frac{4}{\rho^k} - \frac{\beta r}{\rho}\frac{1}{\rho^k} - \frac{r}{\rho}\frac{1}{\rho^k} > \frac{2}{\rho^k} > \frac{(\beta+1)r}{\rho^{k+1}} > d + \delta_i = \eta_i,$$

which proves (4.43).

To prove (4.44) we use a similar procedure as for (4.43) and obtain

$$|u_i| > \frac{1}{r^k} - d - \delta > \frac{1}{r^k} - \frac{\beta r}{\rho}\frac{1}{\rho^k} - \frac{r}{\rho^{k+1}} > \frac{1}{r^k} - \frac{1}{\rho^k} - \frac{1}{\rho^k} > \frac{1}{r^k} - \frac{2}{\beta^k r^k} = \frac{1-2s}{r^k}.$$

Finally, we estimate

$$\eta_i = d + \delta_i < \frac{\beta r}{\rho}\frac{1}{\rho^k} + \frac{r}{\rho^{k+1}} < \frac{1}{\rho^k} + \frac{1}{\rho^k} < \frac{2}{\beta^k r^k} = \frac{2s}{r^k} . \qquad \square$$

From the inclusion

$$Y_i \subseteq \left\{ c_i \; ; \; \frac{k(n-1)}{\rho^{k+1}} \right\} \subseteq \{ c_i \; ; d \} \tag{4.46}$$

(see §4.1), it follows

$$H_k(z_i) - Y_i \subseteq H_k(z_i) - \{c_i \; ; d\} = \{u_i \; ; \eta_i\} .$$

If the disk $\{u_i \; ; \eta_i\}$ does not contain the origin, then obviously also $0 \notin H_k(z_i) - Y_i$. The requirement $0 \notin \{u_i \; ; \eta_i\}$ is equivalent to the inequality (4.43) which means that, under the conditions of Lemma 4.6, the disk $H_k(z_i) - Y_i$ is free of the number 0.

Let

$$\gamma(k) = \begin{cases} \frac{3}{2} & k = 1 \\ 1 & k > 1 \end{cases} , \qquad \theta(k,n) = [\beta(k,n) + 1]\gamma(k) = \begin{cases} \frac{3}{2}(4n+1) & k=1 \\ k(n-1)+1 & k>1 \end{cases} .$$

Sometimes, to simplify the notations, we will write briefly $\gamma$ and $\theta$ instead of $\gamma(k)$ and $\theta(k,n)$.

The convergence order of the iterative method (4.41) is treated in the following theorem ([101]):

**THEOREM 4.5.** *If* $\Delta P(z_i^{(m)}) < |P(z_i^{(m)})|$, $\delta_i^{(m)} \leq r^{(m)}/\rho^{(m)k+1}$ *(i = 1,...,n) and*

$$\rho^{(0)} > \beta(k,n) r^{(0)}, \tag{4.47}$$

*then the convergence order of the iterative method (4.41) is k + 2, that is, the sequence* $(r^{(m)})$ *decreases according to*

$$r^{(m+1)} < \frac{\theta(k,n) r^{(m)k+2}}{(\rho^{(0)} - 2r^{(0)})^{k+1}} .$$

*P r o o f*. Using (2.17), (2.18), (2.19), the inclusion (4.46) and the properties of circular arithmetic, we obtain from (4.41)

$$r_i^{(m+1)} = \text{rad} \frac{1}{[H_k(z_i^{(m)}) - Y_i^{(m)}]_*^{1/k}} \leq \text{rad} \frac{1}{\left\{u_i^{(m)} \; ; \; \eta_i^{(m)}\right\}^{1/k}}$$

$$\leq \text{rad} \left( \frac{1}{\{u_i^{(m)} \; ; \; \eta_i^{(m)}\}} \right)^{1/k} = \text{rad} \left( \frac{\bar{u}_i^{(m)}}{|u_i^{(m)}|^2 - \eta_i^{(m)2}} ; \frac{\eta_i^{(m)}}{|u_i^{(m)}|^2 - \eta_i^{(m)2}} \right)^{\frac{1}{k}}$$

$$= \frac{\eta_i^{(m)}}{\left[|u_i^{(m)}|^2 - \eta_i^{(m)2}\right]^{1/k} \sum_{\lambda=0}^{k-1} |u_i^{(m)}|^{(k-1-\lambda)/k} (|u_i^{(m)}| - \eta_i^{(m)})^{\lambda/k}} .$$

By (4.44) and (4.45) we find for the first iteration

$$r_i^{(1)} \le r^{(1)} < \frac{\eta^{(0)}}{\left[\frac{(1-2s)^2}{r^{(0)2k}} - \frac{4s^2}{r^{(0)2k}}\right]^{1/k} \sum_{\lambda=0}^{k-1} \frac{(1-2s)^{(k-1-\lambda)/k}}{r^{(0)k-1-\lambda}} \cdot \frac{(1-4s)^{\lambda/k}}{r^{(0)\lambda}}}$$

$$= g(k,n)\, r^{(0)k+1} \eta^{(0)},$$

where

$$g(k,n) = \frac{\left(\frac{1-2s}{1-4s}\right)^{1/k} - 1}{2s} .$$

Since

$$\max_{\substack{k \\ n\ge 3}} g(k,n) = \max_{k} g(k,3) = \begin{cases} g(1,3) & k = 1 \\ g(2,3) & k > 1 \end{cases}$$

and

$$g(1,3) = \frac{3}{2}, \quad g(2,3) = 8\left(\sqrt{\tfrac{7}{6}} - 1\right) < 1,$$

we have

$$\max_{\substack{k \\ n\ge 3}} g(k,n) \le \gamma(k).$$

By the last bound and the inequality

$$\eta^{(0)} = d^{(0)} + \delta^{(0)} < \frac{(\beta + 1) r^{(0)}}{\rho^{(0)k+1}},$$

we obtain

$$r^{(1)} < \frac{(\beta + 1)\gamma r^{(0)k+2}}{\rho^{(0)k+1}} = \frac{\theta r^{(0)k+2}}{\rho^{(0)k+1}} . \tag{4.48}$$

Now, we have

$$r^{(1)} < \frac{\theta r^{(0)}}{\left[\frac{\rho^{(0)}}{r^{(0)}}\right]^{k+1}} < \frac{\beta + 1}{\beta} \cdot \frac{\gamma r^{(0)}}{\beta^k} < \frac{5}{4} \cdot \frac{\gamma r^{(0)}}{[\beta(k,3)]^k} ,$$

so that

$$r^{(1)} < \frac{r^{(0)}}{\omega(k)} , \tag{4.49}$$

where

$$\omega(k) = \begin{cases} 6 & k = 1 \\ \frac{1}{2}(2k)^k & k > 1 \end{cases} .$$

For simplicity, we will write only $\omega$ instead of $\omega(k)$.

According to a geometric construction, disjunctivity of the disks produced by (4.41) and the fact that the disks $Z_i^{(m)}$ and $Z_i^{(m+1)}$ must have at least one point in common (the zero $\xi_i$), the following relations can be derived (see [33],[34]):

$$\rho^{(m+1)} \geq \rho^{(m)} - r^{(m)} - 3r^{(m+1)}, \tag{3.50}$$

$$\rho^{(m+1)} \leq \rho^{(m)} + 3r^{(m)} + r^{(m+1)}. \tag{3.51}$$

Using the inequalities (4.49) and (4.50) (for $m = 0$), we find

$$\rho^{(1)} > \beta r^{(0)} - r^{(0)} - \frac{3r^{(0)}}{\omega} > \omega r^{(1)}(\beta - 1 - \frac{3}{\omega}).$$

Since $\beta > (\omega + 3)/(\omega - 1)$ for every k, it follows

$$\rho^{(1)} > \beta(k,n) r^{(1)}. \tag{4.52}$$

Applying complete induction with the argumentation used for derivation of (4.48), (4.49) and (4.52) (which makes the part of the proof with respect to $m = 1$), we prove that the following relations are true for each $m = 0,1,\ldots$:

$$r^{(m+1)} < \frac{\theta(k,n) r^{(m)^{k+2}}}{\rho^{(m)^{k+1}}}, \tag{4.53}$$

$$r^{(m+1)} < \frac{r^{(m)}}{\omega(k)}, \tag{4.54}$$

$$\rho^{(m)} > \beta(k,n) r^{(m)}. \tag{4.55}$$

In regard to (4.50), (4.51) and (4.53) we get the lower and upper bound for $\rho^{(m)}$

$$\rho^{(0)} - 2r^{(0)} < \rho^{(m)} < \rho^{(0)} + 4r^{(0)}. \tag{4.56}$$

According to this and (4.53) we finally find

$$r^{(m+1)} < \frac{\theta(k,n) r^{(m)^{k+2}}}{(\rho^{(0)} - 2r^{(0)})^{k+1}}.$$

Therefore, the convergence order of the iterative method (4.41) remains $k+2$ in the presence of rounding errors when $\delta^{(m)} \le r^{(m)}/\rho^{(m)^{k+1}}$, that is, until the incorporating rounding error remains of the same power order as the radii of inclusion disks.

It is yet to be shown that the iterative process (4.41) is defined in each iteration under the conditions of Theorem 4.4. It can be proved by showing that $0 \notin H_k(z_i^{(m)}) - Y_i^{(m)}$ $(i = 1,\ldots,n;\ m = 0,1,\ldots)$. In view of (4.55) Lemma 4.6 is applicable for each $m = 0,1,\ldots$ so that

$$|\operatorname{mid} H_k(z_i^{(m)}) - c_i^{(m)}| > d^{(m)} + \delta_i^{(m)},$$

which is equivalent to

$$0 \notin H_k(z_i^{(m)}) - \{c_i^{(m)};\eta_i^{(m)}\} \supseteq H_k(z_i^{(m)}) - Y_i^{(m)}.$$

This completes our convergence analysis. □

The explanation for the condition $\delta_i^{(m)} \le r^{(m)}/\rho^{(m)^{k+1}}$ was earlier given in [34] for $k = 2$. The generalized root iterations in ordinary ("point") complex arithmetic, with the convergence order $k+2$, are given by (4.24). The complex number $c_i^{(m)}$ in that formula and the disk $Y_i^{(m)}$ in (4.41) are a kind of correction terms which accelerate the convergence. To preserve the convergence rate, the total rounding error $\delta_i^{(m)}$ must be somewhat smaller than the largest value of $\operatorname{rad} Y_i^{(m)} < d^{(m)} = \beta r^{(m)}/\rho^{(m)^{k+1}}$. Hence, we require

$$\delta_i^{(m)} \le \delta^{(m)} \le \frac{r^{(m)}}{\rho^{(m)^{k+1}}}. \tag{4.57}$$

Using the upper bound for $\rho^{(m)}$, given by (4.56), we can avoid the evaluation of $\rho^{(m)}$ taking more stringent condition

$$\delta_i^{(m)} \le \frac{r^{(m)}}{(\rho^{(0)} + 4r^{(0)})^{k+1}}$$

instead of (4.57).

CONVERGENCE ANALYSIS: THE CASE $\delta \le \min\{1, 1/\rho^{k+1}\}$

In this case the influence of the rounding error, which has the power order as a constant (i.e. $\delta = O(1)$), is sufficient to decrease the accelerating effect of the correction term $Y_i$ to the convergence rate of

Algorithm (4.41) because $\delta_i$ can exceed rad $Y_i$ (decreasing in further iterations).

First, we will give some bounds, which are analogous to those from Lemma 4.6.

**LEMMA 4.7.** *Assume that (4.42) holds. If*

$$\Delta P_i < |P(z_i)|, \quad \delta_i \le \min\{1, 1/\rho^{k+1}\} \quad (i = 1,\dots,n),$$

*then for each $i = 1,\dots,n$ and $k = 1,2,\dots$ we have*

$$|u_i| > \eta_i, \tag{4.58}$$

$$|u_i| > \frac{1-2s}{r^k}, \tag{4.59}$$

$$\eta_i < \frac{2}{\rho^k} < \frac{2s}{r^k}. \tag{4.60}$$

*Proof.* By distinguishing two cases, $\rho < 1 \Rightarrow \min\{1,1/\rho^{k+1}\} = 1$ and $\rho > 1 \Rightarrow \min\{1,1/\rho^{k+1}\} = 1/\rho^{k+1}$, we see that $1/\rho^k > \min\{1,1/\rho^{k+1}\} (\ge \delta_i)$. Using this bound, the inequality

$$d = \frac{\beta r}{\rho}\cdot\frac{1}{\rho^k} < \frac{1}{\rho^k}$$

and the procedure as at the beginning of the proof of Lemma 4.6, we get

$$|u_i| \ge |h_k(z_i) - c_i| - \delta_i > \frac{1}{r^k} - d - \delta_i > \frac{4}{\rho^k} - d - \min\{1,1/\rho^{k+1}\}$$

$$> \frac{4}{\rho^k} - \frac{1}{\rho^k} - \frac{1}{\rho^k} > \frac{1}{\rho^k} + \frac{r}{\rho^{k+1}} > d + \delta_i = \eta_i.$$

Further, in an analogous manner as in Lemma 4.6 we find

$$|u_i| > \frac{1}{r^k} - d - \delta > \frac{1}{r^k} - \frac{1}{\rho^k} - \frac{1}{\rho^k} > \frac{1}{\rho^k} - \frac{2}{\beta^k r^k} = \frac{1-2s}{r^k}$$

and

$$\eta_i < d + \min\{1,1/\rho^{k+1}\} < \frac{1}{\rho^k} + \frac{1}{\rho^k} = \frac{2}{\rho^k} < \frac{2s}{r^k}. \qquad \square$$

Instead of the constant $\theta(k,n)$ from the previously considered case, we will operate with a new constant $\theta_1(k) = 2\gamma(k)$.

**THEOREM 4.6.** *If $\Delta P(z_i^{(m)}) < |P(z_i^{(m)})|$, $\delta_i^{(m)} \le \min\{1, 1/\rho^{(m)k+1}\}$ $(i = 1,\dots,n)$ and (4.47) hold, then the iterative method (4.41) has the convergence order $k+1$, that is, the sequence $(r^{(m)})$ decreases according to*

$$r^{(m+1)} < \theta_1(k) r^{(m)k+1} / (\rho^{(0)} - \tfrac{7}{3} r^{(0)})^k.$$

*Proof*. According to the estimations from Theorem 4.5 and the bounds (4.59) and (4.60), we get

$$r^{(1)} < g(k,n)\, r^{(0)^{k+1}} \eta^{(0)} < \frac{2\gamma r^{(0)^{k+1}}}{\rho^{(0)^k}}$$

or

$$r^{(1)} < \frac{\theta_1(k)\, r^{(0)^{k+1}}}{\rho^{(0)^k}} . \tag{4.61}$$

As in the previous case, from (4.47) and (4.61) we obtain

$$r^{(1)} < \frac{\theta_1(k)\, r^{(0)}}{\left[\dfrac{\rho^{(0)}}{r^{(0)}}\right]^k} < \frac{2\gamma r^{(0)}}{\beta^k} ;$$

hence

$$r^{(1)} < \frac{r^{(0)}}{\omega_1(k)} , \tag{4.62}$$

where

$$\omega_1(k) = \begin{cases} 4 & k = 1 \\ \frac{1}{2}(2k)^k & k > 1 \end{cases} .$$

By (4.50), (4.62) and the inequality $\beta > (\omega_1(k) + 3)/(\omega_1(k) - 1)$, we find

$$\rho^{(1)} > \beta(k,n)\, r^{(1)} . \tag{4.63}$$

Applying again (4.50) and (4.63) we obtain

$$\rho^{(1)} > \rho^{(0)} - \frac{7}{3} r^{(0)} . \tag{4.64}$$

By virtue of (4.61) and (4.63) we establish

$$r^{(1)} < \frac{\theta_1(k)\, r^{(0)^{k+1}}}{\left(\rho^{(0)} - \frac{7}{3} r^{(0)}\right)^k} .$$

By entirely analogous reasoning as for (4.61), (4.63) and (4.64) and by induction, we prove for each $m = 0,1,\ldots$

$$r^{(m+1)} < \frac{\theta_1(k)\, r^{(m)^{k+1}}}{\rho^{(m)^k}} , \tag{4.65}$$

$$\rho^{(m)} > \beta(k,n)\, r^{(m)} , \tag{4.66}$$

$$\rho^{(m)} > \rho^{(0)} - \frac{7}{3} r^{(0)} . \tag{4.67}$$

Finally, on the basis of (4.65) and (4.67), we obtain

$$r^{(m+1)} < \frac{\theta_1(k)\, r^{(m)^{k+1}}}{(\rho^{(0)} - \frac{7}{3} r^{(0)})^k} .$$

Thus, in the case when the total rounding error in the evaluation of $h_k$ is of the power order as a constant, then the convergence order of the iterative method (4.41) is reduced to $k + 1$.

Since the condition (4.66) is valid for each $m$ ( = 0,1,...), we can apply Lemma 4.7 for each $m$ and hence, according to (4.46) and (4.58),

$$|u_i^{(m)}| > \eta_i^{(m)} \iff 0 \notin \{u_i^{(m)} ; \eta_i^{(m)}\} \Rightarrow 0 \notin H_k(z_i^{(m)}) - Y_i^{(m)} .$$

Thus, the iterative method (4.41) is defined in each iterative step.

The proof of Theorem 4.6 is now complete. □

**REMARK 5.** In particular, the iterative interval methods of the third and fourth order follow from (4.14) for $k = 1$ and $k = 2$. The numerical stability of these methods was considered by Gargantini in [33] and [34]. The analysis presented in this section includes these special cases. ®

## 4.4. MULTIPLE ZEROS AND CLUSTERS OF ZEROS

Consider a monic polynomial $P$ of degree $N \geq 3$

$$P(z) = \prod_{i=1}^{n} (z - \xi_i)^{\mu_i}$$

with $n$ ($\leq N$) distinct real or complex zeros $\xi_1, \ldots, \xi_n$ of the respective multiplicities $\mu_1, \ldots, \mu_n$, where $\mu_1 + \cdots + \mu_n = N$. Assume we have found $n$ nonoverlapping initial disks $Z_1^{(0)}, \ldots, Z_n^{(0)}$ such that $\xi_i \in Z_i^{(0)}$ $(i=1,\ldots,n)$. Let us introduce the following notations:

$$r^{(m)} = \max_{1 \leq j \leq n} r_j^{(m)}, \qquad \mu = \min_{1 \leq j \leq n} \mu_j,$$

$$\beta(k,N,\mu) = \begin{cases} 2N & k = 1 \\ k(N-\mu) & k > 1 \end{cases}, \qquad \theta(k,N,\mu) = \begin{cases} 3N & k = 1 \\ k(N-\mu) & k > 1 \end{cases},$$

$$c_i^{(m)} = \sum_{\substack{j=1 \\ j \neq i}}^{n} \mu_j (z_i^{(m)} - z_j^{(m)})^{-k}, \qquad \eta^{(m)} = \frac{\beta(k,N,\mu)\, r^{(m)}}{\rho^{(m)k+1}} .$$

For simplicity, we will write shortly $\beta$ and $\theta$ instead of $\beta(k,\dot{N},\mu)$ and $\theta(k,N,\mu)$.

From the identity

$$h_k(z) = \frac{(-1)^{k-1}}{(k-1)!}\cdot\frac{d^k}{dz^k}(\log P(z)) = \sum_{j=1}^{n} \mu_j(z-\xi_j)^{-k},$$

we find

$$(z-\xi_i)^k = \frac{\mu_i}{h_k(z) - \sum\limits_{\substack{j=1\\ j\neq i}}^{n} \mu_j(z-\xi_j)^{-k}}$$

or

$$\xi_i \equiv z - \frac{1}{\left\{\frac{1}{\mu_i}\left[h_k(z) - \sum\limits_{\substack{j=1\\ j\neq i}}^{n} \mu_j(z-\xi_j)^{-k}\right]\right\}_*^{1/k}}.$$

The value of the k-th root to be chosen is the one which equals $\frac{1}{z-\xi_i}$.

Let $z = z_i^{(0)}$ be the center of the initial disk $Z_i^{(0)}$. Since $\xi_j \in Z_j^{(0)}$ $(j=1,\ldots,n)$, according to the inclusion isotonicity we obtain from the last (fixed point) relation

$$\xi_i \in z_i^{(0)} - \frac{1}{\left\{\frac{1}{\mu_i}\left[h_k(z_i^{(0)}) - \sum\limits_{\substack{j=1\\ j\neq i}}^{n} \mu_j(z_i^{(0)} - Z_j^{(0)})^{-k}\right]\right\}_*^{1/k}} = Z_i^{(1)},$$

where the symbol $*$ denotes the disk which satisfies CCR, i.e., the disk containing the complex number $1/(z_i^{(0)} - \xi_i)$ (see § 4.1),

Therefore, if the disks $Z_i^{(0)}$ contain the corresponding zeros $\xi_i$ $(i=1,\ldots,n)$, then the circular region $Z_i^{(1)}$ is an inclusion region for the zero $\xi_i$. Hence, we construct the generalized interval method for the simultaneous inclusion of multiple complex zeros of the polynomial P:

$$Z_i^{(m+1)} = z_i^{(m)} - \frac{1}{\left\{\frac{1}{\mu_i}\left[h_k(z_i^{(m)}) - \sum\limits_{\substack{j=1\\ j\neq i}}^{n} \mu_j(z_i^{(m)} - Z_j^{(m)})^{-k}\right]\right\}_*^{1/k}}$$

$$(i=1,\ldots,n;\ m=0,1,\ldots;\ k=1,2,\ldots). \qquad (4.68)$$

The following theorem asserts that the sequence $(r^{(m)})$ converges to zero with the convergence order k + 2:

**THEOREM 4.7.** *Let the interval sequences* $(Z_i^{(m)})$ $(i=1,\dots,n)$ *be produced by (4.68). Then, under the condition*

$$\rho^{(0)} > \beta r^{(0)}, \tag{4.69}$$

*we have for each* $i=1,\dots,n$ *and* $m=0,1,\dots$

$1^o$ $\xi_i \in Z_i^{(m)}$;

$2^o$ $r^{(m+1)} < \dfrac{\theta r^{(m)\,k+2}}{(\rho^{(0)}-3r^{(0)})^{k+1}}$.

The assertion $1^\circ$ follows from the construction of Algorithm (4.68) (see Sections 3.1, 3.2 and 4.1). For the proof of the assertion $2^\circ$ it is sufficient to replace the polynomial degree N by the number of different zeros n in all sums, and N-1 by N-$\mu$ in the proof of Theorem 4.1. Using the inclusion (4.10) we obtain

$$\sum_{\substack{j=1\\ j\neq i}}^{n} \mu_j (z_i^{(m)} - Z_j^{(m)})^{-k} \subset \left\{ \sum_{\substack{j=1\\ j\neq i}}^{n} \mu_j (z_i^{(m)} - z_j^{(m)})^{-k} ; \frac{(N-\mu_i)kr^{(m)}}{\rho^{(m)k+1}} \right\}$$

$$\subseteq \{c_i^{(m)} ; \eta^{(m)}\}. \tag{4.70}$$

Further, it is easy to show that (4.69) implies the inequality

$$|h_k(z_i^{(m)}) - c_i^{(m)}| > \eta^{(m)}$$

for multiple zeros, too. From this and (4.70) we conclude that the interval method (4.68) is defined in each iteration, in other words, the radicand in (4.68) does not include the origin so that formula (2.17) can be applied.

**REMARK 6.** In particular, for k = 1 and k = 2, the interval methods for multiple complex zeros presented in [33] and [35], follow from (4.68).®

The total-step method (4.68) can be accelerated if we use the already calculated inclusion disks in the current iterative step. The single-step interval method for multiple zeros is given by the formula

$$z_i^{(m+1)} = z_i^{(m)} - \frac{1}{\left\{ \frac{1}{\mu_i}\left[ h_k(z_i^{(m)}) - \sum_{j=1}^{i-1} \mu_j (z_i^{(m)} - z_j^{(m+1)})^{-k} - \sum_{j=i+1}^{n} \mu_j (z_i^{(m)} - z_j^{(m)})^{-k} \right] \right\}_*^{1/k}}$$

$$(i = 1,\ldots,n;\ m = 0,1,\ldots;\ k = 1,2,\ldots\ ). \qquad (4.71)$$

Its R-order of convergence is given by Theorem 4.3, where the number of different zeros n is the degree of the polynomial $\sigma^n - \sigma - k - 1$.

**EXAMPLE 4.** The polynomial

$$P(z) = z^{13} + (-11+4i)z^{12} + (46-44i)z^{11} + (-74+204i)z^{10} + (-105-516i)z^9$$
$$+ (787+616i)z^8 + (-1564+392i)z^7 + (724-2344i)z^6 + (2351+2616i)z^5$$
$$+ (-4389+980i)z^4 + (430-5148i)z^3 + (4662+540i)z^2 + (-135+2700i)z - 675$$

was taken to illustrate numerically the interval method (4.68) for $k = 3$ (the convergence order equals five). The factorization of P is

$$P(z) = (z+1)^2 (z-3)^3 (z^2-2z+5)^2 (z+i)^4 .$$

The initial disks were selected to be $Z_i^{(0)} = \{z_i^{(0)}; 0.5\}$, where $|z_i^{(0)} - \xi_i| \cong 0.35$ $(i = 1,\ldots,5)$. The distribution of these disks is such that the quotient $\rho^{(0)}/r^{(0)} = 1.83$ is much less than $k(N-\mu) = 33$. Thus, the choice of initial disks was carried out under a weaker condition than (4.69). After the first iterative step we found that the radii $r_i^{(1)}$ were in the range $(7.25\times10^{-4}, 9.68\times10^{-3})$. The following inclusion disks were obtained (Table 4.4):

| i | $Z_i^{(1)}$ | $\mu_i$ |
|---|---|---|
| 1 | $\{-0.99598 - 1.7\times10^{-3}i ; 9.52\times10^{-3}\}$ | 2 |
| 2 | $\{3.00011 - 3.52\times10^{-5}i ; 7.25\times10^{-4}\}$ | 3 |
| 3 | $\{0.99961 + 1.99977\,i ; 1.06\times10^{-3}\}$ | 2 |
| 4 | $\{0.99813 - 2.00337\,i ; 9.68\times10^{-3}\}$ | 2 |
| 5 | $\{-1.25\times10^{-4} - 1.00194\,i ; 4.57\times10^{-3}\}$ | 4 |

Table 4.4

After the second iterative step, the absolute error, bounded by $r^{(2)}$, was inferior to $9\times10^{-13}$.

INTERVAL NEWTON-LIKE METHOD FOR MULTIPLE ZEROS

The simultaneous method

$$z_i^{(m+1)} = z_i^{(m)} - \frac{\mu_i}{h_1(z_i^{(m)}) - \sum_{\substack{j=1 \\ j\neq i}}^{n} \frac{\mu_j}{z_i^{(m)} - z_j^{(m)}}} \quad (i = 1,\ldots,n), \qquad (4.72)$$

which can be obtained from (4.68) for k = 1, was considered extensively by Gargantini [33]. This method has a great practical importance and will be presented here in detail in the original form as in [33]. Beside Algorithm (4.72), where all zeros are simultaneously refined with cubic convergence (refered as *zero distribution (B)*), we will also study the initial geometry (refered as *zero distribution (A)*) which consists of disjoint disks $Z_i$, each disk containing a zero of multiplicity $\mu_i$, with all sets $Z_i$ located inside the disk $\{a ; R\}$. All other zeros are supposed to lie in the region $Z = \{z: |z - a| > R\}$. This kind of distribution corresponds to the case in which one is interested in further improvement of only a certain group of zeros, clustered around a center $a$. The order of convergence is proved to be at least quadratic.

We first deal with the *zero distribution (A)* where the initial disks are represented by the sets $Z_i^{(0)}$ and $Z_0^{(0)}$ which satisfy

(i) $\xi_i \in Z_i^{(0)} = \{z_i^{(0)} ; r_i^{(0)}\}$ $(i = 1,\ldots,k)$;

(ii) $\xi_i \in \text{int}(Z_0^{(0)})$, with $Z_0^{(0)} = \{z: |z - a| \geq R\}$ $(i = k+1,\ldots,n)$;

(iii) $\xi_i \neq z_i^{(0)}$ $(i = 1,\ldots,k)$.

Here int (Z) denotes the interior of Z.

Without loss of generality, one can set $a = 0$, since a translation can map the new onto the original regions. Let us introduce for i = 1, ..., k, k ≥ 2,

$$C_i^{(0)} = \sum_{\substack{j=1 \\ j\neq i}}^{k} \frac{\mu_j}{z_i^{(0)} - z_j^{(0)}}, \qquad D_i^{(0)} = \sum_{j=k+1}^{n} \frac{\mu_j}{z_i^{(0)} - Z_0^{(0)}}.$$

From the fixed point relation

$$\xi_i = z_i^{(0)} - \frac{\mu_i}{h_1(z_i^{(0)}) - \sum_{\substack{j=1\\ j\neq i}}^{n} \frac{\mu_j}{z_i^{(0)} - \xi_j}} \quad (i = 1,\ldots,k),$$

we obtain by the inclusion isotonicity

$$\xi_i \in Z_i^{(0)} - \frac{\mu_i}{h_1(Z_i^{(0)}) - C_i^{(0)} - D_i^{(0)}} = Z_i^{(1)}, \tag{4.73}$$

where $Z_i^{(1)}$ $(i = 1,\ldots,k)$ is the new inclusion approximation to the zero $\xi_i$.

For $k = 1$ formula (4.73) is still valid if we take $C_1^{(0)} = 0$. Also, the zero distribution we have considered requires $n > k$.

At the m-th iteration we define

$$C_i^{(m)} = \sum_{\substack{j=1\\ j\neq i}}^{k} \frac{\mu_j}{z_i^{(m)} - Z_j^{(m)}}, \qquad D_i^{(m)} = \sum_{j=k+1}^{n} \frac{\mu_j}{z_i^{(m)} - Z_0^{(m)}},$$

$$z_i^{(m)} = \text{mid } Z_i^{(m)}, \qquad Z_0^{(m+1)} = Z_0^{(m)}.$$

On the basis of (4.73) we can establish the following algorithm for improving k inclusion circular approximations to the polynomial zeros:

$$Z_i^{(m+1)} = z_i^{(m)} - \frac{\mu_i}{h_1(z_i^{(m)}) - C_i^{(m)} - D_i^{(m)}} \quad (i = 1,\ldots,k;\ m = 0,1,\ldots), \tag{4.74}$$

assuming that $P(z_i^{(m)}) \neq 0$.

Introduce the following notations:

$$\mu = \min_{1 \le i \le k} \mu_i, \qquad \hat{\mu} = \max_{1 \le i \le n} \mu_i,$$

$$\gamma = \frac{\hat{\mu}N}{\mu}, \qquad r^{(m)} = \max_{1 \le i \le k} r_i^{(m)}, \qquad N_1 = \sum_{j=1}^{k} \mu_j,$$

$$\hat{\rho}^{(m)} = \min_{i \neq j} \{|z| : z \in Z_i^{(m)} - Z_j^{(m)}\} \quad (i = 1,\ldots,k;\ j = 0,1,\ldots,k).$$

The order of convergence and the conditions for convergence of the iterative method (4.74) are discussed in the following

**THEOREM 4.8.** *For $\rho^{(0)}$ such that*

$$8\gamma r^{(0)} < \rho^{(0)} \le \hat{\rho}^{(0)}, \tag{4.75}$$

*the sequence $(r^{(m)})$ converges to zero as $m \to \infty$, with*

$$r^{(m+1)} < \frac{4\gamma r^{(m)2}}{\rho^{(0)}}. \tag{4.76}$$

*Proof*. The disk $z_i^{(0)} - Z_j^{(0)}$ $(j \neq 0)$ is the set of points $\zeta$ such that

$$|\zeta - (z_i^{(0)} - z_j^{(0)})| \le r_j^{(0)}, \qquad |\zeta| \ge \rho^{(0)}.$$

Therefore, the inverse of $z_i^{(0)} - Z_j^{(0)}$ is contained in the disk centered at the origin and with radius $1/\rho^{(0)}$. Hence

$$C_i^{(0)} = \sum_{\substack{j=1\\ j\neq i}}^{k} \frac{\mu_j}{z_i^{(0)} - Z_j^{(0)}} \subset \left\{0\ ;\ \frac{N_1}{\rho^{(0)}}\right\}. \tag{4.77}$$

In regard to (2.10) we get

$$(z_i^{(0)} - Z_0^{(0)})^{-1} = \left\{z:\ |z - z_i^{(0)}| \ge R\right\}^{-1} = \left\{-\frac{\bar{z}_i^{(0)}}{R^2 - |z_i^{(0)}|^2}\ ;\ \frac{R}{R^2 - |z_i^{(0)}|^2}\right\}.$$

Because of the initial distribution, we have

$$\frac{R}{R^2 - |z_i^{(0)}|^2} \le \frac{1}{\rho^{(0)}}, \qquad \left|-\frac{\bar{z}_i^{(0)}}{R^2 - |z_i^{(0)}|^2}\right| < \frac{1}{\rho^{(0)}},$$

so that $(z_i^{(0)} - Z_0^{(0)})^{-1} \subset \left\{0\ ;\ \frac{2}{\rho^{(0)}}\right\}$. Therefore

$$D_i^{(0)} \subset \{0\ ;\ 2(N - N_1)/\rho^{(0)}\}$$

and, by (4.77)

$$C_i^{(0)} + D_i^{(0)} \subset \{0\ ;\ 2N/\rho^{(0)}\}. \tag{4.78}$$

We will now prove that $h_1(z_i^{(0)}) - C_i^{(0)} - D_i^{(0)}$ is a disk that does not contain the origin. Using (4.75) we find a bound for $|h_1(z_i^{(0)})|$:

$$|h_1(z_i^{(0)})| \ge \left|\left|\frac{\mu_i}{z_i^{(0)} - \xi_i}\right| - \left|\sum_{\substack{j=1\\ j\neq i}}^{n} \frac{\mu_j}{z_i^{(0)} - \xi_j}\right|\right| \ge \frac{\mu}{r^{(0)}} - \frac{\mu\gamma}{\rho^{(0)}}.$$

In accordance with (4.78), the disk $h_1(z_i^{(0)}) - C_i^{(0)} - D_i^{(0)}$ is enclosed by the disk with center $h_1(z_i^{(0)})$ and radius $2N/\rho^{(0)}$, that is,

$$h_1(z_i^{(0)}) - C_i^{(0)} - D_i^{(0)} = \left\{ h_1(z_i^{(0)}) - \operatorname{mid}(C_i^{(0)} + D_i^{(0)}) \; ; \; \operatorname{rad}(C_i^{(0)} + D_i^{(0)}) \right\}$$

$$\subset \left\{ h_1(z_i^{(0)}) \; ; \; 2N/\rho^{(0)} \right\} .$$

Using (4.79), we obtain

$$|h_1(z_i^{(0)})| - \frac{2N}{\rho^{(0)}} \geq \frac{\mu}{r^{(0)}} - \frac{3\mu\gamma}{\rho^{(0)}} > 0 \tag{4.80}$$

and

$$|h_1(z_i^{(0)}) - \operatorname{mid}(C_i^{(0)} + D_i^{(0)})| > \frac{\mu(\rho^{(0)} - 3\gamma r^{(0)})}{r^{(0)}\rho^{(0)}} . \tag{4.81}$$

By virtue of (4.80) it follows that $0 \notin h_1(z_i^{(0)}) - C_i^{(0)} - D_i^{(0)}$.

To obtain an upper bound for $r^{(1)}$, we derive a convenient upper bound for the radius of $C_i^{(0)} + D_i^{(0)}$. We have

$$\operatorname{rad}(C_i^{(0)} + D_i^{(0)}) = \operatorname{rad} \sum_{\substack{j=1 \\ j\neq i}}^{k} \frac{\mu_j}{Z_i^{(0)} - Z_j^{(0)}} + \operatorname{rad} \frac{N - N_1}{Z_i^{(0)} - Z_0^{(0)}}$$

$$\leq \sum_{\substack{j=1 \\ j\neq i}}^{k} \frac{\mu_j r_j^{(0)}}{\rho^{(0)2}} + \frac{N - N_1}{\rho^{(0)}} < \frac{N}{\rho^{(0)}} . \tag{4.82}$$

From (4.74) there follows

$$r_i^{(1)} = \operatorname{rad}\left( \frac{\mu_i}{h_1(z_i^{(0)}) - C_i^{(0)} - D_i^{(0)}} \right)$$

$$= \frac{\mu_i \operatorname{rad}(C_i^{(0)} + D_i^{(0)})}{|h_1(z_i^{(0)}) - \operatorname{mid}(C_i^{(0)} + D_i^{(0)})|^2 - \left[\operatorname{rad}(C_i^{(0)} + D_i^{(0)})\right]^2}$$

Using (4.75), (4.81) and (4.82), we get

$$r^{(1)} < \frac{4\gamma r^{(0)2}}{\rho^{(0)}} \tag{4.83a}$$

and

$$r^{(1)} < \frac{1}{2} r^{(0)} < \frac{1}{16\gamma} \rho^{(0)} . \tag{4.83b}$$

The intersection $Z_i^{(1)} \cap Z_i^{(0)}$ is not empty (because it contains at least one common point - the zero $\xi_i$), which implies $|z_i^{(1)} - z_i^{(0)}| \le r^{(0)} + r^{(1)}$ (see [33]). The disks $Z_1^{(1)},\ldots,Z_k^{(1)}$ do not intersect if

$$d = \max_{1\le i\le k} (\rho^{(0)} - |z_i^{(0)} - z_i^{(1)}| - 3r^{(1)}) > 0. \tag{4.84}$$

Since $N \ge 3$ we have $\gamma > 1$. According to this and (4.83b) we get

$$|z_i^{(0)} - z_i^{(1)}| + 3r^{(1)} \le r^{(0)} + 4r^{(1)} < \frac{3}{8\gamma}\rho^{(0)},$$

which means that (4.84) holds. Also, $Z_i^{(1)} \cap Z_0^{(1)} = \emptyset$ because of

$$\max_{1\le i\le k} (\rho^{(0)} - |z_i^{(0)} - z_i^{(1)}| - r^{(1)}) > d.$$

Thus, we can choose $\rho^{(1)}$ to satisfy

$$\rho^{(0)} - r^{(0)} - 3r^{(1)} \le \rho^{(1)} \le \hat{\rho}^{(1)}. \tag{4.85}$$

By the first estimate in (4.83b) and by (4.85), we get

$$8\gamma r^{(1)} < \rho^{(0)} - r^{(0)} - 3r^{(1)} < \rho^{(1)} \le \hat{\rho}^{(1)}. \tag{4.86}$$

Using (4.85) and the first estimate in (4.83b), we find

$$\rho^{(1)} > \rho^{(0)} - 6\gamma(r^{(0)} - r^{(1)}). \tag{4.87}$$

The proofs of relations (4.78), (4.80), (4.83), (4.86) and (4.87) can be considered the first part of a proof by induction with $m = 1$. To complete it, we suppose that

$$r^{(m)} < \frac{4\gamma r^{(m-1)^2}}{\rho^{(0)}}, \tag{4.88}$$

$$8\gamma r^{(m)} < \rho^{(m-1)} - r^{(m-1)} - 3r^{(m)} \le \rho^{(m)} \le \hat{\rho}^{(m)} \tag{4.89}$$

and

$$\rho^{(m)} > \rho^{(m-1)} - 6\gamma(r^{(m-1)} - r^{(m)}) \tag{4.90}$$

are true for any m and prove (4.88) - (4.90) for index $m+1$.

From (4.89) (for $j \ne 0$ and $j = 0$) it follows

$$\rho^{(m)} \le \min_{i\ne j} \{|z_i^{(m)} - z_j^{(m)}| - r_j^{(m)}\}$$

and

$$\rho^{(m)} \le R - \max_{1\le i\le k} |z_i^{(m)}|,$$

so that

$$C_i^{(m)} + D_i^{(m)} \subset \{0 \; ; \; 2N/\rho^{(0)}\}.$$

The required bounds are derived essentially in the same way as for $m = 0$. As before, the zero distribution is as follows: $n-k$ zeros are located in the interior of $Z_0^{(m)}$; $k$ zeros are contained in the union of $Z_i^{(m)}$ ($i = 1,\ldots,k$), and no zeros lie in the annular regions $r^{(m)} < |\zeta - z_i^{(m)}| \le \rho^{(m)}$.

Similarly, by the above reasoning as for $m = 0$ we obtain

$$|h_1(z_i^{(m)}) - \mathrm{mid}\,(C_i^{(m)} + D_i^{(m)})| > \frac{\mu(\rho^{(m)} - 3\gamma r^{(m)})}{r^{(m)}\rho^{(m)}},$$

$$\mathrm{rad}\,(C_i^{(m)} + D_i^{(m)}) < \frac{N}{\rho^{(0)}}$$

and

$$r^{(m+1)} < \frac{\gamma r^{(m)2}}{\rho^{(m)} - 6\gamma r^{(m)}}. \tag{4.91}$$

Furthermore, from (4.88), (4.89) and (4.83a) we have

$$r^{(\lambda)} < \frac{1}{2} r^{(\lambda-1)} \qquad (\lambda = 1,\ldots,m),$$

and from (4.91),

$$r^{(m+1)} < \frac{1}{2} r^{(m)}. \tag{4.92}$$

Hence, $r^{(m)} \to 0$ when $m \to \infty$.

By (4.89) and (4.87) we obtain

$$\rho^{(m)} - 6\gamma r^{(m)} > \rho^{(m-1)} - 6\gamma r^{(m-1)} > \cdots$$
$$> \rho^{(1)} - 6\gamma r^{(1)} > \rho^{(0)} - 6\gamma r^{(0)},$$

so that (4.91) becomes

$$r^{(m+1)} < \frac{\gamma r^{(m)2}}{\rho^{(0)} - 6\gamma r^{(0)}} < \frac{4\gamma r^{(m)2}}{\rho^{(0)}},$$

proving that the sequence $(r^{(m)})$ converges to zero at least quadratically.

Applying (4.89) and (4.92), we obtain

$$r^{(m+1)} < \frac{\rho^{(m)}}{16\gamma}.$$

Taking into account this inequality and (4.92), we prove that all intersections $Z_i^{(m+1)} \cap Z_j^{(m+1)}$ $(i = 1,\dots,k;\ j = 0,1,\dots,k)$ are empty, which implies the existence of a $\rho^{(m+1)}$ such that

$$8\gamma r^{(m+1)} < \rho^{(m)} - r^{(m)} - 3r^{(m+1)} \leq \rho^{(m+1)} \leq \hat{\rho}^{(m+1)} .$$

The inequality (4.92) and the last relations give

$$\rho^{(m+1)} > \rho^{(m)} - 6\gamma(r^{(m)} - r^{(m-1)}),$$

which completes the proof of Theorem 4.8. □

**REMARK 7.** The similarity of formula (4.74) with Schröder's formula [113] (modified Newton's method for multiple zeros)

$$z_i^{(m+1)} = z_i^{(m)} - \mu_i \frac{P(z_i^{(m)})}{P'(z_i^{(m)})}$$

becomes quite evident if we write $Z_i^{(m+1)}$ as

$$Z_i^{(m+1)} = z_i^{(m)} - \mu_i \frac{P(z_i^{(m)})}{P'(z_i^{(m)})} \Bigg/ \left(1 - \frac{P(z_i^{(m)})}{P'(z_i^{(m)})}(C_i^{(m)} + D_i^{(m)})\right). \text{ ®}$$

**REMARK 8.** Algorithm (4.74) requires $P(z_i^{(m)}) \neq 0$. If $P(z_i^{(m)}) = 0$ at the m-th iteration, then we take $Z_i^{(m+1)} = \{z_i^{(m)};0\}$, and do not work further for index i. ®

Let us consider now the *zero distribution (B)* where the initial regions are represented by n disks $Z_i^{(0)}$ with center $z_i^{(0)}$ and radius $r_i^{(0)}$, satisfying $z_i^{(0)} \neq \xi_i \;\; Z_i^{(0)}$. The successive approximations to $\xi_i$ are given by (4.72). Algorithm (4.72) provides, at each iteration, a new disk distribution with the properties: (*i*) only the zero of multiplicity $\mu_i$ is contained in $Z_i^{(m)}$, and (*ii*) no zeros lie in the annular regions defined by $r^{(m)} < |\zeta - z_i^{(m)}| \leq \rho^{(m)}$ $(i = 1,\dots,n)$.

Let

$$C_i^{(m)} = \sum_{\substack{j=1 \\ j\neq i}}^{n} \frac{\mu_j}{z_i^{(m)} - z_j^{(m)}}, \quad \mu = \min_{1\leq i\leq n} \mu_i, \quad \hat{\mu} = \max_{1\leq i\leq n} \mu_i, \quad \gamma = \frac{\hat{\mu}(N-\mu)}{\mu},$$

$$r^{(m)} = \max_{1\leq i\leq n} r_i^{(m)}, \quad \hat{\rho}^{(m)} = \min_{i\neq j} \{|z| : z \in Z_i^{(m)} - Z_j^{(m)}\} \quad (i,j = 1,\dots,n).$$

Theorem 4.9 gives the conditions which guarantee that $z_i^{(m)} \to \xi_i$ as $m \to \infty$.

**THEOREM 4.9.** *For $N > \mu$ and for any $\rho^{(0)}$ such that*

$$6\gamma r^{(0)} < \rho^{(0)} \le \hat{\rho}^{(0)},$$

*the sequence $(r^{(m)})$ converges to zero, with*

$$r^{(m+1)} < \frac{3\gamma r^{(m)3}}{\mu\rho^{(0)2}} .$$

*Proof.* We will give only a brief outline of the proof of this theorem. As in the proof of Theorem 4.8 we prove the inclusion

$$C_i^{(0)} \subset \{0 \,;\, (N-\mu)/\rho^{(0)}\}$$

and derive the bounds

$$|h_1(z_i^{(0)}) - \text{mid}\, C_i^{(0)}| \ge \frac{\mu(\rho^{(0)} - 2\gamma r^{(0)})}{r^{(0)}\rho^{(0)}} , \tag{4.93}$$

$$\text{rad}\, C_i^{(0)} \le \frac{(N-\mu)r^{(0)}}{\rho^{(0)2}} . \tag{4.94}$$

Using the inequalities (4.93) and (4.94), from

$$r_i^{(1)} = \mu_i \frac{\text{rad}\, C_i^{(0)}}{|h_1(z_i^{(0)}) - \text{mid}\, C_i^{(0)}|^2 - (\text{rad}\, C_i^{(0)})^2}$$

we find

$$r^{(1)} < \frac{\gamma}{\mu} \cdot \frac{r^{(0)3}}{\rho^{(0)}(\rho^{(0)} - 4\gamma r^{(0)})} < \frac{3\gamma}{\mu} \cdot \frac{r^{(0)3}}{\rho^{(0)2}} < \frac{r^{(0)}}{12} . \tag{4.95}$$

Further, it can be proved that

$$d = \max_{1 \le i \le n} (\rho^{(0)} - |z_i^{(1)} - z_i^{(0)}| - 3r^{(1)}) > 0.$$

According to this we find any $\rho^{(1)}$ such that

$$6\gamma r^{(1)} < \rho^{(0)} - r^{(0)} - 3r^{(1)} \le \rho^{(1)} \le \hat{\rho}^{(1)} \tag{4.96a}$$

and

$$\rho^{(1)} > \rho^{(0)} - 4\gamma(r^{(0)} - r^{(1)}). \tag{4.96b}$$

As in the previous case, we assume that (4.95), (4.96a) and (4.96b) are true for any m and we prove that these relations are also true for m + 1. From the analogy to (4.95), that is

$$r^{(m+1)} < \frac{3\gamma}{\mu} \cdot \frac{r^{(m)3}}{\rho^{(0)2}} < \frac{r^{(m)}}{12},$$

we conclude that the sequence $(r^{(m)})$ converges to zero at least cubically. □

**EXAMPLE 5.** ( [33]) The polynomial

$$P(z) = z^9 + (-2+3i)z^8 + (48-6i)z^7 + (-94+152i)z^6 + (522-298i)z^5 + (-950+1974i)z^4$$
$$+ (-1400-3650i)z^3 + (3750+1200i)z^2 + (-1875+1250i)z - 625i$$

was chosen to illustrate numerically the algorithms (4.72) and (4.74). The factorization of $P(z)$ is

$$P(z) = (z-1)^2(z+i)^3(z-5i)^2(z+5i)^2.$$

For Algorithm (4.74) the initial approximations were selected to be

$$Z_1^{(0)} = \{1.2 + 0.1i\ ;\ 0.3\}, \qquad Z_2^{(0)} = \{0.2 - 0.9i\ ;\ 0.3\},$$
$$Z_0^{(0)} = \{z\colon |z| \geq 4\}.$$

As results we obtained

$$Z_1^{(1)} = \{0.993 + 7.8\times 10^{-3}i\ ;\ 3.9\times 10^{-2}\},$$
$$Z_2^{(1)} = \{-5.6\times 10^{-3} - 0.998i\ ;\ 2.3\times 10^{-2}\},$$

$$Z_1^{(2)} = \{0.999998 - 2.4\times 10^{-5}i\ ;\ 6.4\times 10^{-5}\},$$
$$Z_2^{(2)} = \{7.4\times 10^{-7} - 0.999999i\ ;\ 1.3\times 10^{-5}\},$$

For Algorithm (4.72) the initial approximations $Z_1^{(0)}$ and $Z_2^{(0)}$ were taken to be the same as before, while $Z_3^{(0)}$ and $Z_4^{(0)}$ were given by

$$Z_3^{(0)} = \{0.1 - 5.1i\ ;\ 0.3\}, \qquad Z_4^{(0)} = \{-0.1 + 5.2i\ ;\ 0.3\}.$$

After the first iterative step we found that the radii $r_i^{(1)}$ were in the range $[8\times 10^{-4}, 10^{-2}]$. For the second iteration we obtained the following disks:

$$Z_1^{(2)} = \{0.99999983 - 2.7\times 10^{-7}i\ ;\ 5.7\times 10^{-7}\},$$
$$Z_2^{(2)} = \{-3.4\times 10^{-8} - 1.0000000i\ ;\ 5.9\times 10^{-8}\},$$
$$Z_3^{(2)} = \{7.6\times 10^{-11} - 5.0000000i\ ;\ 3.\times 10^{-10}\},$$
$$Z_4^{(2)} = \{1.2\times 10^{-10} + 5.0000000i\ ;\ 6.4\times 10^{-10}\}.$$

## CLUSTERS OF ZEROS

We have mainly considered interval algorithms that require relatively well-separated initial disks. The problem concerning one isolated *cluster* (i.e., the set of several zeros that either coincide or are very close) was treated in [33] (see this section) and [93] (§5.3). Nevertheless, the problem of multiple zeros and clusters of zeros remains a very difficult one and it requires a robust software, sometimes including even an interactive approach (see [25]). It is well known that most algorithms break down as soon as the distance between any zeros becomes comparable in size to errors of the corresponding approximations or if any zero is multiple. Most frequently, a distinction between multiple zero and zero cluster by numerical computations is very cumbersome or even impossible. The question of distinguishing a zero cluster from a multiple zero has been tackled theoretically by Ostrowski [80] and Hoffmann [53]. As an illustration of the above problem we give the following example (based on an example by Wavrik [133]).

The zeros of the polynomial

$$\begin{aligned} P(z) &= 70z^5 - (350+99i)z^4 + (980+396i)z^3 - (1540+990i)z^2 \\ &\quad +(1470+1188i)z - (630+891i) \\ &= 70(z^2+2z+3)^2\left(z-1-\frac{99}{70}i\right) \end{aligned} \tag{4.97}$$

are

$\xi_1 = 1 - i\sqrt{2} = 1 - i\,1.41421356\ldots$ of the multiplicity 2,

$\xi_2 = 1 + i\sqrt{2} = 1 + i\,1.41421356\ldots$ of the multiplicity 2,

$\xi_3 = 1 + i\frac{99}{70} = 1 + i\,1.41428571\ldots$ of the multiplicity 1.

Applying a standard numerical method to the equations $P(z) = 0$ and $P'(z) = 0$ in arithmetic with five decimal digits, we have found the following approximations to the zeros:

$z_1 = 1.00002 - 1.41421\,i,$

$z_2 = 1.00003 + 1.41425\,i,$

$z_3 = 0.99997 + 1.41423\,i.$

Obviously, it is quite difficult to determine from the above results whether the polynomial P has a triple zero near 1. + 1.4142 i, three closely spaced simple zeros, or (correctly) a double zero and a nearby simple zero.

Any method for finding numerical values of the zeros of a polynomial P depends upon information provided by computed values of P(z). Since the magnitude of P(z) decreases rapidly near a zero of high multiplicity, the band in which the error is significant will be especially large near a multiple zero. The machine error may radically change the situation and must be appropriately taken into account. For example, the equation $Az^n = 0$ may be represented in a computing machine as $(A + \varepsilon_1)z^n + \varepsilon_2 = 0$; thus, the zero of multiplicity n "explodes" into a cluster of n very close zeros."The problem does not stem from the particular zero-finding method used, but rather from the inability of a machine using fixed precision floating-point arithmetic to compute accurately the value of P(z)" (Wavrik [133]). To avoid the effects of rounding errors, numerical examples should be done in multiple-precision arithmetic. M. R. Farmer and G. Loizou, whose contribution in theoretical research of zero-finding algorithms and their practical implementation on digital machines is remarkable, have concluded that "there is still a lot of work to be done ... on the problem involved when clusters are present" ([25]). The aim of the following part of this section is to present and study a class zero-finding methods which meet some of the above requirements.

We begin our discussion of zero clusters with the problem of close zeros clustered around a center $z_0$, which was presented in [37] (see also [51, § 6.6]). Although the authors have not solved this problem completely, their results have permitted the construction of an algorithm for isolating zeros in cases where the Weyl algorithm merely furnishes clusters of zeros.

Let m and n be integers, $1 < m < n$, where n is the degree of a polynomial P whose zeros are $\xi_1,\dots,\xi_n$. Let there be given $n - m$ circular regions $Z_{m+1},\dots,Z_n$ such that

$$\xi_j \in Z_j \quad (j = m+1,\dots,n) \tag{4.98}$$

and let $z_0$ be a complex number satisfying $z_0 \notin Z_j$ and $0 \notin (z_0 - Z_j)^k$ ($j = m+1,\ldots,n$; $k = 1,\ldots,m$). The values of the functions

$$q_k(z) := \frac{1}{(k-1)!}\frac{d^k}{dz^k}(\log P(z)) = \frac{1}{(k-1)!}\left(\frac{P'(z)}{P(z)}\right)^{(k-1)} \qquad (4.99)$$

$$(k = 1,\ldots,m)$$

at the point $z = z_0$ will be denoted by $q_k(z_0) = c_k$ ($k = 1,\ldots,m$). In order to localize the zeros $\xi_1,\ldots,\xi_m$, Gargantini and Henrici [37] have constructed the circular disks $B_1,\ldots,B_m$ so that these zeros are of the form

$$\xi_k = z_0 + w_k \quad (k = 1,\ldots,m), \qquad (4.100)$$

where the numbers $w_1,\ldots,w_m$ are the zeros of a polynomial

$$g(w) = b_m w^m + \cdots + b_2 w^2 + b_1 w + 1 \qquad (4.101)$$

such that

$$b_k \in B_k \quad (k = 1,\ldots,m). \qquad (4.102)$$

The construction of the disks $B_k$ is such that their radii tend to zero if the radii of all disks $Z_{m+1},\ldots,Z_n$ tend to zero. However, apart from the exception of the case $m = 1$, it is not claimed that *every* polynomial $g$ of the form (4.101) has zeros $w_k$ such that the resulting $\xi_k$ belongs to a polynomial $P$.

The construction is based on the following corollary of Newton's identities, easily proved by means of generating functions:

**LEMMA 4.8.** *Let $z_0, \xi_1,\ldots,\xi_m$ be complex numbers, $z_0 \neq \xi_k$ ($k = 1,\ldots,m$), and let*

$$s_k = -\sum_{j=1}^{m} \frac{1}{(\xi_j - z_0)^k} \quad (k = 1,\ldots,m). \qquad (4.103)$$

*If the numbers $b_1,\ldots,b_m$ are defined recursively by*

$$b_k = \frac{1}{k}(s_k + b_1 s_{k-1} + \cdots + b_{k-1}s_1) \quad (k = 1,\ldots,m), \qquad (4.104)$$

*then $\xi_j = z_0 + w_j$ ($j = 1,\ldots,m$), where $w_1,\ldots,w_m$ are the zeros of the polynomial (4.101).*

Since

$$q_1(z) = \frac{P'(z)}{P(z)} = -\sum_{j=1}^{n} \frac{1}{\xi_j - z},$$

there follows upon differentiation

$$q_k(z) = -\sum_{j=1}^{n} \frac{1}{(\xi_j - z)^k} \quad (k = 1,2,\ldots).$$

Hence, in view of (4.99), the numbers $s_k$ defined by (4.103) satisfy

$$s_k = c_k + \sum_{j=m+1}^{n} \frac{1}{(\xi_j - z_0)^k},$$

so that, by virtue of (4.98),

$$s_k \in S_k \quad (k = 1,\ldots,m),$$

where

$$S_k = c_k + \sum_{j=m+1}^{n} \frac{1}{(Z_j - z_0)^k}\,.^{*)} \tag{4.105}$$

The power of $(Z_j - z_0)^{-1}$ may be computed by (2.12). From (4.104) it follows that the coefficients of the polynomial (4.101) (with the zeros $w_j := \xi_j - z_0$) satisfy (4.102), where the sets $B_k$ are defined by

$$B_k = \frac{1}{k}(S_k + S_{k-1}B_1 + \cdots + S_1B_{k-1}), \quad B_0 = 1 \quad (k = 1,\ldots,m). \tag{4.106}$$

According to the above, the following theorem was stated in [37]:

**THEOREM 4.10.** *Let $z_0, c_1, \ldots, c_m$ be complex numbers, and let $Z_{m+1},\ldots,Z_n$ be disks such that $z_0 \notin Z_k$ $(k = m+1,\ldots,n)$. If the conditions (4.98) and (4.99) are valid, then the zeros $\xi_1,\ldots,\xi_m$ of a polynomial $P$ are of the form $z_0 + w_k$ $(k = 1,\ldots,m)$, where $w_k$ are the zeros of a polynomial (4.101) whose coefficients $b_k$ satisfy (4.102), the sets $B_k$ being defined by (4.105) and (4.106).*

Let us consider the case $m = 2$. This requires the value $c_2$ of

$$q_2(z) = \frac{P(z)P''(z) - P'(z)^2}{P(z)^2}$$

at $z_0$ in addition to $c_1 = P'(z_0)/P(z_0)$. We have

---

*) The disks $1/(Z_j - z_0)^k$ were originally used in the paper [37]. But, the circular extensions $(1/(Z_j - z_0))^k$ would provide a better result compared to $1/(Z_j - z_0)^k$ appearing in (4.105) (see Remark 1 in this section or (2.20')). Besides, the use of the latter circular term requires much stronger condition $0 \notin (z_0 - Z_j)^k$ in relation to $0 \notin z_0 - Z_j$, which is sufficient for the disk $(1/(Z_j - z_0))^k$.

$$S_1 = c_1 + D, \quad S_2 = c_2 + E,$$

where

$$D = \sum_{j=3}^{n} \frac{1}{Z_j - z_0}, \qquad E = \sum_{j=3}^{n} \frac{1}{(Z_j - z_0)^2}.$$

Since $B_1 = S_1$, $B_2 = \frac{1}{2}(S_1^2 + S_2)$, the zeros $w_1$ and $w_2$ of $g(w) = b_2 w^2 + b_1 w + 1$ are contained in the sets

$$W_{1,2} = \frac{2}{-S_1 \pm \sqrt{-S_1^2 - 2S_2}} = \frac{2}{-c_1 - D \pm \sqrt{-(c_1 + D)^2 - 2c_2 - 2E}},$$

where the square root is computed according to (2.17). The zeros $\xi_1$ and $\xi_2$ are then contained in $Z_{1,2} := z_0 \pm W_{1,2}$.

The above formula can be used to separate or refine two close zeros contained in the same disk produced by the Weyl algorithm or some other method. The following example is taken from [51, p. 484]:

**EXAMPLE 6.** For the polynomial

$$P(z) = (z+4.1)(z+3.8)(z+2.05)(z+1.85)(z-1.95)(z-2.15)(z-3.9)(z-4.05),$$

working with $z_0 = -4$ and the disks

$$Z_3 = Z_4 = \{-2\,;\,0.3\}, \quad Z_5 = Z_6 = \{2\,;\,0.3\}, \quad Z_7 = Z_8 = \{4\,;\,0.3\},$$

the formula presented above yields

$$Z_1 = \{-4.0999\,;\,0.0013\}, \qquad Z_2 = \{-3.7985\,;\,0.0052\}.$$

With the data $z_0 = -2$ the double inclusion disks $Z_3 = Z_4$ are separated into the disks

$$Z_3 = \{-2.0500\,;\,0.00011\}, \qquad Z_4 = \{-1.8499\,;\,0.00096\};$$

$z_0 = 2$ yields

$$Z_5 = \{1.9500\,;\,0.0003\}, \qquad Z_6 = \{2.1502\,;\,0.0027\}$$

and with $z_0 = 4$ we finally get

$$Z_7 = \{3.9000\,;\,7\times 10^{-6}\}, \qquad Z_8 = \{4.0500\,;\,2\times 10^{-6}\}.$$

The presented formulas are simplified in a special situation where it is known that a disk $\{z_0\,;\,r\}$ contains at most m zeros of the given polynomial. We then may apply Theorem 4.10 where $Z_k = \{z : |z - z_0| \geq r\}$ (k =

$m+1,\dots,n$). By the rules of circular arithmetic we find

$$\sum_{j=m+1}^{n} \frac{1}{(Z_j - z_0)^k} = \{0 \,;\, (n-m)/r^k\}$$

so that the disks $S_k$ have the particularly simple form

$$S_k = \{c_k \,;\, (n-m)/r^k\} \quad (k = 1,2,\dots).$$

For $m = 2$ the zero sets $W_1$ and $W_2$ are given by

$$W_{1,2} = \frac{2}{\left\{-c_1 \,;\, \frac{n-2}{r}\right\} \pm \left\{-2c_2 - c_1^2 \,;\, \frac{n-2}{r^2}(n + 2r|c_1|)\right\}^{1/2}},$$

where $c_k = q_k(z_0)$.

The proposed procedure is based on the circular extension of the explicit formula for the zeros of a polynomial $g(w) = b_m w^m + \cdots + b_1 w + 1$. But, as it is well known, such formulas exist only for the polynomial whose degree is less than five, while the corresponding explicit formulas for polynomials of degree three and four are very complicated and practically useless. Therefore, the presented algorithm for isolating close zeros is applicable only for $m \le 2$, which is the main deficiency of this algorithm.

Using the results presented in [37] Glatz continued the study of the previous problem. His algorithm, proposed in [40], was constructed with special regard to multiple zeros and zero clusters.

Suppose again that we have found the circular regions $Z_{k+1},\dots,Z_n$ $(1 < k < n)$ such that $\xi_j \in Z_j$ and $z_0 \notin Z_j$ $(j = k+1,\dots,n)$, where $z_0$ is an arbitrary complex number. Glatz [40] proved the following assertion:

**THEOREM 4.11.** *The circular region*

$$Z = z_0 - \frac{\{1 \,;\, k-1\}}{\dfrac{P'(z_0)}{P(z_0)} - \displaystyle\sum_{j=k+1}^{n} (z_0 - Z_j)^{-1}} \qquad (4.107)$$

*contains at least one of the zeros* $\xi_1,\dots,\xi_k$.

If sufficiently good initial approximations $z_1^{(0)},\dots,z_k^{(0)}$ to the zeros $\xi_1,\dots,\xi_k$ are known, then from (4.107) it follows the iterative method for the inclusion of the remaining zeros $\xi_1,\ldots,\xi_k$,

$$Z_i^{(m+1)} = z_i^{(m)} - \frac{\{1\,;\,k-1\}}{\dfrac{P'(z_i^{(m)})}{P(z_i^{(m)})} - \sum_{j=k+1}^{n} \dfrac{1}{z_i^{(m)} - Z_j}} \tag{4.108}$$

$$(z_i^{(m)} = \text{mid}\; Z_i^{(m)};\; i = 1,\ldots,k;\; m = 0,1,\ldots).$$

The main disadvantages of Algorithm (4.108) are: 1) the produced circular regions $Z_1^{(m)},\ldots,Z_k^{(m)}$ can be overlapping; 2) there is no information about the number of zeros in each of these disks; 3) the convergence is only linear (as it will be shown later).

As in the case of the problem considered by Gargantini and Henrici [37], we will consider again a special distribution of initial inclusion regions, where it is known that a disk $Z_k^{(0)} = \{z^{(0)}\,;\,r\}$ includes k zeros $\xi_1,\ldots,\xi_k$ and the open circular region $Z = \{z: |z - z^{(0)}| \geq r\}$ contains the remaining zeros $\xi_{k+1},\ldots,\xi_n$. This case is of interest when the zeros $\xi_1$, $\ldots,\xi_k$ are very close and clustered around the point $z_0$. Using (2.10) we simplify (4.107) as follows

$$Z_k = z^{(0)} - \frac{\{1\,;\,k-1\}}{\left\{\dfrac{P'(z^{(0)})}{P(z^{(0)})}\,;\,\dfrac{n-k}{r}\right\}},$$

while the iterative method (4.108) is reduced to

$$Z_k^{(m+1)} = z^{(m)} - \frac{\{1\,;\,k-1\}}{\dfrac{P'(z^{(m)})}{P(z^{(m)})} - (n-k)\dfrac{\{\bar{z}^{(0)} - \bar{z}^{(m)}\,;\,r\}}{r^2 - |z^{(0)} - z^{(m)}|^2}} \tag{4.109}$$

$$(z^{(m)} = \text{mid}\; Z_k^{(m)};\; m = 0,1,\ldots).$$

The subscript "k" in the last formula points to the possibility of enclosure of some zeros from the cluster containing the zeros $\xi_1,\ldots,\xi_n$. In the sequel, such set of very close zeros $\xi_1,\ldots,\xi_n$ will be refered to as a *cluster* $\{\xi_1,\ldots,\xi_n\}$.

We note that the convergence of the algorithms (4.108) and (4.109) is merely *linear* due to the multiplication by the disks $\{1\,;\,k-1\}$; namely, from (4.108) or (4.109) we obtain

$$r^{(m+1)} = O(\text{rad}(\{1\,;\,k-1\}\cdot\{wr^{(m)}\,;\,ar^{(m)2}\}))$$

$$= O(|w|r^{(m)} + kar^{(m)2}) = O(r^{(m)}),$$

where $r^{(m)} = \text{rad}\ Z^{(m)}$, $w \in \mathbb{C}$, $a \in \mathbb{R}^+$.

In the following we will assume that Algorithm (4.109) is converging, that is, $r^{(m)} = \text{rad}\ Z_k^{(m)} \to 0$ when $m \to \infty$. For that reason, Algorithm (4.109) improves inclusion disks. Unfortunely, in the course of the iterative process it may happen that one or more zeros from the cluster "drop out" from the disk $Z_k^{(m)}$ due to the fact that formula (4.107) does not guarantee the inclusion of more than one zero. Nevertheless, certain control of safe enclosure could be provided by an information about the number of zeros in the disk $Z_k^{(m)}$ in each iteration (applying some of the "counting" methods, see the references in § 2.2). When this number becomes less than k, the iterative process should be stopped and the disk from the preceding iteration should be taken as a circular approximation of the cluster. Obviously, such control is very costly. Moreover, the iterative process (4.109) may be broken off although the disk $Z_k^{(m)}$ is not sufficiently improved, which is an additional deficiency.

Another approach consists of the continuation of the iterative process (4.109) in a situation when the obtained disks still do not contain all zeros from the cluster. According to Theorem 4.11, the produced disks will contain at least one zero. Under the assumption $r^{(m)} = \text{rad}\ Z_k^{(m)} \to 0$, after a sufficiently great number of iterations (because of the slow convergence), we will obtain the disk including *only one* (simple or multiple) zero. If the center $z^{(m)}$ of this resulting disk $Z^{(m)}$ was determined with a great precision (in other words, the radius $r^{(m)}$ is small enough), then the multiplicity $\mu$ of the included zero could be found using the limiting formula (2.2) (by means of an approximation improved by (2.4), eventually). To complete the zero-finding algorithm, the corresponding factor $(z - z^{(m)})^{\mu}$ is then removed from the original polynomial and the procedure is employed again to find the zeros of the deflated polynomial whose degree is now lowered by $\mu$. The iterative method (4.109) starts again with the inclusion disk defined by (4.107), and so forth. To avoid the round-off errors and the effects of the deflation, the described procedure should be realized in multiple-precision arithmetic.

**REMARK 9.** Numerous examples showed that, realizing the iterative process (4.109), the number of zeros $N^{(m)}$ contained in the produced disk $Z_k^{(m)}$, may vary (see Example 7). But, if the method is convergent, then $N^{(m)} \to 1$ when $m \to \infty$. ●

As a numerical illustration of the above observation, we present

**EXAMPLE 7.** We considered the polynomial (4.97) whose zeros $\xi_2 = 1+i\sqrt{2}$ of the multiplicity 2 and $\xi_3 = 1 + \frac{99}{70}i$ of the multiplicity 1 are very close. We isolated these zeros by the disk $Z^{(0)} = \{1.2 + 1.5i \,;\, 0.5\}$, the remaining double zero $\xi_1$ belongs to ext $Z^{(0)}$. To improve the circular approximation of $\xi_2$ and $\xi_3$, we applied Algorithm (4.109) (with $k = 3$, $z^{(0)} = 1.2 + 1.5i$ and $r = 0.5$). This algorithm furnished the inclusion disk approximations $Z_3^{(m)} = \{z^{(m)} \,;\, r^{(m)}\}$, displayed in Table 4.5.

| m | $z^{(m)}$ | $r^{(m)}$ |
|---|---|---|
| 1 | 1.1306 + 1.4666 i | 0.2196 |
| 2 | 1.0851 + 1.4469 i | 0.1280 |
| 3 | 1.0557 + 1.4350 i | 0.0761 |
| 4 | 1.0366 + 1.4276 i | 0.0464 |
| 5 | 1.0242 + 1.4230 i | 0.0290 |
| 6 | 1.0160 + 1.4200 i | 0.0180 |
| 7 | 1.0106 + 1.4180 i | 0.0119 |
| 8 | 1.0071 + 1.4168 i | 0.0070 |
| 9 | 1.0047 + 1.4159 i | 0.0051 |
| 10 | 1.0031 + 1.4153 i | 0.0030 |
| 11 | 1.0021 + 1.4150 i | 0.0022 |
| 12 | 1.0014 + 1.4147 i | 0.0015 |

Table 4.5

The disks $Z_3^{(1)},\ldots,Z_3^{(12)}$ contain both zeros $\xi_2$ and $\xi_3$, but $Z_3^{(13)} = \{1.0009 + 1.4146\,i \,;\, 0.001\}$ contains only $\xi_3 = 1 + \frac{99}{70}i$. Therefore, the disk

$$Z_3^{(12)} = \{1.0014 + 1.4147\,i \,;\, 0.0015\}$$

is taken as a circular approximation of the cluster $\{\xi_2,\xi_3\}$. In the continuation of the iterative process we obtain $Z_3^{(13)},\ldots,Z_3^{(19)}$ containing only the zero $\xi_3$, but

$$Z_3^{(20)} = \{1.000045 + 1.41425\,i \,;\, 6.4 \times 10^{-5}\}$$

again includes the cluster $\{\xi_2,\xi_3\}$! The diameter $2r^{(12)} = 0.003$ of the disk $Z_3^{(12)}$ is considerably larger than the distance $|\xi_3 - \xi_2| = 7.2\times 10^{-5}$ between these zeros, wherefrom we conclude that $Z_3^{(12)}$ is insufficiently good approximation of the cluster $\{\xi_2,\xi_3\}$. On the other hand, the value $2r^{(20)} = 1.28\times 10^{-4}$ shows that $Z_3^{(20)}$ is a quite satisfactory including approximation, but this disk is an accidental enclosure rather than an expected result. Finally, we note that the applied iterative method converges rather slowly.

We shall now describe an algorithm for the determination of the zeros belonging to an isolated cluster. It combines iterative processes in circular interval and ordinary complex arithmetic and is based on „splitting" zeros from a cluster. This „splitting" method will be called the *S-method* for brevity.

The S-method requires the following distribution of initial disks: the zeros $\xi_1,\ldots,\xi_k$, which form a cluster $\{\xi_1,\ldots,\xi_k\}$, are isolated in a disk $Z^{(0)} = \{z^{(0)}; r\}$, the remaining zeros $\xi_{k+1},\ldots,\xi_n$ are contained in well-separated nonintersecting disks $Z_{k+1}^{(0)},\ldots, Z_n^{(0)}$ such that

$$\xi_j \in Z_j^{(0)}, \quad Z_i^{(0)} \cap Z_j^{(0)} = \emptyset \ (i \neq j), \quad Z^{(0)} \cap Z_j^{(0)} = \emptyset \quad (i,j = k+1,\ldots,n).$$

The improved inclusion disks of the zeros $\xi_{k+1},\ldots,\xi_n$ are obtained by the following version of the iterative method (4.22):

$$Z_i^{(m+1)} = z_i^{(m)} - \cfrac{1}{\cfrac{P'(z_i^{(m)})}{P(z_i^{(m)})} - \sum_{\substack{j=k+1\\ j\neq i}}^{n} (Z_i^{(m)} - Z_j^{(m)})^{-1} - k\,(Z_i^{(m)} - Z^{(0)})^{-1}}$$

$$(z_i^{(m)} = \operatorname{mid} Z_i^{(m)};\ i = k+1,\ldots,n;\ m = 0,1,\ldots). \qquad (4.110)$$

The order of convergence of the last method is *two*.

**REMARK 10.** If the given polynomial P has multiple zeros $\xi_{k+1},\ldots,\xi_q$ ($q < n$) of the multiplicity $\mu_{k+1},\ldots,\mu_q$ ($\mu_{k+1} + \cdots + \mu_q + k = n$), then their inclusion disks are improved by the algorithm

$$Z_i^{(m+1)} = z_i^{(m)} - \cfrac{\mu_i}{\cfrac{P'(z_i^{(m)})}{P(z_i^{(m)})} - \sum_{\substack{j=k+1\\ j\neq i}}^{q} \mu_j (Z_i^{(m)} - Z_j^{(m)})^{-1} - k\,(Z_i^{(m)} - Z^{(0)})^{-1}}$$

$$(i = k+1,\ldots,q;\ m = 0,1,\ldots), \qquad (4.110a)$$

which is a modifiaction of the parallel algorithm (4.74), proposed by Gargantini [33]. ®

Assume that the resulting inclusion disks $Z_{k+1}^{(m)},\ldots,Z_n^{(m)}$ have sufficiently small radii. Then we may take their centers $z_{k+1}^{(m)},\ldots,z_n^{(m)}$ as the approximations to the zeros $\xi_{k+1},\ldots,\xi_n$. If the wanted accuracy (defined by $r^{(m)} = \max_{k+1\le i\le n} r_i^{(m)}$) has not been attained, then these point approximations can be employed as the starting values in some of the (local convergent) iterative processes with the undeflated, original polynomial for further refinement. Forming the polynomial

$$Q(z) = \prod_{i=k+1}^{n} (z - z_i^{(m)})$$

and dividing the original polynomial P(z) by Q(z), as a result we obtain a deflated polynomial

$$T(z) = b_k z^k + b_{k-1} z^{k-1} + \cdots + b_1 z + b_0 \quad \left(= \frac{P(z)}{Q(z)}\right), \tag{4.111}$$

whose zeros $\xi_1,\ldots,\xi_k$ lie in the isolated disk $Z^{(0)} = \{z^{(0)}; r\}$ and belong to the cluster $\{\xi_1,\ldots,\xi_k\}$. This polynomial will be called the *cluster polynomial*, *C-polynomial* for short, emphasizing that it has all zeros in one cluster. In order to alleviate unfavourable effects of deflation, the performed procedure should be carried out in multiple-precision arithmetic. In such case we may hope that the coefficents $b_0, b_1, \ldots, b_k$ of the C-polynomial T will be determined with a sufficiently great accuracy.

The S-method is based on the substitution of the complex argument z in the polynomial (4.111) by a new argument w using the linear transformation

$$z = -\frac{b_{k-1}}{k b_k} + w\cdot 10^{-h}, \tag{4.112}$$

where $h > 1$ is a suitably chosen integer. In this manner we obtain a new polynomial, say

$$G(w) = s_k w^k + s_{k-1} w^{k-1} + \cdots + s_1 w + s_0. \tag{4.113}$$

The zeros of this polynomial will be denoted by $\eta_1,\ldots,\eta_k$.

The argumentation for the substitution (4.112) is as follows: By Vieta's formulas we find

$$\xi_B = \frac{\xi_1 + \cdots + \xi_k}{k} = -\frac{b_{k-1}}{k\,b_k},$$

where $\xi_B$ denotes the „barycenter" of the zeros $\xi_1,\ldots,\xi_k$ from the cluster. The complex point $\xi_B$ can also be regarded as the barycenter of the considered cluster (with regard to its geometric sense). From (4.112) there follows

$$w = (z - \xi_B)\cdot 10^h.$$

If $z = \xi_i$ $(i \in \{1,\ldots,k\})$, where $\xi_i$ is a zero from the cluster $\{\xi_1,\ldots,\xi_k\}$, then $d_i = |\xi_i - \xi_B|$ is the distance from this zero to the barycenter $\xi_B$. The quantities $d_i$ are very small in the case of very close zeros, but, multiplying $(\xi_i - \xi_B)$ by $10^h$, the mentioned distances become enlarged. Since

$$\eta_i = (\xi_i - \xi_B)\cdot 10^h$$

(in view of (4.112)), the zeros $\eta_1,\ldots,\eta_k$ of the new polynomial (4.113) *are diverging* from each other, namely

$$|\eta_i - \eta_j| = |\xi_i - \xi_j|\cdot 10^h \quad (i \neq j).$$

Hence, the zeros of the polynomial G are separated (with the exception when a cluster is reduced to a multiple zero). The use of the barycentric point ensures that the „explosion" of the zeros from the cluster is *uniform*. Clearly, the distances between the new zeros are larger if the value of h is greater.

Let sep (P) be a measure of separation of zeros $\xi_1,\ldots,\xi_n$ of a polynomial P (generally speaking), defined by

$$\operatorname{sep}(P) = \min_{\substack{i,j\\ i\neq j}} |\xi_i - \xi_j|$$

(see, e.g., Güting [43]). Evidently, the exponent h in (4.112) should be chosen so that the value of $10^{-h}$ is of the same order of magnitude as sep (T). Unfortunately, we do not have any information about the value of sep (T). The estimations of sep (T) in general, given in the literature (e.g. Mahler [66], Mignote [68]), are rather rough ones and practically inapplicable for our problem. For that reason, let us consider a disk $Y_j = \{\xi_B\,;\,r_j\}$ with the center at the barycentric point $\xi_B$ such that it contains all zeros from the cluster $\{\xi_1,\dots,\xi_k\}$, and let

$$r = \inf_j \left\{ r_j:\ \xi_i \in \{\xi_B ; r_j\}\ (i = 1,\dots,k) \right\} .$$

Then $Y = \{\xi_B ; r\}$ is the smallest disk with this inclusion property. Let the *radius* of the cluster $\{\xi_1,\dots,\xi_k\}$ be defined by

$$R(\{\xi_1,\dots,\xi_k\}) := r.$$

It is convenient to choose h so that $10^{-h}$ is of the same order as R. To find an upper bound of R, it is useful to apply the result (2.1) given in §2.1. Since (2.1) defines an inclusion disk (for all polynomial zeros) centered at the origin, to obtain a disk with a radius as small as possible, it is necessary to *translate* the barycenter $\xi_B$ into the origin. Therefore, in the *first* step of the application of the S-method, we represent the C-polynomial T(z) by the powers of $(z - \xi_B)$, that is,

$$T(z) = c_k(z-\xi_B)^k + c_{k-1}(z-\xi_B)^{k-1} + \cdots + c_1(z-\xi_B) + c_0.$$

The coefficients $c_0, c_1, \dots, c_k$ can be obtained by the Horner scheme. Substituting $y = z - \xi_B$, we get the polynomial

$$F(y) = T(y+\xi_B) = c_k y^k + c_{k-1} y^{k-1} + \cdots + c_1 y + c_0 , \tag{4.114}$$

whose zeros are $\zeta_i = \xi_i - \xi_B$ $(i = 1,\dots,k)$. It is obvious that $c_{k-1} = 0$ because of

$$\frac{c_{k-1}}{c_k} = - \sum_{i=1}^{k} \zeta_i = - \sum_{i=1}^{k} (\xi_i - \xi_B) = - \sum_{i=1}^{k} \xi_i + k\xi_B = 0.$$

This fact can be usefully employed in controling the influence of rounding error.

The zeros $\zeta_1,\dots,\zeta_k$ of the polynomial (4.114) are clustered around the origin, which is their barycenter. According to (2.1) these zeros are contained in the inclusion disk $\{0 ; r\}$, where

$$r = 2 \max_{1 \le \lambda \le k} |c_{k-\lambda}/c_k|^{1/\lambda} = R(\{\zeta_1,\dots,\zeta_k\}). \tag{4.115}$$

We are now able to estimate the exponent h appearing in (4.112). In view of the preceding discussion, from the equation $10^{-h} = R$ and (4.115) we find

$$h = I\left(-\log_{10}(2 \max_{1 \le \lambda \le k} |c_{k-\lambda}/c_k|^{1/\lambda})\right), \tag{4.116}$$

where $I(a)$ denotes the integer which is closest to a real positive number $a$.

In the *second* step we set $y = w\cdot 10^{-h}$ in (4.114) and obtain the polynomial $G(w)$ in the form of (4.113). The coefficients in (4.113) are given by $s_m = c_m\cdot 10^{-mh}$ $(m = 0,1,\dots,k)$ and so, they are very small. For this reason, in the *third* step we carry out a normalization of these coefficients multiplying $G(w)$ by $10^{kh}$. The resulting polynomial is

$$G^*(w) = s_k^* w^k + s_{k-1}^* w^{k-1} + \cdots + s_1^* w + s_0^* ,$$

where $s_m^* = c_m\cdot 10^{(k-m)h}$ $(m = 0,1,\dots k)$. This polynomial has the well-separated zeros $\eta_1,\dots,\eta_k$ and finally, we can apply some of the numerous zero-finding algorithms. A great number of numerical examples showed that the initial point $w^{(0)} = 0$ (the new barycenter of cluster) can often be satisfactory for running a lot of algorithms. If $w_i^{(M)}$ are the resulting approximations to the zeros $\eta_i$ $(i = 1,\dots,k)$ (solving the polynomial equation $G^*(w) = 0$), then the zeros $\xi_1,\dots,\xi_k$ of the C-polynomial $T(z)$ are determined approximately from (4.112) as

$$\xi_i \cong -\frac{b_{k-1}}{k\,b_k} + w_i^{(M)}\cdot 10^{-h} \qquad (i = 1,\dots,k).$$

The zeros $\xi_1,\dots,\xi_k$ can be further improved by using their approximations, obtained by the last formula, as the initial values in some of the iterative processes employing the original polynomial $P$ (the so-called "purification" procedure). This is the *fourth* (final) step of the S-method. For example, we can apply the modified Newton method

$$z^{(m+1)} = z^{(m)} - \frac{P(z^{(m)})P'(z^{(m)})}{P'(z^{(m)})^2 - P(z^{(m)})P''(z^{(m)})} \quad (m = 0,1,\dots),$$

which has a quadratic convergence although the multiplicity of the requested zero is unknown.

As it could be expected, and as numerous examples have shown, the S-method is less stable if the zeros of a cluster are closer (that is, if its radius $R(\{\xi_1,\dots,\xi_k\})$ is smaller). The presence of multiple zeros leads to further instability of the described zero cluster procedure. In fact, as it is well known, the enormous closeness of these zeros makes the problem of clusters a very difficult one. As an illustration we discuss the results from Example 9. Although we used the approximations with at least 23 exact decimal digits in the deflation procedure, the S-method produced the approximations of the zeros belonging to a cluster (with the radius $R(\{\xi_1,\xi_2,\xi_3\}) = 4.8\times 10^{-5}$) with only 9 exact decimal digits. In this process multiple-precision arithmetic with 34 significant digits was employed.

**REMARK 11.** From a practical point of view, the situations with polynomials which have zero clusters are rare. In addition, if such polynomial is a mathematical model (or its part) of a certain, for instance, real physical device, the presence of clusters is often connected with the unstable performances of this device. Therefore, the main attention has to be paid to the reconstruction and improvement of the considered device; the problem of clusters is of a secondary importance because its appearance is most frequently only a consequence of a badly-built device. ®

The presented S-method is not necessarily the best possible, but we believe that it can solve, partially or even completely, the problem of one cluster in numerous nondegenerative situations. To improve this method, a lot of study is still to be done; first of all, it is necessary to provide that the coefficients of C-polynomial are determined with a great precision, the problem of choosing the exponent h should be aimed toward finding very sharp bounds for sep (T) and, especially, all significant digits of the coefficients $s_0,s_1,\dots,s_k$ (in (4.113)) should be preserved. In any case, it is desirable to employ multiple-precision arithmetic.

**EXAMPLE 8.** We demonstrate the S-method by considering the C-polynomial

$$T(z) = z^3 - 0.79z^2 + 0.2078z - 0.0182 = (z-0.25)(z-0.26)(z-0.28)$$

in arithmetic with eight decimal digits.

Since $\xi_B = \dfrac{0.79}{3}$, substituting $y = z - \dfrac{0.79}{3}$ in $T(z)$, we obtain

$$F(y) = y^3 - 2.3333333\times 10^{-4}\,y - 7.4074\times 10^{-7}.$$

According to (4.116) it follows

$$h = I(-\log_{10}(2\max\{(2.3333333\times 10^{-4})^{1/2},\ (7.4074\times 10^{-7})^{1/3}\})) = I(1.816) = 2.$$

Putting $y = w\cdot 10^{-2}$ in $F(y)$, we obtain

$$G(w) = 10^{-6}w^3 - 2.3333333\times 10^{-6}w - 7.4074\times 10^{-7},$$

or, after normalization

$$G^*(w) = w^3 - 2.3333333\,w - 0.74074.$$

For finding a zero of the polynomial $G^*(w)$ we apply the modified Newton method (2.4) (presented also on p. 132). Starting with $w^{(0)} = 0$, after the third iteration we obtain

$$w_1 = w^{(3)} = -0.33333298.$$

Solving the deflated quadratic equation $\frac{G^*(w)}{w - w_1} = 0$, that is,

$$w^2 - 0.33333298\,w - 2.222222424$$

by the usual formula, we find

$$w_2 = 1.6666665, \qquad w_3 = -1.3333336.$$

Finally, using the substitution $z = \frac{0.79}{3} + w\cdot 10^{-2}$, we calculate the corresponding zeros of the polynomial $T$,

$$\xi_1 = 0.26, \qquad \xi_2 = 0.28, \qquad \xi_3 = 0.25.$$

**EXAMPLE 9.** A complete procedure of finding zeros of a polynomial which has one cluster, consisting of the interval method (4.110a) and the S-method, was demonstrated by means of the polynomial

$$P(z) = 70z^6 - (280 + 99i)z^5 + (630 + 297i)z^4 - (560 + 594i)z^3 + (-70 + 198i)z^2$$
$$+ (840 + 297i)z - (630 + 891i)$$
$$= 70(z^2 - 2z + 3)^2(z - 1 - \tfrac{99}{70}i)(z + 1).$$

The polynomial $P$ is formed by multiplying the polynomial (4.97) by $z + 1$.

We found that the disks $Z_4^{(0)} = \{-1.2 + 0.3i\ ;\ 0.8\}$ and $Z_5^{(0)} = \{1.4 - 1.2i\ ;\ 0.8\}$ contain the zeros $\xi_4$ and $\xi_5$ of the multiplicity 1 and 2, respectively. The remaining zeros $\xi_1$, $\xi_2$ and $\xi_3$ form a cluster $\{\xi_1, \xi_2, \xi_3\}$ and they were included in the isolated circular region $Z^{(0)} = \{1.3 + 1.6i\ ;\ 1\}$.

Applying the interval method (4.110a) with $k = 3$, after the fourth iterative step as results we obtained (using quad precision arithmetic - about 34 significant decimal digits):

$Z_4^{(4)} = \{-0.99999999999999999999999999999979 + 1.57 \times 10^{-30}\, i\ ;\ 9.35 \times 10^{-28}\}$,

$Z_5^{(4)} = \{1.000000000000000000000000002615235 - 1.414213562373095048801689743902\, i\ ;\ 6.12 \times 10^{-24}\}$.

Dividing the original polynomial $P(z)$ by $(z - z_4^{(4)})(z - z_5^{(4)})$ ($z_i^{(4)} = \text{mid}\ Z_i^{(4)}$, $i = 4,5$), we furnished the deflated polynomial

$$T(z) = b_3 z^3 + b_2 z^2 + b_1 z + b_0$$

with the coefficients listed below:

| k | $\text{Re}\{b_k\}$ | $\text{Im}\{b_k\}$ |
|---|---|---|
| 0 | 350.01428534987281966273041511 81961 | -98.98989873223330683223249429 87094 |
| 1 | -210.01428534987281966273590723 06745 | 593.97979746446661366447105709 86516 |
| 2 | -209.99999999999999999999963386 55458 | -296.98989873223330683223656414 62439 |
| 3 | 70. | 0. |

The zeros $\xi_1, \xi_2, \xi_3$ of the polynomial $T$ lie in the isolated disk $Z^{(0)} = \{1.3 + 1.6i\ ;1\}$ and form the cluster $\{\xi_1, \xi_2, \xi_3\}$. Their barycenter is $\xi_B = -b_2/(3b_3)$. Substituting $y = z - \xi_B$ in $T(z)$, this barycenter was translated into the origin; the coefficients of the corresponding polynomial of the form

$$F(y) = T(y + \xi_B) = c_3 y^3 + c_2 y^2 + c_1 y + c_0 ,$$

where

| k | $\text{Re}\{c_k\}$ | $\text{Im}\{c_k\}$ |
|---|---|---|
| 0 | -0.00000000000000000002929111612473 | 0.000000000001947636099672486461678 |
| 1 | 0.000000121470964874326235656325122 | -0.00000000000000000001035591603190 |
| 2 | 0. | $4.9 \times 10^{-32}$ |
| 3 | 70. | 0. |

According to the value of $c_2 = 0 - 4.9 \times 10^{-32}\, i$ we conclude that the influence of rounding errors in the previously performed procedure is neglectable. In the following we calculated with $c_2 = 0$.

The exponent h appearing in the relation of substitution was calculated by (4.116) using the coefficients of polynomial F,

$$h = I\left( -\log_{10}\left( 2 \max \left\{ \left| \frac{c_1}{c_3} \right|^{1/2}, \left| \frac{c_0}{c_3} \right|^{1/3} \right\} \right) \right) = I(4.073) = 4.$$

In the next step we set $y = w \cdot 10^{-4}$ in $F(y)$ and, after multiplication with $10^{12}$ $(= 10^{kh})$, we obtained the polynomial

$$G^*(w) = s_3^* w^3 + s_2^* w^2 + s_1^* w + s_0^*$$

with the following coefficients:

| k | $\mathrm{Re}\{s_k^*\}$ | $\mathrm{Im}\{s_k^*\}$ |
|---|---|---|
| 0 | $0.29291116124734927824497787379977 \times 10^{-8}$ | $1.9476360996724864616783008070656$ |
| 1 | $12.147096487432623565632512152353 2$ | $-0.10355916031899070202853658277 36 \times 10^{-12}$ |
| 2 | 0. | 0. |
| 3 | 70. | 0. |

In order to find one zero of $G^*$, we applied the modified Newton method (2.4) with the initial approximation $w^{(0)} = 0$ (the barycenter of the zeros of $G^*$). In the first seven iterative steps this method showed a very slow (only linear) convergence and then, the convergence rate became quadratic. After 10 iterations we obtained the following value

$$w_1 = w^{(10)} = 0.0000065139045016047530376366697 16 - 0.2405019237220188933341335498709 89i,$$

with $|G^*(w^{(10)})| = 1.63 \times 10^{-29}$. Since the absolute value of rounding error was about $10^{-34}$, we could not expect any further improvement of the above approximations $w^{(10)}$ and we terminated the iterative process (according to the stopping criterion described in [33] (see, also, [101]).

The deflated quadratic equation $\dfrac{G^*(w)}{w - w^{(10)}} = 0$, solved by the usual formula, gave the solutions

$$w_2 = 0.65138241240483628226 \times 10^{-5} - 0.24051082710423370253\, i,$$

$$w_3 = 0.80377556390215021989 \times 10^{-10} + 0.48101275082625259586\, i.$$

Using the relation of substitution

$$z_i = -\frac{b_2}{3b_3} + w_i \cdot 10^{-4} \quad (i = 1,2,3),$$

we found the following approximations $z_1$, $z_2$ and $z_3$ to the zeros $\xi_1$, $\xi_2$ and $\xi_3$ of P:

$$z_1 = 1.000000000\underline{6}51390 + 1.414213562\underline{8}18262\, i,$$

$$z_2 = 0.999999999\underline{3}48617 + 1.414213561\underline{9}27924\, i,$$

$$z_3 = 0.99999999999999\underline{1} + 1.41428571428571\underline{7}\, i.$$

The first incorrect digit is underlined.

The approximations $z_1$, $z_2$ and $z_3$ were improved by the Newton-like method (2.4) (starting with $z_i^{(0)} = z_i$, $i=1,2,3$). Applying only one iteration, we obtained

$$z_1^{(1)} = 1.0000000000000000000\underline{4}51749 + 1.41421356237309\underline{6}0133543576\,i\ ,$$

$$z_2^{(1)} = 1.00000000000000\underline{5}1705966511 + 1.4142135623730\underline{8}90840876960\,i\ ,$$

$$z_3^{(1)} = 0.99999999999999999999999\underline{7}6 + 1.41428571428571428571428\underline{7}9\,i\ .$$

Since $|P(z_1^{(1)})| = 1.97 \times 10^{-31}$, $|P(z_2^{(1)})| = 8.14 \times 10^{-30}$ and $|P(z_3^{(1)})| = 2.3 \times 10^{-29}$, for the above mentioned remark concerning rounding errors, further improvements were not possible.

The approximations $z_1^{(1)}$ and $z_2^{(1)}$ are correct to 13 decimal places. The approximation $z_3^{(1)}$ is much better: it has even 23 correct decimal digits. Since the first digits of $z_1^{(1)}$ and $z_2^{(1)}$ are identical, it is logical to conjecture that $z_1^{(1)}$ and $z_2^{(1)}$ are approximations of the *same* zero, say $\xi_1$, of the multiplicity $\mu_1 = 2$. This explains why very small values of P at the points $z_1^{(1)}$ and $z_2^{(1)}$ were obtained although these approximations are correct to 13 decimal digits ($|P(z_i^{(1)})| < 10^{-29}$, $i=1,2$); namely, the polynomial P contains the factor $(z_1^{(1)} - \xi_1)(z_2^{(1)} - \xi_1)$ which practically means that $z_1^{(1)} - \xi_1$ or $z_2^{(1)} - \xi_1$ (of the order $10^{-14}$) is squared. The above conjecture was verified by calculating $P'(z_1^{(1)})$ and $P'(z_2^{(1)})$. In both cases we found that $|P'(z_i^{(1)})|$ $(i=1,2)$ were of the order $10^{-14}$. The refinement of the approximations ($z_1^{(1)}$ or $z_2^{(1)}$) to the zero $\xi_1$ was attained by solving the equation $P'(z) = 0$ using ordinary Newton's method

$$z^{(m+1)} = z^{(m)} - \frac{P'(z^{(m)})}{P''(z^{(m)})}\ ,$$

starting with $z^{(0)} = z_1^{(1)}$ or $z_2^{(1)}$. We found

$$z^{(2)} = 1.00000000000000000000000000000\underline{3} + 1.414213562373095048801688724207\underline{2}\,i\ .$$

Finally, we obtained the following approximate values of the zeros $\xi_1$, $\xi_2$, $\xi_3$ and $\xi_4$ of the multiplicity $\mu_1 = 2$, $\mu_2 = 1$, $\mu_3 = 1$ and $\mu_4 = 2$, respectively:

$$\xi_1 \cong z^{(2)} = 1.00000000000000000000000000000\underline{3} + 1.414213562373095048801688724207\underline{2}\,i$$ (by the S-method and the equation $P'(z) = 0$),

$$\xi_2 \cong z_3^{(1)} = 0.99999999999999999999999\underline{7}6 + 1.41428571428571428571428\underline{7}9\,i$$ (by the S-method),

$$\xi_3 \cong z_4^{(4)} = -\,0.99999999999999999999999999999979 + 1.57\times 10^{-30} i\,,$$

(by the interval method (4.110a).

$$\xi_4 \cong z_5^{(4)} = 1.00000000000000000000000026 - 1.41421356237309504880168897\,i$$

The underlined digit indicates the first incorrect digit.

POINT ROOT ITERATIONS FOR MULTIPLE ZEROS

For a sufficiently small radii $r_j$ $(j = 1,\ldots,n)$, from (4.68) we obtain the following expression for the center $z_i^{(m+1)}$ of the disk $Z_i^{(m+1)}$:

$$z_i^{(m+1)} = z_i^{(m)} - \frac{1}{\left\{\frac{1}{\mu_i}\left[h_k(z_i^{(m)}) - \sum_{\substack{j=1\\ j\neq i}}^{n} \mu_j (z_i^{(m)} - z_j^{(m)})^{-k}\right]\right\}_*^{1/k}} \tag{4.117}$$

$$(i = 1,\ldots,n;\ m = 0,1,\ldots;\ k = 1,2,\ldots\ ).$$

As before, the symbol $*$ denotes that the "appropriate" k-th root (among k values of the denominator in (4.117) has to be chosen in accordance with CCR. The above formula may also be derived from the fixed point relation given at the beginning of § 4.4. By (4.117) the generalized iterative procedure for improving, simultaneously, approximations to multiple complex zeros (in standard complex arithmetic) is formulated. The single-step version of (4.117) is given by

$$z_i^{(m+1)} = z_i^{(m)} - \frac{1}{\left\{\frac{1}{\mu_i}\left[h_k(z_i^{(m)}) - \sum_{j=1}^{i-1} \mu_j (z_i^{(m)} - z_j^{(m+1)})^{-k} - \sum_{j=i+1}^{n} \mu_j (z_i^{(m)} - z_j^{(m)})^{-k}\right]\right\}_*^{\frac{1}{k}}}$$

$$(i = 1,\ldots,n;\ m = 0,1,\ldots;\ k = 1,2,\ldots\ ). \tag{4.118}$$

The last iterative formula may also be obtained from (4.71). The lower bound of $O_R((4.118),\xi)$ is determined by Theorem 4.3.

**EXAMPLE 10.** The simultaneous iterative method of the fifth order, defined by (4.117) for $k = 3$, was applied for finding complex zeros of the polynomial

$$\begin{aligned} P(z) = {} & z^{15} + (6-6i)z^{14} + (-12-36i)z^{13} + (-134+8i)z^{12} + (-55+420i)z^{11} + (1140+330i)z^{10} \\ & + (1356-2872i)z^{9} + (-5692-4616i)z^{8} + (-11069+8808i)z^{7} + (12030+21870i)z^{6} \\ & + (39204-12020i)z^{5} + (-2750-61336i)z^{4} + (-77325-31500i)z^{3} + (-75000+41550i)z^{2} \\ & + (-22500+50000i)z + 15000\,i\,, \end{aligned}$$

whose factorization is

$$P(z) = (z-3)(z+1)^3(z-2i)^3(z^2+4z+5)^2(z^2-4z+5)^2 .$$

As the initial approximations to the zeros of P, the following complex numbers were taken:

$z_1^{(0)} = -3.4 + 0.2i$, $z_2^{(0)} = -0.7 - 0.3i$, $z_3^{(0)} = 0.3 + 2.4i$, $z_4^{(0)} = -2.3 + 0.6i$,
$z_5^{(0)} = -1.7 - 0.7i$, $z_6^{(0)} = 2.3 + 1.4i$, $z_7^{(0)} = 1.6 - 0.7i$.

The results of the first and second iterative step are given in Table 4.6.

| i | $z_i^{(1)}$ | $z_i^{(2)}$ | $\mu_i$ |
|---|---|---|---|
| 1 | $-3.00991 - 6.57\times10^{-3}i$ | $-3.00000000005241638 - 2.2\times10^{-10}i$ | 1 |
| 2 | $-0.99751 - 2.5\times10^{-3}i$ | $-0.99999999999976589 - 5.04\times10^{-13}i$ | 3 |
| 3 | $-8.53\times10^{-4} + 2.00023\,i$ | $3.94\times10^{-17} + 2.00000000000000041\,i$ | 3 |
| 4 | $-1.98889 + 1.01753\,i$ | $-2.00000000037126743 + 1.00000000004327441\,i$ | 2 |
| 5 | $-2.02133 - 0.99856\,i$ | $-2.00000000035192846 - 1.00000000052464129\,i$ | 2 |
| 6 | $1.99751 + 0.99922\,i$ | $2.00000000000000554 + 0.99999999999999786\,i$ | 2 |
| 7 | $1.99955 - 0.99829\,i$ | $2.00000000000000112 - 0.99999999999999801\,i$ | 2 |

Table 4.6

The greatest distances of the zeros,

$$\varepsilon^{(m)} = \max_{1\le i\le n} |z_i^{(m)} - \xi_i| \qquad (m = 0,1,\ldots),$$

were $\varepsilon^{(1)} \cong 2.1\times10^{-2}$ and $\varepsilon^{(2)} \cong 6.32\times10^{-10}$.

From a practical point of view, the iterative formulas

$$\hat{z}_i = z_i - \frac{\mu_i}{h_1(z_i) - \sum_{\substack{j=1\\ j\ne i}}^{n} \mu_j (z_i - z_j)^{-1}} \qquad (i = 1,\ldots,n) \tag{4.119}$$

and

$$\hat{z}_i = z_i - \frac{\sqrt{\mu_i}}{\left[h_2(z_i) - \sum_{\substack{j=1\\ j\neq i}}^{n} \mu_j (z_i - z_j)^{-2}\right]_*^{1/2}} \quad (i = 1,\ldots,n), \tag{4.120}$$

obtained from (4.117) for k = 1 and k = 2, have a great importance. In the two last formulas the iteration index is omitted for brevity. We recall that the functions $h_1$ and $h_2$ are given by

$$h_1(z) = \frac{P'(z)}{P(z)} \quad \text{and} \quad h_2(z) = \frac{P'(z)^2 - P(z)P''(z)}{P(z)^2}.$$

The iterative methods (4.119) and (4.120), as well as their modifications, were studied in [96]. As in § 4.1, we will only give a list of these methods, marked by the same abbreviations as in the case of simple zeros (see § 4.1). For example, the methods (4.119) and (4.120) are marked (TS) (total-step).

Let $\mu$ be the multiplicity of the zero $\xi$ of P. By means of $h_1(z)$ and $s(z) = P''(z)/P'(z)$, we define

$$h_2(z) = h_1(z)[h_1(z) - s(z)] \qquad \text{(Ostrowski's function)},$$

$$N(z) = \mu/h_1(z) \qquad \text{(Newton's correction)},$$

and

$$H(z) = 2[(1+\tfrac{1}{\mu})h_1(z) - s(z)]^{-1} \qquad \text{(Halley's correction)}.$$

We recall that the correction terms N(z) and H(z) appear in the iterative formulas

$\hat{z} = z - N(z)$ (Schröder's modification of Newton's method for multiple zeros [113])

and

$\hat{z} = z - H(z)$ (modification of Halley's method, introduced by Hansen and Patrick [45] for multiple zeros),

with the convergence orders 2 and 3, respectively. We note that the order of multiplicity in these iterative formulas takes the values $\mu_1,\ldots,\mu_n$. Ostrowski's function has already been mentioned concerning the square root interval method for simple zeros (§ 4.1). Newton's and Halley's corrections will be used in § 5.5 for constructing some improved Halley-like algorithms.

The modified methods are presented below:

$$\text{(SS)}: \quad \hat{z}_i = z_i - \frac{(\mu_i)^{1/k}}{\left\{ h_k(z_i) - \sum_{j<i} \mu_j (z_i - \hat{z}_j)^{-k} - \sum_{j>i} \mu_j (z_i - z_j)^{-k} \right\}_*^{1/k}}$$

$(i = 1,\ldots,n;\ k = 1,2)$, (4.121)

$$\text{(TSN)}: \quad \hat{z}_i = z_i - \frac{(\mu_i)^{1/k}}{\left\{ h_k(z_i) - \sum_{j\neq i} \mu_j [z_i - z_j + N(z_j)]^{-k} \right\}_*^{1/k}} \tag{4.122}$$

$(i = 1,\ldots,n;\ k = 1,2)$,

$$\text{(SSN)}: \quad \hat{z}_i = z_i - \frac{(\mu_i)^{1/k}}{\left\{ h_k(z_i) - \sum_{j<i} \mu_j (z_i - \hat{z}_j)^{-k} - \sum_{j>i} \mu_j [z_i - z_j + N(z_j)]^{-k} \right\}_*^{1/k}}$$

$(i = 1,\ldots,n;\ k = 1,2)$, (4.123)

$$\text{(TSH)}: \quad \hat{z}_i = z_i - \frac{\sqrt{\mu_i}}{\left\{ h_2(z_i) - \sum_{j\neq i} \mu_j [z_i - z_j + H(z_j)]^{-2} \right\}_*^{1/2}} \tag{4.124}$$

$(i = 1,\ldots,n)$,

$$\text{(SSH)}: \quad \hat{z}_i = z_i - \frac{\sqrt{\mu_i}}{\left\{ h_2(z_i) - \sum_{j<i} \mu_j (z_i - \hat{z}_j)^{-2} - \sum_{j>i} \mu_j [z_i - z_j + H(z_j)]^{-2} \right\}_*^{1/2}}$$

$(i = 1,\ldots,n)$. (4.125)

**REMARK 12.** An application of Halley's correction in the basic iterative formula (4.119) would decrease the efficiency of the new modified method. Namely, to evaluate $H(z_j)$ it is necessary to evaluate the second derivative of a polynomial, which is not required for the basic method (4.119). Therefore, Halley's correction has been used only in the case of the square root method (4.120), where the functions $h_2(z)$ and the correction $H(z)$ are evaluated employing the same functions $h_1(z)$ and $s(z)$. ®

The total-step method (TSN) has the order of convergence equals k+3 (k = 1,2), while the order of the iterative method (TSH) is even six. The convergence rate of the single-step methods listed above was considered in the following theorem ([96]), which is very similar to Theorem 4.2:

**THEOREM 4.12.** *The R-order of convergence of the single-step methods (SS), (SSN) and (SSH) is given by*

$$O_R((SS),\xi) \geq k + 1 + t_n \quad (k = 1,2),$$

$$O_R((SSN),\xi) \geq k + 1 + x_n \quad (k = 1,2),$$

$$O_R((SSH),\xi) \geq 3 + y_n,$$

*where $t_n$, $x_n$ and $y_n$ are the unique positive roots of the equations*

$$t^n - t - k - 1 = 0 \quad (k = 1,2),$$

$$x^n - 2^{n-1}x - (k+1)\cdot 2^{n-1} = 0 \quad (k = 1,2),$$

$$y^n - 3^{n-1}y - 3^n = 0,$$

*respectively.*

**EXAMPLE 11.** In order to illustrate numerically the efficiency of the modified methods TS (4.120), SS (4.121), TSN (4.122), SSN (4.123), TSH (4.124) and SSH (4.125) of square root type (k = 2), these methods were applied for the determination of zeros of the polynomial

$$P(z) = z^9 - 7z^8 + 20z^7 - 28z^6 - 18z^5 + 110z^4 - 92z^3 - 44z^2 + 345z + 225.$$

The zeros of this polynomial are $\xi_1 = 1 + 2i$, $\xi_2 = 1 - 2i$, $\xi_3 = -1$ and $\xi_4 = 3$, with the multiplicities $\mu_1 = 2$, $\mu_2 = 2$, $\mu_3 = 3$ and $\mu_4 = 2$. As initial approximations to these zeros the following complex numbers were taken:

$$z_1^{(0)} = 1.8 + 2.7i, \quad z_2^{(0)} = 1.8 - 2.7i, \quad z_3^{(0)} = -0.3 - 0.8i, \quad z_4^{(0)} = 2.3 - 0.7i.$$

In realizing the (TSN), (SSN), (TSH) and (SSH) methods with Newton's and Halley's corrections, before calculating new approximations $z_i^{(m+1)}$ the values $h_1(z_i^{(m)})$ as well as $s(z_i^{(m)})$ $(m = 0,1,\ldots)$ were calculated. The same values were used for calculating Newton's corrections

$$N(z_i^{(m)}) = \mu_i / h_1(z_i^{(m)}) \quad (i = 1,\ldots,n),$$

Ostrowski's function

| | i | $\mathrm{Re}\{z_i^{(2)}\}$ | $\mathrm{Im}\{z_i^{(2)}\}$ |
|---|---|---|---|
| (TS) (4.120) | 1 | 0.999999853800923892 | 2.000000112716998844 |
| | 2 | 0.999999826741999847 | -2.000000351383949125 |
| | 3 | -0.999999859207295616 | $-8.18 \times 10^{-7}$ |
| | 4 | 3.000000527270300803 | $-3.48 \times 10^{-8}$ |
| (SS) (4.121) | 1 | 0.999999939617346251 | 1.999999964305993363 |
| | 2 | 1.000000861310650873 | -2.000000509862992614 |
| | 3 | -0.999999999709498985 | $1.35 \times 10^{-9}$ |
| | 4 | 3.000000000000030662 | $7.16 \times 10^{-14}$ |
| (TSN) (4.122) | 1 | 0.999999455077856744 | 2.000000212961094747 |
| | 2 | 1.000000018147137107 | -2.000000068835695135 |
| | 3 | -0.999999974528732211 | $3.43 \times 10^{-8}$ |
| | 4 | 3.000000722708680682 | $-9.58 \times 10^{-8}$ |
| (SSN) (4.123) | 1 | 0.999999894885117145 | 2.000000042747320793 |
| | 2 | 0.999999994177457521 | -2.000000000709903145 |
| | 3 | -1.000000000007845003 | $3.82 \times 10^{-11}$ |
| | 4 | 2.999999999999997525 | $-6.58 \times 10^{-15}$ |
| (TSH) (4.124) | 1 | 1.000000000098386276 | 1.999999999890580897 |
| | 2 | 1.000000000450329186 | -2.000000000521585396 |
| | 3 | -0.999999999986166747 | $-2.93 \times 10^{-12}$ |
| | 4 | 3.000000000368704406 | $-6.92 \times 10^{-10}$ |
| (SSH) (4.125) | 1 | 1.000000000032764666 | 2.000000000002146278 |
| | 2 | 1.000000000000674921 | -1.999999999997025086 |
| | 3 | -1.000000000000001322 | $2.74 \times 10^{-15}$ |
| | 4 | 3.000000000000000383 | $-2.01 \times 10^{-16}$ |

Table 4.7

$$h_2(z_i^{(m)}) = h_1(z_i^{(m)})[h_1(z_i^{(m)}) - s(z_i^{(m)})] \quad (i = 1,\ldots,n)$$

and Halley's corrections

$$H(z_i^{(m)}) = 2\left[(1+\frac{1}{\mu_i})h_1(z_i^{(m)}) - s(z_i^{(m)})\right]^{-1} \quad (i = 1,\ldots,n).$$

Thus, the proposed iterative methods with Newton's and Halley's correction terms require slightly more numerical operations in relation to the basic methods (4.119) and (4.120). Taking into account the significantly increased order of convergence, it is obvious that the presented methods have a great efficiency.

In spite of rough initial approximations ( $\min_i |z_i^{(0)} - \xi_i| \cong 1$), the modified methods demonstated very fast convergence. Numerical results, obtained in the second iteration, are given in Table 4.7.

MAEHLY'S COMBINED METHOD

Because of the great computational cost of most interval methods, in this book we have established a number of combined methods in order to preserve the great *efficiency* of point iterative processes and the *inclusion property* of interval methods (that is, the enclosure of zeros). As the result, we have obtained the efficient iterative methods which possess (*i*) a small computational cost and (*ii*) the ability of automatic determination of the upper error bounds concerning the produced approximations to polynomial zeros.

Among all combined methods considered the greatest attention is paid to combined methods which are based on Maehly's interval method (4.72). The reason is the great efficiency of this iterative scheme (see §6.3). The analysis of computational amount of work and a great numerical examples have shown that the following point iterative methods are suitable for a combination with Maehly's interval method:

- Weierstrass' method (3.3);
- Maehly's method (4.119);
- Maehly's method with Newton's correction (4.122) (k = 1).

In the case of multiple zeros, Schröder's modification of Newton's method

$$\hat{z}_i = z_i - \mu_i \frac{P(z_i)}{P'(z_i)} \tag{4.126}$$

may be applied instead of Weierstrass' method (3.3). In this manner we construct the following four combined methods:

$(K_W)$: ((3.3),(4.72)),

$(K_N)$: ((4.126),(4.72)),

$(K_M)$: ((4.119),(4.72)),

$(K_{MN})$: ((4.122),(4.72)).

The subscripts W,N,M,MN indicate the used point iterative methods.

Let $Z_1^{(0)} = \{z_1^{(0)}; r_1^{(0)}\}, \ldots, Z_n^{(0)} = \{z_n^{(0)}; r_n^{(0)}\}$ be the initial inclusion disks including the zeros $\xi_1, \ldots, \xi_n$, and let $z_1^{(0)}, \ldots, z_n^{(0)}$ be the starting approximations for a point iterative method of the order $k_P$. The improved approximations generated by this method after M iterative steps will be denoted with $z_1^{(M)}, \ldots, z_n^{(M)}$. Applying Maehly's interval method in the final iteration, we find the improved circular approximations given by

$$Z_i^{(M,1)} = z_i^{(M)} - \frac{\mu_i}{\dfrac{P'(z_i^{(M)})}{P(z_i^{(M)})} - \sum\limits_{\substack{j=1 \\ j\neq i}}^{n} \mu_j (z_i^{(M)} - Z_j^{(0)})^{-1}} \quad (i = 1,\ldots,n). \quad (4.127)$$

Let $r^{(m)} = \max\limits_{1\le i\le n} \operatorname{rad} Z_i^{(m)}$ be the largest radius of disks produced by Maehly's interval method at the m-th iteration. The notations $r_N^{(M,1)}$ and $r_M^{(M,1)}$ have a similar meaning for the combined methods $(K_N)$ and $(K_M)$, where Newton's and Maehly's point methods are applied M times. We have the following estimation of the circular approximations generated by (4.127):

$$r^{(M,1)} = O\left((r^{(0)})^{2k_P^M + 1}\right),$$

where $r^{(0)} = \max\limits_{1\le i\le n} r_i^{(0)}$ and $r^{(M,1)} = \max\limits_{1\le i\le n} \operatorname{rad} Z_i^{(M,1)}$, assuming that $r^{(0)}$ is reasonable small.

The efficiency and the convergence properties of Maehly's combined methods are discussed by means of some numerical examples. For comparison, Maehly's interval method (4.72) is also considered in the same examples.

**EXAMPLE 12.** Consider the polynomial

$$P(z) = z^{11} + (-7+2i)z^{10} + (11-14i)z^9 + (19+24i)z^8 + (-70+24i)z^7$$
$$+(42-116i)z^6 + (198+108i)z^5 + (-234+280i)z^4 + (-491-360i)z^3$$
$$+(45-702i)z^2 + (351-270i)z + 135,$$

whose factorization is

$$P(z) = (z+1)^4(z-3)^3(z+i)^2(z^2-2z+5).$$

The initial inclusion disks were selected to be

$$Z_1^{(0)} = \{-0.7 + 0.3i \; ; \; 0.7\}, \quad Z_2^{(0)} = \{2.7 + 0.2i \; ; \; 0.7\}, \quad Z_3^{(0)} = \{0.2 - 1.2i \; ; \; 0.7\},$$
$$Z_4^{(0)} = \{1.1 - 2.1i \; ; \; 0.7\}, \quad Z_5^{(0)} = \{1.1 + 2.1i \; ; \; 0.7\}.$$

Applying Maehly's interval method, after the first step we found $r^{(1)} = 9.\times 10^{-2}$. For the second iteration we obtained the following disks:

$$Z_1^{(2)} = \{-1.000000039 - 3.6\times 10^{-7}i \; ; \; 8.\times 10^{-7}\},$$
$$Z_2^{(2)} = \{2.999999906 + 5.1\times 10^{-8}i \; ; \; 4.48\times 10^{-7}\},$$
$$Z_3^{(2)} = \{1.79\times 10^{-5} - 0.999977837i \; ; \; 1.27\times 10^{-4}\},$$
$$Z_4^{(2)} = \{1.000078637 - 1.999943391i \; ; \; 2.86\times 10^{-4}\},$$
$$Z_5^{(2)} = \{0.999998794 + 1.999998722i \; ; \; 9.25\times 10^{-6}\}.$$

For the combined method $(K_N)$, after one Newton's point iterative step and one Maehly's interval iteration, the value $r_N^{(1,1)}$ was less than $5\times 10^{-3}$. Using two Newton's point iterations, from (4.127) we obtained the inclusion disks as listed below:

$$Z_1^{(2,1)} = \{-1.0000014 - 8.17\times 10^{-7}i \; ; \; 1.38\times 10^{-5}\},$$
$$Z_2^{(2,1)} = \{2.9999953 - 1.19\times 10^{-6}i \; ; \; 1.73\times 10^{-5}\},$$
$$Z_3^{(2,1)} = \{5.24\times 10^{-6} - 0.9999955i \; ; \; 1.29\times 10^{-5}\},$$
$$Z_4^{(2,1)} = \{0.9999781 - 2.0000475i \; ; \; 7.66\times 10^{-5}\},$$
$$Z_5^{(2,1)} = \{1.0000069 + 2.0000033i \; ; \; 1.61\times 10^{-5}\}.$$

The combined method $(K_M)$ produced slightly better results compared to $(K_N)$, although only one point iterative step was employed! By applying $(K_M)$ we furnished the following disks:

$$Z_1^{(1,1)} = \{-0.999999971 + 1.39 \times 10^{-8} i \,;\, 2.78 \times 10^{-7}\},$$
$$Z_2^{(1,1)} = \{3.000000064 - 1.42 \times 10^{-7} i \,:\, 5.53 \times 10^{-7}\},$$
$$Z_3^{(1,1)} = \{-1.91 \times 10^{-5} - 1.000012832 i \,;\, 4.31 \times 10^{-5}\},$$
$$Z_4^{(1,1)} = \{0.999997748 - 2.000004602 i \,;\, 7.54 \times 10^{-6}\},$$
$$Z_5^{(1,1)} = \{1.000000026 + 2.000000206 i \,;\, 4.34 \times 10^{-7}\}.$$

We note that these circular approximations are even better than those obtained by Maehly's interval method applying two iterative steps. Since the combined method $(K_M)$ requires a smaller number of operations, its efficiency is evident (see § 6.4).

**EXAMPLE 13.** Maehly's interval method (4.72) and the combined methods $(K_N)$ and $(K_M)$ were applied for solving the polynomial equation

$$z^9 + z^8 + 46z^7 + 46z^6 + 436z^5 + 420z^4 - 1950z^3 - 2750z^2 + 6875z - 3125 = 0, \qquad (4.128)$$

whose roots are $\xi_1 = 1$, $\xi_{2,3} = -2 \pm i$ and $\xi_{4.5} = \pm 5i$ with the multiplicities $\mu_1 = 3$, $\mu_2 = \mu_3 = 1$ and $\mu_4 = \mu_5 = 2$. The following initial disks for these zeros were taken:

$$Z_1^{(0)} = \{1.2 + 0.3i \,;\, 1.5\}, \quad Z_2^{(0)} = \{-2.2 + 1.2i \,;\, 1.5\}, \quad Z_3^{(0)} = \{-2.2 - 1.2i \,;\, 1.5\},$$
$$Z_4^{(0)} = \{0.3 + 4.7i \,;\, 1.5\}, \quad Z_5^{(0)} = \{0.3 - 4.7i \,;\, 1.5\}.$$

We analyzed the behavior of the applied methods changing only the position of the center of the disk $Z_1^{(0)}$. Starting with the above given approximations, by Maehly's interval method (4.72) we obtained:

$$Z_1^{(2)} = \{1.000000044 + 2.97 \times 10^{-8} i \,;\, 1.19 \times 10^{-7}\},$$
$$Z_2^{(2)} = \{-1.99994983 + 0.99992107 i \,;\, 2.16 \times 10^{-4}\},$$
$$Z_3^{(2)} = \{-1.99991838 - 0.99997813 i \,;\, 1.99 \times 10^{-4}\},$$
$$Z_4^{(2)} = \{2.29 \times 10^{-7} + 5.00000006 i \,;\, 7.08 \times 10^{-7}\},$$
$$Z_5^{(2)} = \{-1.9 \times 10^{-8} - 5.000000021 i \,;\, 7.4 \times 10^{-8}\}.$$

The combined method $(K_N)$ produced:

$$Z_1^{(2,1)} = \{0.9999999996 + 1.14 \times 10^{-9} i \,;\, 1.18 \times 10^{-8}\},$$
$$Z_2^{(2,1)} = \{-2.0000105 + 0.9999789 i \,;\, 8.44 \times 10^{-5}\},$$
$$Z_3^{(2,1)} = \{-1.9999993 - 0.9999725 i \,;\, 8.05 \times 10^{-5}\},$$

$Z_4^{(2,1)} = \{3.75 \times 10^{-6} + 4.9999969i \; ; \; 1.44 \times 10^{-5}\}$,

$Z_5^{(2,1)} = \{1.6 \times 10^{-6} - 4.9999996i \; ; \; 1.26 \times 10^{-5}\}$.

The combined method $(K_M)$ furnished the following circular approximations by only one point iteration:

$Z_1^{(1,1)} = \{1.0000000004 + 1.41 \times 10^{-9} i \; ; \; 1.44 \times 10^{-8}\}$,

$Z_2^{(1,1)} = \{-1.999998432 + 0.999999961i \; ; \; 5.73 \times 10^{-6}\}$,

$Z_3^{(1,1)} = \{-1.999998874 - 0.999998957i \; ; \; 4.56 \times 10^{-6}\}$,

$Z_4^{(1,1)} = \{1.76 \times 10^{-8} + 4.999999985i \; ; \; 6.87 \times 10^{-8}\}$,

$Z_5^{(1,1)} = \{1.74 \times 10^{-8} - 4.999999996i \; ; \; 1.36 \times 10^{-7}\}$.

For the combined method $(K_N)$ we found $r_N^{(1,1)} = 8.61 \times 10^{-3}$, while Maehly's interval method produced in the first iterative step $r^{(1)} = 0.12$. On the other hand, we observe that the disks $Z_i^{(1,1)}$, generated by $(K_M)$, are smaller that those determined by $(K_N)$ after two Newton's point iterations as well as by Maehly's interval method in two iterations.

As noticed from this example, interval methods may converge very slowly at the beginning of the iterative process (for instance, $r^{(1)} = 0.12$) because of the large initial disks or their "bad separation". But, the situation may be even worse; namely, the size of initial disks and their distribution can be such that the applied interval method does not converge. The following example illustrates the described situation.

Dislocating only the center of the initial disk $Z_1^{(0)} = \{1.2 + 0.3i \; ; \; 1.5\}$ to be $z_1^{(0)} = -0.2 - 0.7i$ (and, consequently, changing the position of the new disk $Z_1^{(0)} = \{-0.2 - 0.7i \; ; \; 1.5\}$ in reference to the other disks), we obtain such distribution of the initial disks which leads to the break of Maehly's interval method because the denominator in (4.72) includes the number 0. At the same time, the combined method $(K_M)$ converges very fast with the same initial approximations. We found $r_M^{(1,1)} = 8.76 \times 10^{-4}$, while two point iterative steps of Maehly's type produced the disks

$Z_1^{(2,1)} = \{1.00000000000000000058 - 8.26 \times 10^{-19} i \; ; \; 3.73 \times 10^{-18}\}$,

$Z_2^{(2,1)} = \{-2.000000000000000296 + 1.000000000000000063i \; ; \; 4.23 \times 10^{-16}\}$,

$Z_3^{(2,1)} = \{-2.000000000000000520 - 1.000000000000000120i \; ; \; 6.34 \times 10^{-16}\}$,

$Z_4^{(2,1)} = \{1.21 \times 10^{-20} + 4.99999999999999999996i \; ; \; 8 \times 10^{-20}\}$,

$Z_5^{(2,1)} = \{2.13 \times 10^{-20} - 4.99999999999999999996i \; ; \; 6.79 \times 10^{-20}\}$.

This example as well as many others of polynomial equations which were tested, points to the advantage of combined methods consisting of the convergence of these methods under weaker initial conditions compared to interval methods.

We also solved the equation (4.128) with the unchanged disks $Z_2^{(0)}$, $Z_3^{(0)}$, $Z_4^{(0)}$, $Z_5^{(0)}$ and (again) the new disk $Z_1^{(0)} = \{0.2 + 0.7i \; ; \; 1.5\}$. The largest radii concerning the three applied methods are displayed in Table 4.8.

| | m = 1 | m = 2 | m = 3 | m = 4 |
|---|---|---|---|---|
| $r^{(m)}$ | $6.16 \times 10^{-1}$ | $7.13 \times 10^{-2}$ | $1.37 \times 10^{-6}$ | |
| $r_N^{(m,1)}$ | $2.53 \times 10^{-2}$ | $1.94 \times 10^{-4}$ | $1.83 \times 10^{-8}$ | $2.58 \times 10^{-16}$ |
| $r_M^{(m,1)}$ | $1.17 \times 10^{-4}$ | $1.97 \times 10^{-18}$ | | |

Table 4.8

## 4.5. SOME APPLICATIONS OF CIRCULAR INTERVAL ALGORITHMS

In Sections 4.3 and 5.3 we have presented the algorithms for the simultaneous inclusion of k ( < n ) zeros of a polynomial, where n is the number of different zeros. Special case when k = 1 is considered in §5.4 (the iterative formula (5.86) and, particularly, (5.87) and (5.88)). Such procedures could also be applied to the root iteration methods (4.14) and (4.68). From a practical point of view the case k = 1, considered in this section, is particularly interesting for the following reasons: (*i*) it provides the construction of an iterative process of the order k + 1 for finding a simple zero of a polynomial; (*ii*) zero-finding algorithm can be extended from polynomials to entire functions with infinitely many zeros and exponent of convergence < 1; (*iii*) a computational test for the existence of a simple polynomial zero may be established. The aforementioned applications of circular algorithms are considered at the end of this chapter as some kind of consecutive results obtained from the root iteration methods.

### INCLUSION OF ZEROS OF ENTIRE FUNCTIONS

The results concerning the inclusion of zeros of an entire function, introduced by Gargantini and Henrici [37], are here presented in their original form.

Let f be an entire function. The *order* of f is the greatest lower bound of all numbers $\alpha$ such that

$$|f(z)| < e^{|z|^{\alpha}}$$

for all sufficiently large $|z|$. If $\xi_1, \xi_2, \ldots$ are the zeros of f different from zero, the *exponent of convergence* of f is the greatest lower bound of all numbers $\beta > 0$ such that

$$\sum \frac{1}{|\xi_i|^{\beta}} < +\infty .$$

It can be shown (see, e.g. [120]) that the exponent of convergence never exceeds the order.

Here we will consider entire functions with infinitely many zeros and exponent of convergence < 1. Such functions necessarily have the form

$$f(z) = z^{\lambda} e^{g(z)} \prod_{i=1}^{\infty} \left(1 - \frac{z}{\xi_i}\right) , \tag{4.129}$$

where $\lambda$ is a nonnegative integer, and where g is an entire function. Using the logarithmic derivative we obtain

$$u(z) := \frac{f'(z)}{f(z)} = g'(z) + \sum_{i=0}^{\infty} \frac{1}{z - \xi_i} , \tag{4.130}$$

where the sequence $(\xi_i)$ has been renumbered to include any possible zeros at the origin.

To approximate one zero of f, say $\xi_0$, we claim disks $Z_i$ such that $\xi_i \in Z_i$ $(i = 1,2,\ldots)$, and the values of the functions $u(z)$ and $g'(z)$ at a point $z_0$ not contained in any of the disks $Z_1, Z_2, \ldots$ . From (4.130) we obtain

$$\xi_0 = z_0 - \frac{1}{u(z_0) - \left[ g'(z_0) + \sum_{i=1}^{\infty} (z_0 - \xi_i)^{-1} \right]} . \tag{4.131}$$

Under the assumption that the radii of the disks $Z_i$ form a bounded sequence, the series of disks

$$C = \sum_{i=1}^{\infty} \frac{1}{z_0 - Z_i}$$

converges and defines a disk C such that

$$\sum_{i=1}^{\infty} \frac{1}{z_0 - \xi_i} \in C.$$

Using the last relation and the inclusion isotonicity, from (4.131) we obtain that $\xi_0 \in Z_0$, where

$$Z_0 = z_0 - \frac{1}{u(z_0) - [g'(z_0) + C]} . \tag{4.132}$$

The following assertion was proved in [37]:

**THEOREM 4.13.** *Let $f$ be an entire function with infinitely many zeros $\xi_0, \xi_1, \ldots$ of exponent of convergence < 1. If $\xi_i \in Z_i$ $(i = 1,2,\ldots)$, where $Z_1, Z_2, \ldots$ is a sequence of disks whose radii are bounded, then for any $z_0$ not contained in $Z_1, Z_2, \ldots$ the zero $\xi_0$ lies in the circular region $Z_0$ defined by (4.132).*

Theorem 4.13 can be applied to obtain rigorous bounds for the solution of certain transcendental equations. For a numerical illustration, we present an example taken from [37].

**EXAMPLE 14.** Let $k(z)$ be an even entire function of order one whose infinitely many zeros are real, $k(0) \neq 0$. Then $f(z) := k(\sqrt{z})$ is an entire function of order $\frac{1}{2}$ which can be represented in the form (4.129) where $\lambda = 0$, $g(z) = \text{const.}$, and all zeros $\xi_i$ are positive. We assume that the positive zeros of $k$ with the exception of the smallest one are known to lie in the intervals $[n+\alpha, n+\beta]$, where $0 < \alpha < \beta$ and $n = 1,2,\ldots$ . The corresponding zeros of $f$ are then contained in the disks

$$Z_n = \left\{ \tfrac{1}{2}[(n+\alpha)^2 + (n+\beta)^2] ;\ \tfrac{1}{2}[(n+\beta)^2 - (n+\alpha)^2] \right\} .$$

To apply Theorem 4.12 with $z_0 = 0$, we find $u(0) = f'(0)/f(0)$ and the disk

$$C = \{c\,;\,r\} = -\sum_{n=1}^{\infty} \frac{1}{Z_n} .$$

Since

$$\frac{1}{Z_n} = \left\{ \frac{(n+\alpha)^2 + (n+\beta)^2}{2(n+\alpha)^2(n+\beta)^2} ;\ \frac{(n+\beta)^2 - (n+\alpha)^2}{2(n+\alpha)^2(n+\beta)^2} \right\} ,$$

we have

$$c = -\tfrac{1}{2}[\psi(1+\alpha) + \psi(1+\beta)], \qquad r = \tfrac{1}{2}[\psi(1+\alpha) - \psi(1+\beta)] ,$$

where $\psi(1+\sigma) = \sum_{n=1}^{\infty} (n+\sigma)^{-2}$ is the Digamma function, tabulated, for instance, in [2].

In particular, let us determine the smallest positive zero of the Bessel function $J_\mu$. The function $k(z) = \text{const}\cdot z^{-\mu} J_\mu(\pi z)$ satisfies the above hypotheses for $-\frac{1}{2} < \mu \leq \frac{1}{2}$, where

$$\alpha = \frac{3}{4} + \frac{\mu}{2}, \qquad \beta = \frac{7}{8} + \frac{\mu}{4}$$

(see [132, p. 490]). For a suitable choice of the constant we have

$$f(0) = 1, \qquad f'(0) = -\frac{\pi^2}{4(\mu+1)} \quad .$$

There follows

$$Z_0 = \frac{\pi^2}{4(\mu+1)} - \left\{ \frac{\psi(\frac{7}{4}+\frac{\mu}{2}) + \psi(\frac{15}{8}+\frac{\mu}{4})}{2} \;;\; \frac{\psi(\frac{7}{4}+\frac{\mu}{2}) - \psi(\frac{15}{8}+\frac{\mu}{4})}{2} \right\}^{-1} .$$

The following inclusion intervals were obtained for $\mu = 0, \frac{1}{3}$ and $\frac{1}{2}$:

| $\mu$ | $Z_0$ | Exact value of $\xi_0$ |
|---|---|---|
| 0 | {0.57637 ; 0.01071} | 0.58596 |
| $\frac{1}{3}$ | {0.84878 ; 0.00677} | 0.85363 |
| $\frac{1}{2}$ | {1.00000 ; 0.00000} | 1.00000 |

Table 4.9

## SOME COMPUTATIONAL TESTS FOR THE EXISTENCE OF POLYNOMIAL ZEROS

The problem of finding the zeros of any analytic function

$$f(z) = f(x+iy) = u(x,y) + i\,v(x,y)$$

can be regarded as the problem of solving the pair of simultaneous equations

$$u(x,y) = 0,$$
$$v(x,y) = 0.$$

Suppose that it is required to find those zeros of f lying in a region S in the complex plane. Most frequently, S is a disk or a rectangle with sides parallel to the coordinate axes. In order to examine whether S may contain more than one zero, we can use the following result given in [41]:

**THEOREM 4.14.** *Let S be a closed convex region and let $f(z) = u(x,y) + iv(x,y)$ be an analytic function. If f(z) has two or more zeros in S, then the partial derivatives $u_x$, $u_y$, $v_x$ and $v_y$ all take the value zero somewhere in S, though not necessarily simultaneosly.*

Since f is analytic, by the Cauchy-Riemann relations we have $u_x = v_y$ and $u_y = -v_x$ and so, it is sufficient to consider only two partial derivatives, say $u_x$ and $v_x$, with $f'(z) = u_x + i v_x$.

Let $Z = \{z ; r\} = \{w: |w-z| \leq r\}$ be a given disk with center $z = x + iy = \text{mid } Z$ and radius $r = \text{rad } Z$, and let $S = X + iY = is(Z)$ be the inclusion square for the disk Z, where $X = [x-r, x+r]$, $Y = [y-r, y+r]$. Further, let $F'(S) = F'(X+iY) \supseteq \{f'(x+iy) = u_x(x,y) + iv_x(x,y) : x+iy \in S\}$ be rectangular interval extension of f' over S. Since F'(S) is a rectangle with sides parallel to the coordinate axes, the condition

$$0 \notin F'(S) \tag{4.133}$$

implies that $u_x$ and $v_x$ do not both vanish anywhere in S. Therefore, according to Theorem 4.14, a sufficient condition for S to include at most one zero of f is given by (4.133). This condition is, obviously, computationally verifiable.

As a particular case we will consider a polynomial P having real or complex coefficients. As above, let $Z = \{z ; r\}$ be a given disk and $S = X + iY = is(Z)$ its inclusion square. If $P(z) = P(x+iy) = u(x,y)+iv(x,y)$ and $\mathbb{P}'(is(Z))$ denotes an interval extension of the derivatives P' over the square $S = is(Z)$, then, in view of (4.133), the polynomial P will contain at most one zero in the disk Z if

$$0 \notin \mathbb{P}'(is(Z)). \tag{4.134}$$

Rectangular extension $\mathbb{P}'(is(Z))$ can be computed, for instance, applying the Horner scheme in rectangular arithmetic.

For simple complex zeros $\xi_1, \ldots, \xi_n$ of the polynomial of degree n we have derived in § 4.1 the fixed point relation

$$\xi_i = z - \frac{1}{\left[h_k(z) - \sum_{\substack{j=1 \\ j \neq i}}^{n} (z - \xi_j)^{-k}\right]_*^{1/k}} \quad (i = 1, \ldots, n). \tag{4.135}$$

Assume that we have found an inclusion disk $\{z: |z-a| \leq R\} = \{a ; R\}$ with center a and radius R containing only one simple zero $\xi_i$ of P. All other zeros are supposed to lie in the region $W = \{w: |w-a| > R\}$, that is, in the exterior of the disk $\{a ; R\}$. Using the inclusion isotonicity we get

$$\frac{1}{(z-\xi_j)^k} \in \left(\frac{1}{z-W}\right)^k \quad (j \neq i;\ z \in \{a;R\}). \tag{4.136}$$

Since $z \notin W$, that is $|z-a| < R$, according to (2.10) we find

$$(z-W)^{-1} = \left\{ w:\ \left| w - \frac{\bar{a}-\bar{z}}{R^2-|a-z|^2} \right| \leq \frac{R}{R^2-|a-z|^2} \right\} = \{c ; d\},$$

where

$$c = \text{mid}\ (z-W)^{-1} = \frac{\bar{a}-\bar{z}}{R^2-|a-z|^2},$$

$$d = \text{rad}\ (z-W)^{-1} = \frac{R}{R^2-|a-z|^2}.$$

Therefore, from (4.136) there follows

$$\frac{1}{(z-\xi_j)^k} \in \{c ; d\}^k. \tag{4.137}$$

By virtue of (4.135) and (4.137) we obtain the inclusion relation

$$\xi_i \in z - \frac{1}{\left[h_k(z) - (n-1)\{c ; d\}^k\right]_*^{1/k}}. \tag{4.138}$$

The k-th root of a disk is the union of k disks (formula (2.17)) and the symbol $*$ in (4.138) denotes that only one of the k disks is chosen in accordance with CCR (see §4.1). We recall that the selected disks should contain the value $(z-\xi_i)^{-1}$.

Let $Z^{(0)} = Z = \{a ; R\}$. The relation (4.138) suggests the following iterative process for the inclusion of simple complex zeros $\xi_i$ of P:

$$Z^{(m+1)} = Y_k(z^{(m)}, Z) = z^{(m)} - \frac{1}{\left[h_k(z^{(m)}) - (n-1)\{c^{(m)} ; d^{(m)}\}^k\right]_*^{1/k}}$$

$$(m = 0,1,\ldots), \tag{4.139}$$

where

$$z^{(m)} = \text{mid}\ Z^{(m)}, \quad c^{(m)} = \frac{\bar{a}-\bar{z}^{(m)}}{R^2-|a-z^{(m)}|^2}, \quad d^{(m)} = \frac{R}{R^2-|a-z^{(m)}|^2}$$

and $Y_k$ is the interval operator. From (4.138), it follows that $\xi_i \in Z^{(m)}$

for each $m = 0,1,\ldots$ if the inequality $|z^{(m)} - a| < R$ is satisfied. The validity of this inequality can be provided under suitable initial condition depending on $h(a)$ and $Z$, similarly as in [92].

Similarly to the convergence analysis presented by Gargantini [33] (see § 4.4, Theorem 4.8), it is easy to show that the convergence order of the iterative interval method (4.139) is $k+1$. Because of the great computational cost, the next two special cases, following from (4.139) for $k = 1$ and $k = 2$, have a particular importance:

$$Z^{(m+1)} = Y_1(z^{(m)}, Z) = z^{(m)} - \frac{1}{\dfrac{P'(z^{(m)})}{P(z^{(m)})} - (n-1)\{c^{(m)}; d^{(m)}\}} \qquad (4.140)$$

$$(m = 0,1,\ldots),$$

and

$$Z^{(m+1)} = Y_2(z^{(m)}, Z) = z^{(m)} - \frac{1}{\left[h_2(z^{(m)}) - (n-1)\{c^{(m)}; d^{(m)}\}^2\right]_*^{1/2}}$$

$$(m = 0,1,\ldots), \qquad (1.141)$$

where $h_2(z) = \dfrac{P'(z)^2 - P(z)P''(z)}{P(z)^2}$.

We note that (4.141) is a special case of the iterative formula (5.87) for a multiple complex zero. The convergence order of the iterative methods (4.140) and (4.141) is *two* and *three*, respectively.

**REMARK 13.** Let $Q_k^{(m)} = h_k(z^{(m)}) - (n-1)\{c^{(m)}; d^{(m)}\}^k$ and let

$$Q_k^{(m)\,1/k} = \bigcup_{\lambda=1}^{k} \{q_\lambda^{(m)}; \varepsilon^{(m)}\}.$$

With regard to CCR, described in § 4.1, the disk to be chosen in (4.139) is that whose center minimizes

$$\left| \frac{P'(z^{(m)})}{P(z^{(m)})} - q_\lambda^{(m)} \right| \qquad (\lambda = 1,\ldots,k). \ ®$$

Let $Z^{(0)} = Z = \{a; R\}$. Then $z^{(0)} = a$, $c^{(0)} = 0$, $d^{(0)} = 1/R$ and formula (4.139) is simplified to

$$Y_k(a,Z) = a - \frac{1}{\left\{h_k(a);\ (n-1)/R^k\right\}_*^{1/k}}. \qquad (4.142)$$

The following theorem gives a computational test for the existence of a simple zero of a polynomial P in a given disk Z.

**THEOREM 4.15.** *Let $Z = \{a\,;R\}$ be a given disk and let $F'(is(Z))$ be a rectangular extension of the derivative P' over the inclusive square $S = is(Z)$ for which (4.134) is satisfied. In addition, assume that the inequality*

$$|h_k(a)| > \frac{n-1}{R^k} \tag{4.143}$$

*is valid. Then we have the following statements:*

*$1^o$ If Z contains a zero of P, then so does $Y_k(a,Z)$;*

*$2^o$ If $Y_k(a,Z) \cap Z = \emptyset$, then Z is free of any zeros of P;*

*$3^o$ If $Y_k(a,Z) \subseteq Z$, then there exists a unique simple zero of P in Z.*

*Proof.* With regard to (4.143), the disk $\{h_k(a)\,;\,(n-1)/R^k\}$ does not contain the origin and, therefore $0 \notin \{h_k(a)\,;\,(n-1)/R^k\}^{1/k}$ (see § 2.2). Consequently, there exists the inverse disk in (4.142) and $Y_k(a,Z)$ is also a disk.

From (4.134) we conclude that there are at least n-1 zeros of P outside of the disk Z. Without loss of generality, we will denote these zeros with $\xi_2,\ldots,\xi_n$. Furthermore, evidently,

$$|a - \xi_j| > R \quad (j = 2,\ldots,n). \tag{4.144}$$

Suppose that the remaining zero $\xi_1$ belongs to the disk Z and let W be the exterior of Z, i.e., $W = \{w\colon |w-a| > R\}$. Since $\xi_j \in W$ $(j = 2,\ldots,n)$ and

$$(a - W)^{-1} = \{w'\colon |w'| > R\}^{-1} = \left\{0\,;\,\frac{1}{R}\right\},$$

we have

$$\sum_{j=2}^{n} (a-\xi_j)^{-k} \in \sum_{j=2}^{n} (a-W)^{-k} = (n-1)(a-W)^{-k} = \left\{0\,;\,\frac{n-1}{R^k}\right\}. \tag{4.145}$$

Using (4.145) and the identity

$$h_k(z) = \sum_{j=1}^{n} (z-\xi_j)^{-k} \tag{4.146}$$

(see § 4.1, p. 70), we obtain

$$\xi_1 = a - \frac{1}{(a-\xi_1)^{-1}} = a - \frac{1}{\left[\sum_{j=1}^{n}(a-\xi_j)^{-k} - \sum_{j=2}^{n}(a-\xi_j)^{-k}\right]_*^{1/k}}$$

$$\in a - \frac{1}{\left[h_k(a) - \{0\,;\,(n-1)/R^k\}\right]_*^{1/k}} = Y_k(a,Z),$$

which proves $1^\circ$.

The assertion $2^\circ$ follows immediately from $1^\circ$. Namely, we have just proved that, if the zero $\xi_1$ lies in $Z$, then also $\xi_1 \in Y_k(a,Z)$. Therefore, $Y_k(a,Z) \cap Z \neq \emptyset$ (because there exists at least one common point for the disks $Z$ and $Y_k(a,Z)$ - the zero $\xi_1$). Otherwise, if this intersection is empty, then there is no zero of $P$ in $Z$.

To prove $3^\circ$ we start with the assumption that the inclusion $Y_k(a,Z) \subseteq Z$ is valid, which is equivalent to the inequality

$$|a - \operatorname{mid} Y_k(a,Z)| \leq R - \operatorname{rad} Y_k(a,Z). \tag{4.147}$$

In view of (4.142) and (2.17) we find

$$\operatorname{mid} Y_k(a,Z) = a - \frac{\overline{[h_k(a)]^{1/k}}}{|h_k(a)|^{2/k} - v^2},$$

$$\operatorname{rad} Y_k(a,Z) = \frac{v}{|h_k(a)|^{2/k} - v^2},$$

where

$$v = |h_k(a)|^{1/k} - \left(|h_k(a)| - \frac{n-1}{R^k}\right)^{1/k}.$$

The inequality (4.147) now becomes

$$\frac{|h_k(a)|^{1/k}}{|h_k(a)|^{2/k} - v^2} < R - \frac{v}{|h_k(a)|^{2/k} - v^2},$$

or

$$\frac{1}{|h_k(a)|^{1/k} - v} < R,$$

whence

$$\frac{1}{\left(|h_k(a)| - \frac{n-1}{R^k}\right)^{1/k}} < R. \tag{4.148}$$

From the condition (4.143) we conclude that either the disk $Z$ does not contain any zero of $P$ or $Z$ contains at most one zero, say $\xi_1$. Using (4.146) and (4.144) we estimate

$$|h_k(a)| = \left|\sum_{j=1}^{n}(a-\xi_j)^{-k}\right| < |a-\xi_1|^{-k} + \sum_{j=2}^{n}|a-\xi_j|^{-k} < |a-\xi_1|^{-k} + \frac{n-1}{R^k}.$$

Taking into account the last bound and the inequality (4.148), we can write

$$\frac{1}{\left[\left(|a-\xi_1|^{-k} + \frac{n-1}{R^k}\right) - \frac{n-1}{R^k}\right]^{1/k}} < \frac{1}{\left[|h_k(a)| - \frac{n-1}{R^k}\right]^{1/k}} < R,$$

that is,

$$|a-\xi_1| < R,$$

which means that $\xi_1 \in \{a\,;\,R\} = Z$. □

Let $\mathbb{P}_c(Z)$ be a circular extension of $P$ over the given disk $Z$. At the beginning of the test of existence we also check whether $0 \in \mathbb{P}_c(Z)$, which is a necessary condition for $Z$ to contain any zero of $P$. According to the previous facts, if for the initial disk $Z^{(0)}$ we have

$$0 \in \mathbb{P}_c(Z^{(0)}) \wedge 0 \notin \mathbb{P}'(is(Z^{(0)})) \wedge Y_k(z^{(0)}, Z^{(0)}) \subseteq Z^{(0)},$$

then there exists a unique simple zero of $P$ in $Z^{(0)}$.

The test of the existence of a polynomial zero, presented in Theorem 4.15, is computationally verifiable and can be usefully combined with some of the methods from the family (4.139) for the improvement of circular approximation of the requested zero. Obviously, it is much simpler to check the inclusion $Y_1(a,Z) \subseteq Z$ rather than $Y_k(a,Z) \subseteq Z$ $(k > 1)$, but the iterative methods (4.139) for $k > 1$ have a fast convergence. Moreover, the calculated disk $Y_k(a,Z)$ has already been used as the first improved circular approximation in the iterative process. The presented tests of existence and the corresponding iterative methods were illustrated by two examples for $k = 1$ and $k = 2$.

**EXAMPLE 15.** Consider the polynomial

$$P(z) = z^4 - 5z^3 - 5z^2 - 5z - 6$$

whith the roots -1, ±i, 6, and the initial region [3.2 , 7.2] + i[-1 , 3]. We divide this square into four subsquares

$$S_1 = [3.2\ ,\ 5.2] + i[-1\ ,\ 1],\qquad S_2 = [3.2\ ,\ 5.2] + i[1\ ,\ 3],$$

$$S_3 = [5.2\ ,\ 7.2] + i[1\ ,\ 3],\qquad S_4 = [5.2\ ,\ 7.2] + i[-1\ ,\ 1]$$

and, using the Horner scheme $\mathbb{P}'(S) = ((4S-15)\cdot S-10)\cdot S-5$ in rectangular arithmetic, we calculate

$$\mathbb{P}'(S_1) = [-163.888\ ,\ 147.232] + i[-163.76\ ,\ 163.76]\ ,$$

$$\mathbb{P}'(S_2) = [-543.088\ ,\ 72.832] + i[-152.48\ ,\ 463.44]\ ,$$

$$\mathbb{P}'(S_3) = [-502.448\ ,\ 582.992] + i[90.8\ ,\ 1176.24]\ ,$$

$$\mathbb{P}'(S_4) = [36.432\ ,\ 709.792] + i[-400.08\ ,\ 400.08]\ .$$

Since $0 \notin \mathbb{P}'(S_3)$ and $0 \notin \mathbb{P}'(S_4)$, the squares $S_3$ and $S_4$ may contain at most one zero of P. Let $Z_3$ be the disk enclosed by the square $S_3$, that is, $S_3 = is(Z_3)$. The inequality (4.143) is not valid at the point $a = \text{mid } Z_3$ (with $R = 1$ and $n = 4$) and hence, $Y_1(\text{mid } Z_3\ ,\ Z_3)$ is not a closed circular region. For this reason and the fact that $0 \in \mathbb{P}(S_3)$, we divide $S_3$ into four subsquares

$$S_{3,1} = [5.2\ ,\ 6.2] + i[1\ ,\ 2],\qquad S_{3,2} = [5.2\ ,\ 6.2] + i[2\ ,\ 3],$$

$$S_{3,3} = [6.2\ ,\ 7.2] + i[2\ ,\ 3],\qquad S_{3,4} = [6.2\ ,\ 7.2] + i[1\ ,\ 2].$$

By the Horner scheme $\mathbb{P}(S) = (((S-5)\cdot S-5)\cdot S-5)\cdot S-6$ we calculate

$$\mathbb{P}(S_{3,1}) = [-715.7824\ ,\ -25.8144] + i[-104.672\ ,\ 585.296]\ ,$$

$$\mathbb{P}(S_{3,2}) = [-1351.6624\ ,\ -221.0624] + i[-336.\ ,\ 764.6]\ ,$$

$$\mathbb{P}(S_{3,3}) = [-1661.8704\ ,\ -44.6944] + i[36.4\ ,\ 1653.576]\ ,$$

$$\mathbb{P}(S_{3,4}) = [-703.6704\ ,\ 372.0656] + i[157.448\ ,\ 1233.184]\ .$$

Thus, $0 \notin \mathbb{P}(S_{3,j})$ for each $j = 1,2,3,4$ and, consequently, the square $S_3$ does not contain any zero of P.

The remaining square $S_4 = [5.2\ ,\ 7.2] + i[-1\ ,\ 1]$ is the inclusion square for the disk $Z^{(0)} = \{6.2\ ;\ 1\}$, that is, $S_4 = S = is(Z^{(0)})$. Let $z^{(m)} = \text{mid } Z^{(m)}$ $(m = 0,1,\ldots)$. Since

$$|h_1(z^{(0)})| = \left|\frac{P'(z^{(0)})}{P(z^{(0)})}\right| \cong 5.46 > \frac{n-1}{R} = 3\ ,$$

the conditions mentioned in Theorem 4.15 (for $k=1$) are satisfied. Using (4.142) for $k=1$, we find

$$Y_1(z^{(0)}, Z^{(0)}) = \{5.937\ ;\ 0.145\},$$

which means that

$$Y_1(z^{(0)}, Z^{(0)}) \subset Z^{(0)}.$$

Therefore, there exists a unique simple zero of P in $Z^{(0)}$, which is also contained in $Z^{(1)} = Y_1(z^{(0)}, Z^{(0)}) = \{5.938\ ;\ 0.145\}$.

Applying the iterative formula (4.140) we find the improved disks

$$Z^{(2)} = Y_1(z^{(1)}, Z^{(0)}) = \{6.001\ ;\ 0.013\}$$

and

$$Z^{(3)} = Y_1(z^{(2)}, Z^{(0)}) = \{5.9999993\ ;\ 5.1\times 10^{-7}\}$$

which contain the exact zero $\xi = 6$.

**EXAMPLE 16.** The polynomial

$$P(z) = z^4 - 3z^3 - 6z^2 - 12z - 40$$

and the initial region $[2.2, 6.2] + i[-1, 3]$ were chosen to illustrate the test of existence and the iterative method (4.141). We divide the initial square into four subsquares

$$S_1 = [2.2, 4.2] + i[-1, 1], \qquad S_2 = [2.2, 4.2] + i[1, 3],$$

$$S_3 = [4.2, 6.2] + i[1, 3], \qquad S_4 = [4.2, 6.2] + i[-1, 1].$$

Using the Horner scheme $\mathbb{P}(S) = (((S-3)\cdot S-6)\cdot S-12)\cdot S-40$ in rectangular arithmetic, we calculate

$$\mathbb{P}(S_1) = [-328.87, -21.11] + i[-199.68, 199.68],$$

$$\mathbb{P}(S_2) = [-816.63, 93.19] + i[-648., 261.82],$$

$$\mathbb{P}(S_3) = [-1661.86, 367.29] + i[-519.69, 1509.46],$$

$$\mathbb{P}(S_4) = [-321.14, 587.45] + i[-542.75, 542.75].$$

Since $0 \notin \mathbb{P}(S_1)$, the square $S_1$ is free of any zero of the given polynomial.

By the Horner scheme $\mathbb{P}'(S) = ((4S-9)\cdot S-12)\cdot S-12$ we evaluate

$$\mathbb{P}'(S_2) = [-438.53\,,\,50.19] + i[-128.48\,,\,360.24]\,,$$

$$\mathbb{P}'(S_3) = [-471.89\,,\,471.55] + i[57.6\,,\,1001.04]\,,$$

$$\mathbb{P}'(S_4) = [17.79\,,\,586.35] + i[-341.68\,,\,341.68]\,.$$

Since $0 \notin \mathbb{P}'(S_3)$ and $0 \notin \mathbb{P}'(S_4)$, the squares $S_3$ and $S_4$ may contain at most one zero of P. Dividing $S_3$ into four subsquares

$$S_{3,1} = [4.2\,,\,5.2] + i[1\,,\,2]\,, \qquad S_{3,2} = [4.2\,,\,5.2] + i[2\,,\,3]\,,$$

$$S_{3,3} = [5.2\,,\,6.2] + i[2\,,\,3]\,, \qquad S_{3,4} = [5.2\,,\,6.2] + i[1\,,\,2]$$

and applying the Horner scheme, we find

$$\mathbb{P}(S_{3,1}) = [-562.19\,,\,-23.18] + i[-81.19\,,\,457.74]\,,$$

$$\mathbb{P}(S_{3,2}) = [-1077.32\,,\,-180.63] + i[-326.16\,,\,570.53]\,,$$

$$\mathbb{P}(S_{3,3}) = [-1388.85\,,\,-52.23] + i[-11.36\,,\,1325.26]\,,$$

$$\mathbb{P}(S_{3,4}) = [-585.57\,,\,293.37] + i[123.36\,,\,1002.3]\,.$$

We observe that $0 \notin \mathbb{P}(S_{3,j})$ for each $j = 1,2,3,4$ and, therefore, there is no zeros of P in $S_3$.

The remaining square $S_4$ is the inclusion square for the disk $Z^{(0)} = \{z^{(0)}\,;\,r^{(0)}\} = \{5.2\,;\,1\}$, $S_4 = is\,(Z^{(0)})$. In addition,

$$|h_2(z^{(0)})| \cong 25.067 > \frac{n-1}{R} = 3$$

so that the conditions from Theorem 4.15 are fulfiled. By the interval formula (4.142) for $k = 2$ we find

$$Y_2(z^{(0)}, Z^{(0)}) = \{4.9995\,;\,0.0124\} \subset Z^{(0)}.$$

Thus, with regard to the statement $3^{\circ}$ of Theorem 4.15, there exists a unique simple zero of P in $Z^{(0)}$, which is also included in $Z^{(1)} = Y_2(z^{(0)}, Z^{(0)}) = \{4.9995\,;\,0.0124\}$. The iterative formula (4.141) yields the improved disks

$$Z^{(2)} = Y_2(z^{(1)}, Z^{(0)}) = \{5.0000000000037\,;\,2.79 \times 10^{-10}\}$$

and

$$Z^{(3)} = Y_2(z^{(2)}, Z^{(0)}) = \{5.0000000000000000000000000000000000\,;\,1.16 \times 10^{-34}\},$$

which contain the exact zero $\xi = 5$.

The presented numerical example was realized in quad precision arithmetic (about 34 significant decimal digits).

**REMARK 14.** In practical realization of any iterative formula on a digital computer, where floating-point arithmetic of fixed lenght is used, the round-off errors should be taken into consideration. It is well known that the rounded complex rectangular arithmetic, constructed by the rounded real interval arithmetic (see, e.g. [109]), takes into account rounding errors, but complex circular arithmetic does not. Consequently, the iterative formula (4.139) ( in terms of circular regions) may appear to be inefficient in a computer implementation. It is possible for us to obtain improved rectangular approximations (instead of circular approximations defined by (4.140)) of the requested zero applying the iterative formula

$$S^{(m+1)} = \left\{ z^{(m)} - \frac{P(z^{(m)})}{P'(z^{(m)}) - (n-1)P(z^{(m)})(z^{(m)} - S^{(0)})^{-1}} \right\} \cap S^{(m)}$$

$$(m = 0,1,\dots),$$

where $S^{(m)}$ is the rectangle and $z^{(m)}$ is its center, with $S^{(0)} = S$. The operations of rectangular arithmetic are defined as in [109]. Next advantage of rectangular arithmetic is the use of intersection of rectangles. When the last formula is applied, then we have the chain of inclusions $S^{(0)} \supseteq S^{(1)} \supseteq S^{(2)} \supseteq \cdots$ which provides the convergence of interval sequence $(S^{(m)})$ in certain cases when the initial region $S^{(0)} = S$ is not small enough. We note that the k-th root iteration methods (4.139) for $k > 1$ are not suitable for the application in rectangular arithmetic because they require the k-th root of a rectangle, which is rather complicated. ®

# CHAPTER 5

# BELL'S POLYNOMIALS AND PARALLEL DISK ITERATIONS

The fixed point relation based on Bell's polynomials and introduced by Wang and Zheng [127], can be applied effectively to construct a family of interval iterations for finding all zeros of a polynomial simultaneously with rapid convergence. The order of convergence of the basic method is k+2, where k is the order of the highest derivative of polynomial used in the iterative formula. But, in contrast to the family presented in Chapter 4, these procedures do not use the extraction of a root. The basic iterative formula and some of its modifications developed from the proposed family will be investigated in this chapter. In particular, Newton's method with q steps and Halley-like algorithms, which are of special practical interest, will also be discussed in detail.

## 5.1. BELL'S POLYNOMIALS AND FIXED POINT RELATIONS

Let k be a natural number and let $(z_1,\ldots,z_k) \in \mathbf{C}^k$. The sum of the products of powers of $z_1,\ldots,z_k$

$$B_k(z_1,\ldots,z_k) := \sum_{\nu=1}^{k} \sum \prod_{\lambda=1}^{k} \frac{1}{q_\lambda!}\left(\frac{z_\lambda}{\lambda}\right)^{q_\lambda}, \quad B_0 = 1 \tag{5.1}$$

is called *Bell's polynomial* (cf. [7],[108,Ch.5]). The second sum on the right-hand side runs over all nonnegative integers $(q_1,\ldots,q_k)$ satisfying the pair of equations

$$q_1 + 2q_2 + \cdots + kq_k = k,$$

$$q_1 + q_2 + \ldots + q_k = \nu \quad (1 \le \nu \le k).$$

For example, the expressions for $B_k = B_k(z_1,\ldots,z_k)$ $(k=1,2,3,4)$ are as follows:

$$B_1 = z_1,$$

$$B_2 = \frac{1}{2}z_2 + \frac{1}{2}z_1^2,$$

$$B_3 = \frac{1}{3}z_3 + \frac{1}{2}z_2z_1 + \frac{1}{6}z_1^3,$$

$$B_4 = \frac{1}{4}z_4 + \frac{1}{3}z_3z_1 + \frac{1}{8}z_2^2 + \frac{1}{4}z_2z_1^2 + \frac{1}{24}z_1^4 .$$

Consider now a monic polynomial P of degree N with n $(\leq N)$ distinct real or complex zeros $\xi_1,\ldots,\xi_n$ of the multiplicities $\mu_1,\ldots,\mu_n$ respectively , where $\mu_1+\cdots+\mu_n = N$. The factorization of P is

$$P(z) = \prod_{i=1}^{n}(z-\xi_i)^{\mu_i} \qquad (z \in \mathbb{C}). \tag{5.2}$$

Associate to P(z) the polynomial

$$f_i(z) = \prod_{\substack{j=1\\ j\neq i}}^{n}(z-\xi_j)^{\mu_j}.$$

Let $D_z^{(\nu)}f$ denote the $\nu$-th derivative of a function f, that is,

$$D_z^{(\nu)}f = \frac{d^\nu f}{dz^\nu},$$

and let

$$\Delta_{\nu,i}(z) = \frac{(-1)^\nu}{\nu!}P(z)^{1/\mu_i}D_z^{(\nu)}P(z)^{-1/\mu_i},$$

$$F_{\nu,i}(z) = \frac{(-1)^\nu}{\nu!}f_i(z)^{1/\mu_i}D_z^{(\nu)}f_i(z)^{-1/\mu_i},$$

with $\Delta_{0,i}(z) = F_{0,i}(z) = 1$.

Starting from

$$f_i(z)^{-1/\mu_i} = (z-\xi_i)P(z)^{-1/\mu_i},$$

we obtain

$$(z-\xi_i)\frac{(-1)^k}{k!}f_i(z)^{1/\mu_i}D_z^{(k)}f_i(z)^{-1/\mu_i} = \frac{(-1)^k}{k!}P(z)^{1/\mu_i}D_z^{(k)}f_i(z)^{-1/\mu_i}. \tag{5.3}$$

The left in (5.3) is equal to $(z-\xi_i)F_{k,i}(z)$. Applying the Leibniz formula to the expression on the right, we find

$$\frac{(-1)^k}{k!}P(z)^{1/\mu_i}D_z^{(k)}f_i(z)^{-1/\mu_i} = \frac{(-1)^k}{k!}P(z)^{1/\mu_i}D_z^{(k)}((z-\xi_i)P(z)^{-1/\mu_i})$$

$$= \frac{(-1)^k}{k!}P(z)^{1/\mu_i}[(z-\xi_i)D_z^{(k)}P(z)^{-1/\mu_i} + kD_z^{(k)}P(z)^{-1/\mu_i}]$$

$$= (z-\xi_i)\Delta_{k,i}(z) - \Delta_{k-1,i}(z).$$

Therefore, with regard to (5.3) it follows

$$(z-\xi_i)F_{k,i}(z) = (z-\xi_i)\Delta_{k,i}(z) - \Delta_{k-1,i}(z), \tag{5.4}$$

wherefrom

$$\xi_i = z - \frac{\Delta_{k-1,i}(z)}{\Delta_{k,i}(z) - F_{k,i}(z)}. \tag{5.5}$$

The fixed point relation (5.5) is the base for developing a family of parallel iterations for finding the zeros of polynomial $P(z)$ simultaneously (Wang and Zheng [127]). In order to realize any iterative process from this family, it is necessary to find some expressions or a computational method for $\Delta_{k,i}(z)$ and $F_{k,i}(z)$.

The generating function of $\Delta_{k,i}(z)$ is $\frac{P(z+x)}{P(z)}$, that is

$$\Delta_{k,i}(z) = \frac{(-1)^k}{k!}\left[D_x^{(k)}\left(\frac{P(z+x)}{P(z)}\right)^{-1/\mu_i}\right]_{x=0}.$$

Using the rule of derivation for a compound function (see Bell [7]), we obtain

$$\Delta_{k,i}(z) = \sum_{\nu=1}^{k}(-1)^{k-\nu}\frac{1}{\mu_i}\left(\frac{1}{\mu_i}+1\right)\cdots\left(\frac{1}{\mu_i}+\nu-1\right)\sum\prod_{\lambda=1}^{k}\frac{1}{q_\lambda!}\left(\frac{P^{(\lambda)}(z)}{\lambda!P(z)}\right)^{q_\lambda}, \tag{5.6}$$

where the second sum on the right-hand side runs over all nonnegative integers $(q_1,\ldots,q_k)$ which satisfy $q_1+2q_2+\cdots+kq_k=k$, $q_1+q_2+\cdots+q_k=\nu$ (see [131]).

In particular, if $\mu_i=1$, then $\Delta_{k,i}(z)$ can be computed recurrently as

$$\Delta_{k,i}(z) = \sum_{\nu=1}^{k}\frac{(-1)^{\nu+1}}{\nu!}\cdot\frac{P^{(\nu)}(z)}{P(z)}\Delta_{n-\nu,i}(z).$$

Furthermore, starting from

$$F_{k,i}(z) = \frac{1}{k!} f_i(z)^{1/\mu_i} \left[ D_x^{(k)} \exp(-\frac{1}{\mu_i} \ln f_i(z-x)) \right]_{x=0} ,$$

one obtains

$$F_{k,i}(z) = \sum_{\nu=1}^{k} \sum \prod_{\lambda=1}^{k} \frac{1}{q_\lambda !} \left( \frac{s_{\lambda,i}(z)}{\lambda} \right)^{q_\lambda} , \tag{5.7}$$

where

$$s_{\lambda,i}(z) = \frac{1}{\mu_i} \sum_{\substack{j=1 \\ j\neq i}}^{n} \frac{\mu_j}{(z-\xi_j)^\lambda} \tag{5.8}$$

and for the second sum the same is valid as in (5.6).

Besides, since

$$F_{k,i}(z) = \frac{(-1)^{k-1}}{\mu_i k!} f_i(z)^{-1/\mu_i} D_z^{(k-1)} \left( f_i(z)^{-1/\mu_i} \frac{f_i'(z)}{f_i(z)} \right) ,$$

we obtain the following recursion relation

$$F_{k,i}(z) = \frac{1}{k} \sum_{\nu=1}^{k} s_{\nu,i}(z) F_{k-\nu,i}(z) . \tag{5.9}$$

**REMARK** 1. According to the definition (5.1) and the relation (5.9), we have

$$F_{k,i}(z) = B_k(s_{1,i}(z), \ldots, s_{k,i}(z)) . \tag{5.10}$$

Therefore, in view of (5.9), Bell's polynomials can be computed recurrently as

$$B_k(z_1, \ldots, z_k) = \frac{1}{k} \sum_{\nu=1}^{k} z_\nu B_{k-\nu}(z_1, \ldots, z_k) , \quad B_0 = 1. \circledR$$

## 5.2. A FAMILY OF PARALLEL INTERVAL ITERATIONS

Let P be a monic polynomial with the factorization (5.2). Assume that we have found n disjoint disks $Z_1, \ldots, Z_n$ which contain zeros $\xi_1, \ldots, \xi_n$ of P. Let

$$S_{\nu,i} = \frac{1}{\mu_i} \sum_{\substack{j=1 \\ j\neq i}}^{n} \mu_j (z - Z_j)^{-\nu} \quad (z \notin Z_j \text{ for each } j \neq i)$$

and

$$U_{k,i} = B_k(S_{1,i},\dots,S_{k,i}).$$

For $s_{\nu,i}(z)$, defined by (5.8), it is obvious that $s_{\nu,i}(z) \in S_{\nu,i}$ so that from (5.10) it follows

$$F_{k,i}(z) = B_k(s_{1,i}(z),\dots,s_{k,i}(z)) \in B_k(S_{1,i},\dots,S_{k,i}) = U_{k,i}.$$

According to this and the fixed point relation (5.5) we obtain

$$\xi_i \in \hat{Z}_i = z - \frac{\Delta_{k-1,i}(z)}{\Delta_{k,i}(z) - U_{k,i}}, \tag{5.11}$$

where $\hat{Z}_i$ is a new circular approximation to the exact zero $\xi_i$.

Based on the relation (5.11), Wang and Zheng have established in [127] a family of interval iterations for simultaneous inclusion of complex multiple zeros of a polynomial using circular arithmetic.

Let $Z_1^{(0)},\dots,Z_n^{(0)}$ be disjoint disks containing the exact zeros $\xi_1,\dots,\xi_n$ respectively, and let $Z_j^{(m)} = \{z_j^{(m)}; r_j^{(m)}\}$ be a circular approximation to $\xi_j$ $(j \in \{1,\dots,n\})$, obtained in the m-th iteration, with center $z_j^{(m)} = \text{mid } Z_j^{(m)}$ and radius $r_j^{(m)} = \text{rad } Z_j^{(m)}$. For a natural number k $(\le n)$ the presented family of iteration interval methods has the form

$$Z_i^{(m+1)} = z_i^{(m)} - \frac{\Delta_{k-1,i}(z_i^{(m)})}{\Delta_{k,i}(z_i^{(m)}) - B_k(S_{1,i}^{(m)},\dots,S_{k,i}^{(m)})} \tag{5.12}$$

$$(i = 1,\dots,n;\ m = 0,1,\dots),$$

where

$$S_{\nu,i}^{(m)} = \frac{1}{\mu_i}\sum_{\substack{j=1\\ j\neq i}}^{n} \mu_j \left(\frac{1}{z_i^{(m)} - Z_j^{(m)}}\right)^{\nu}.$$

It has been proved in [127] that the convergence order of the family (5.12) is equal to k+2. For example, if k=1, from (5.12) it follows

$$Z_i^{(m+1)} = z_i^{(m)} - \frac{\mu_i}{\dfrac{P'(z_i^{(m)})}{P(z_i^{(m)})} - \displaystyle\sum_{\substack{j=1\\ j\neq i}}^{n} \frac{\mu_j}{z_i^{(m)} - Z_j^{(m)}}} \tag{5.13}$$

$$(i = 1,\dots,n;\ m = 0,1,\dots),$$

which is the method proposed by Gargantini [33] with *cubic* convergence (see formula (4.72) in § 4.4).

In the following, we will describe some modifications of (5.12), proposed in [131], which deal with two parameters k and q ($k,q \in \mathbb{N}$). The basic feature of these modifications is based on a combination of interval arithmetic and the principle given by Ehrmann [20] (see, also, Alefeld and Herzberger [6], D. Wang and Wu [125]). Namely, in forming one iterative step, only the circular interval $B_k$ in (5.12) is calculated while the same values $\Delta_{k-1,i}(z_i^{(m)})$ and $\Delta_{k,i}(z_i^{(m)})$ are used several times (q times exactly; so these methods are often called *q step methods*). In that way, the number of numerical operations is considerably reduced. Besides, the corresponding algorithms are suitable for programming and occupy little storage space at digital computers. The abbreviations TS(k,q) and SS(k,q) will be used to denote *total-step* and *single-step methods* with parameters k and q.

**ALGORITHM TS(k,q):** For each $m = 0,1,2,\ldots,$ let for $i = 1,\ldots,n$ and $\lambda = 0,1,\ldots,q-1$

$$z_i^{(m)} = \operatorname{mid} Z_i^{(m)},$$

$$S_{\nu,i}^{(m+\frac{\lambda+1}{q})} = \frac{1}{\mu_i}\sum_{\substack{j=1\\ j\neq i}}^{n} \mu_j \left(\frac{1}{z_i^{(m)} - Z_j^{(m+\frac{\lambda}{q})}}\right)^{\nu} \qquad (\nu = 1,\ldots,k),$$

$$U_{k,i}^{(m+\frac{\lambda+1}{q})} = B_k\left(S_{1,i}^{(m+\frac{\lambda+1}{q})},\ldots,S_{k,i}^{(m+\frac{\lambda+1}{q})}\right).$$

Then the successive circular approximations to the zeros $\xi_1,\ldots,\xi_n$ are given by

$$Z_i^{(m+\frac{\lambda+1}{q})} = z_i^{(m)} - \frac{\Delta_{k-1,i}(z_i^{(m)})}{\Delta_{k,i}(z_i^{(m)}) - U_{k,i}^{(m+\frac{\lambda+1}{q})}} \qquad (\lambda = 0,1,\ldots,q-1). \quad (5.14)$$

The following assertion has been proved in [131]:

**THEOREM 5.1.** *Let* $(Z_i^{(m)})$ *$(i = 1,\ldots,n)$ be the sequences of disks produced by TS(k,q). If* $\xi_i \in Z_i^{(0)}$ *for all* $i = 1,\ldots,n$, *then:* $1^o$ $\xi_i \in Z_i^{(m)}$ *for all* $m = 1,2,\ldots$ ; $2^o$ *the convergence order of the iterative method TS(k,q) is at least* $(k+1)q+1$.

The total-step method (5.14) can be accelerated applying the Gauss-Seidel approach, that is, using new circular approximations as soon as they become available (the single-step procedure). With the previous notations, we have

**ALGORITHM SS(k,q):** The successive circular approximations to the zeros are given by

$$Z_i^{(m+\frac{\lambda+1}{q})} = z_i^{(m)} - \frac{\Delta_{k-1,i}(z_i^{(m)})}{\Delta_{k,i}(z_i^{(m)}) - \tilde{U}_{k,i}^{(m+\frac{\lambda+1}{q})}} \quad (i = 1,\ldots,n), \tag{5.15}$$

where

$$\tilde{U}_{k,i}^{(m+\frac{\lambda+1}{q})} = B_k\left(\tilde{S}_{1,i}^{(m+\frac{\lambda+1}{q})},\ldots,\tilde{S}_{k,i}^{(m+\frac{\lambda+1}{q})}\right),$$

$$\tilde{S}_{\nu,i}^{(m+\frac{\lambda+1}{q})} = \frac{1}{\mu_i}\left[\sum_{j=1}^{i-1}\mu_j\left(z_i^{(m)} - Z_j^{(m+\frac{\lambda+1}{q})}\right)^{-\nu} + \sum_{j=i+1}^{n}\mu_j\left(z_i^{(m)} - Z_j^{(m+\frac{\lambda}{q})}\right)^{-\nu}\right]$$

$$(\nu = 1,\ldots,k) .$$

In finding the R-order of convergence of the single-step method SS(k,q), denoted by $O_R(SS(k,q),\xi)$ ( $\xi = (\xi_1 \cdots \xi_n)$ ), a technique proposed by Alefeld and Herzberger [4] may be applied (see Section 2.3). The following assertion has been stated in [131]:

**THEOREM 5.2.** *Let* $(Z_i^{(m)})$ *(i = 1,...,n) be the sequences of disks produced by SS(k,q). If* $\xi_i \in Z_i^{(0)}$ *for all i = 1,...,n, then:* $1^o$ $\xi_i \in Z_i^{(m)}$ *for all m = 1,2,... ;* $2^o$ *the R-order of convergence* $O_R(SS(k,q),\xi)$ *of the iterative method SS(k,q) is at least equal to the spectral radius ρ(A) of the n × n matrix A given by*

$$A = \begin{bmatrix} \overbrace{k+1 \;\ldots\; k+1}^{q} & 1 & & & & & & \\ & k+1 \;\ldots\; k+1 & 1 & & & & O & \\ & & \cdot & \cdot & \cdot & & & \\ & O & & \cdot & \cdot & \cdot & & \\ & & & & \cdot & \cdot & \cdot & \\ & & & & & k+1 \;\ldots\; k+1 & 1 & \\ k+1 & 1 & & & & & k+1 \;\ldots\; k+1 & \\ \vdots & & \cdot & & O & & \cdot & \vdots \\ \vdots & & & \cdot & & & & \vdots \\ k+1 \;\ldots\; k+1 & 1 & & & & & & k+1 \end{bmatrix} \left.\begin{matrix} \\ \\ \\ \\ \end{matrix}\right\} q .$$

For $k,q = 1(1)3$ the values of $\rho(A)$, calculated by the power method, are shown in Table 5.1.

| | k = 1 | | | | | | | | | |
|---|---|---|---|---|---|---|---|---|---|---|
| q \ n | 2 | 3 | 4 | 5 | 6 | 7 | 8 | 9 | 10 | n → ∞ |
| 1 | 4. | 3.521 | 3.353 | 3.267 | 3.215 | 3.180 | 3.154 | 3.135 | 3.121 | 3. |
| 2 | 8. | 6.372 | 5.938 | 5.696 | 5.557 | 5.464 | 5.397 | 5.348 | 5.309 | 5. |
| 3 | 12. | 9.617 | 8.531 | 8.182 | 7.942 | 7.776 | 7.667 | 7.582 | 7.516 | 7. |
| | k = 2 | | | | | | | | | |
| q \ n | 2 | 3 | 4 | 5 | 6 | 7 | 8 | 9 | 10 | n → ∞ |
| 1 | 5.303 | 4.672 | 4.453 | 4.341 | 4.274 | 4.229 | 4.196 | 4.172 | 4.153 | 4. |
| 2 | 11.42 | 9. | 8.347 | 8. | 7.798 | 7.664 | 7.568 | 7.497 | 7.441 | 7. |
| 3 | 17.45 | 13.92 | 12.27 | 11.75 | 11.38 | 11.14 | 10.98 | 10.85 | 10.75 | 10. |
| | k = 3 | | | | | | | | | |
| q \ n | 2 | 3 | 4 | 5 | 6 | 7 | 8 | 9 | 10 | n → ∞ |
| 1 | 6.561 | 5.796 | 5.534 | 5.401 | 5.321 | 5.268 | 5.230 | 5.201 | 5.179 | 5. |
| 2 | 14.84 | 11.62 | 10.75 | 10.30 | 10.04 | 9.860 | 9.736 | 9.643 | 9.570 | 9. |
| 3 | 22.90 | 18.22 | 16. | 15.31 | 14.82 | 14.51 | 14.29 | 14.12 | 14. | 13. |

Table 5.1 The spectral radius of matrix A

For $k = 1$ and $q = 2$ from (5.15) we obtain the iterative method SS(1,2):

$$z_i^{(m+\frac{\lambda+1}{2})} = z_i^{(m)} - \left[ \frac{P'(z_i^{(m)})}{P(z_i^{(m)})} - \sum_{j=1}^{i-1} \left( z_i^{(m)} - z_j^{(m+\frac{\lambda+1}{2})} \right)^{-1} - \sum_{j=i+1}^{N} \left( z_i^{(m)} - z_j^{(m+\frac{\lambda}{2})} \right)^{-1} \right]^{-1}$$

$$(i = 1,\dots,n;\ m = 0,1,\dots;\ \lambda = 0,1).$$

This method for simple zeros was proposed by D. Wang and Wu [125] and its convergence analysis was given in [95].

As one particular case from the family (5.14), we will consider in detail the iterative method TS(1,q), which is a generalization of the total-step procedure TS(1,1) presented by Gargantini [33]. X. Wang and S. Zheng [131] have suggested Newton's method with q steps:

**ALGORITHM TS(1,q):** For any natural number q the inclusion disks for the zeros $\xi_1,\ldots,\xi_n$ are calculated by

$$Z_i^{(m+\frac{\lambda+1}{q})} = z_i^{(m)} - \frac{\mu_i}{\dfrac{P'(z_i^{(m)})}{P(z_i^{(m)})} - \displaystyle\sum_{\substack{j=1\\ j\neq i}}^{n} \frac{\mu_j}{z_i^{(m)} - Z_j^{(m+\frac{\lambda}{q})}}} \tag{5.16}$$

$$(i=1,\ldots,n;\ m=0,1,\ldots;\ \lambda=0,1,\ldots,q-1).$$

Introduce the notations

$$r_i^{(m+\frac{\lambda}{q})} = \operatorname{rad} Z_i^{(m+\frac{\lambda}{q})}, \quad r^{(m+\frac{\lambda}{q})} = \max_{1\le i\le n} r_i^{(m+\frac{\lambda}{q})},$$

$$z_i^{(m+\frac{\lambda}{q})} = \operatorname{mid} Z_i^{(m+\frac{\lambda}{q})},$$

$$\rho_{ij}^{(m+\frac{\lambda}{q})} = \left| z_i^{(m+\frac{\lambda}{q})} - z_j^{(m+\frac{\lambda}{q})} \right| - r_j^{(m+\frac{\lambda}{q})},$$

$$\rho^{(m+\frac{\lambda}{q})} = \min_{1\le i,j\le n} \rho_{ij}^{(m+\frac{\lambda}{q})},$$

$$\theta^{(m+\frac{\lambda}{q})} = r^{(m+\frac{\lambda}{q})} / \rho^{(m+\frac{\lambda}{q})}, \quad \mu = \min_{1\le i\le n} \mu_i,$$

$$\alpha^{(m+\frac{\lambda}{q})} = \sqrt{\frac{N-\mu}{\mu}}\cdot\theta^{(m+\frac{\lambda}{q})},$$

$$W_i^{(m+\frac{\lambda}{q})} = \frac{1}{\mu_i}\left(\sum_{j=1}^{n} \frac{\mu_j}{z_i^{(m)} - \xi_j} - \sum_{\substack{j=1\\ j\neq i}}^{n} \frac{\mu_j}{z_i^{(m)} - Z_j^{(m+\frac{\lambda}{q})}}\right).$$

For simplicity, whenever it is possible all superscripts $(m+\frac{\lambda}{q})$ are abbreviated to $(\lambda/q)$. In particular, if $\lambda = 0$ the whole superscript will be omitted, and for $\lambda = q$ it will be written as asterisk $(*)$. For example, we will write $z_i$, $z_i^*$, $z_i^{(\lambda/q)}$ instead of $z_i^{(m)}$, $z_i^{(m+1)}$, $z_i^{(m+\frac{\lambda}{q})}$ respectively, and so on.

To establish a precise convergence theorem for the method TS(1,q), the following two lemmas are necessary.

**LEMMA 5.1.** *If m is fixed and* $\alpha \le \frac{1}{4}$,[*] *then*

$$r^{(\lambda/q)} \le \frac{\alpha^{2\lambda}}{1 - 2\sum_{\nu=1}^{2\lambda} \alpha^{\nu}}\, r \qquad (\lambda \in \{1,\ldots,q\}). \tag{5.17}$$

*P r o o f.* Using the properties of circular arithmetic, we obtain

$$\left| \operatorname{mid}\left(\frac{1}{z_i - \xi_j} - \frac{1}{Z_i - Z_j^{(\lambda/q)}}\right)\right| = \left| \frac{1}{z_i - \xi_j} - \frac{\bar{z}_i - \bar{z}_j^{(\lambda/q)}}{|z_i - z_j^{(\lambda/q)}|^2 - (r_j^{(\lambda/q)})^2}\right|$$

$$= \left| \frac{(\bar{z}_i - \bar{z}_j^{(\lambda/q)})(\xi_j - z_j^{(\lambda/q)}) - (r_j^{(\lambda/q)})^2}{(z_i - \xi_j)\left[|z_i - z_j^{(\lambda/q)}|^2 - (r_j^{(\lambda/q)})^2\right]}\right|$$

$$\le \frac{|z_i - z_j^{(\lambda/q)}|\, r_j^{(\lambda/q)} + (r_j^{(\lambda/q)})^2}{|z_i - \xi_j|\left[|z_i - z_j^{(\lambda/q)}|^2 - (r_j^{(\lambda/q)})^2\right]}$$

$$= \frac{r_j^{(\lambda/q)}}{|z_i - \xi_j|\left[|z_i - z_j^{(\lambda/q)}| - r_j^{(\lambda/q)}\right]},$$

wherefrom

$$\left| \operatorname{mid}\left(\frac{1}{z_i - \xi_j} - \frac{1}{Z_i - Z_j^{(\lambda/q)}}\right)\right| \le \frac{r^{(\lambda/q)}}{\rho(\rho - 2r^{(\lambda/q)})} \qquad (j \ne i). \tag{5.18}$$

Similarly, we estimate

$$\operatorname{rad} \frac{1}{z_i - Z_j^{(\lambda/q)}} \le \frac{r^{(\lambda/q)}}{\rho(\rho - 2r^{(\lambda/q)})} \qquad (j \ne i). \tag{5.19}$$

Applying (5.18) and (5.19) we find

*) In the original paper [131] the authors have used the condition $\alpha \le 1/\sqrt{6}$ which however leads to certain difficulties in the beginning of the inductive proof of Lemma 6.2 (see [131,p.341]). To overcome that problem, the (slightly stronger) condition $\alpha \le 1/4$ is introduced in the presented analysis of Algorithm TS(1,q).

$$\left|\operatorname{mid} W_i^{(\lambda/q)}\right| \geq \frac{1}{|z_i - \xi_i|} - \sum_{j\neq i} \frac{\mu_j}{\mu_i}\left|\operatorname{mid}\left(\frac{1}{z_i - \xi_j} - \frac{1}{z_i - Z_j^{(\lambda/q)}}\right)\right|$$

$$\geq \frac{1}{r} - \frac{N-\mu}{\mu}\cdot\frac{r^{(\lambda/q)}}{\rho(\rho - 2r^{(\lambda/q)})} \tag{5.20}$$

and

$$\operatorname{rad} W_i^{(\lambda/q)} = \sum_{j\neq i} \frac{\mu_j}{\mu_i} \operatorname{rad} \frac{1}{z_i - Z_j^{(\lambda/q)}} \leq \frac{N-\mu}{\mu}\cdot\frac{r^{(\lambda/q)}}{\rho(\rho - 2r^{(\lambda/q)})} . \tag{5.21}$$

Using the inequality $\frac{r}{\rho} \leq \alpha$, for $\lambda = 0$ we obtain from (5.20) and (5.21)

$$\left|\operatorname{mid} W_i\right| \geq \frac{1}{r}\left(1 - \frac{\alpha^2}{1-2\alpha}\right) \tag{5.22}$$

and

$$\operatorname{rad} W_i \leq \frac{1}{r}\cdot\frac{\alpha^2}{1-2\alpha} . \tag{5.23}$$

Since

$$Z_i^{(\frac{\lambda+1}{q})} = z_i - \frac{1}{W_i^{(\lambda/q)}} , \tag{5.24}$$

applying the estimations (5.22) and (5.23) we find for $\lambda = 1$

$$r_i^{(1/q)} = \operatorname{rad} \frac{1}{W_i} = \frac{\operatorname{rad} W_i}{\left|\operatorname{mid} W_i\right|^2 - (\operatorname{rad} W_i)^2}$$

$$\leq \frac{\frac{1}{r}\cdot\frac{\alpha^2}{1-2\alpha}}{\frac{1}{r^2}\left(1 - \frac{\alpha^2}{1-2\alpha}\right)^2 - \frac{1}{r^2}\left(\frac{\alpha^2}{1-2\alpha}\right)^2} = \frac{\alpha^2}{1 - 2\alpha - 2\alpha^2} r$$

for each $i \in \{1,\dots,n\}$, which means that

$$r^{(1/q)} = \max_{1\leq i\leq n} r_i^{(1/q)} \leq \frac{\alpha^2}{1 - 2(\alpha+\alpha^2)} r .$$

Therefore, the inequality (5.17) holds for $\lambda = 1$. Suppose that it holds for a certain $\lambda \in \{1,\dots,q\}$ and show that the inequality (5.17) is also valid for $\lambda+1$.

First, assuming that (5.17) is true for any $\lambda$, we estimate

$$\sqrt{\frac{N-\mu}{\mu}}\cdot\frac{r^{(\lambda/q)}}{\rho - 2r^{(\lambda/q)}} \le \sqrt{\frac{N-\mu}{\mu}}\cdot\frac{r\alpha^{2\lambda}/(1-2\sum_{\nu=1}^{2\lambda}\alpha^{\nu})}{\rho - 2r\alpha^{2\lambda}/(1-2\sum_{\nu=1}^{2\lambda}\alpha^{\nu})} = \frac{\alpha^{2\lambda+1}}{1-2\sum_{\nu=1}^{2\lambda+1}\alpha^{\nu}} .$$

Therefore, the inequalities (5.20) and (5.21) become

$$|\operatorname{mid} W_i^{(\lambda/q)}| \ge \frac{1}{r}\left(1 - \frac{\alpha^{2\lambda+2}}{1-2\sum_{\nu=1}^{2\lambda+1}\alpha^{\nu}}\right)$$

and

$$\operatorname{rad} W_i^{(\lambda/q)} \le \frac{1}{r}\cdot\frac{\alpha^{2\lambda+2}}{1-2\sum_{\nu=1}^{2\lambda+1}\alpha^{\nu}} .$$

Taking into account the last two inequalities and (5.24), we find

$$r_i^{(\frac{\lambda+1}{q})} = \operatorname{rad}\frac{1}{W_i^{(\lambda/q)}} = \frac{\operatorname{rad} W_i^{(\lambda/q)}}{|\operatorname{mid} W_i^{(\lambda/q)}|^2 - (\operatorname{rad} W_i^{(\lambda/q)})^2} \le \frac{\alpha^{2(\lambda+1)}}{1-2\sum_{\nu=1}^{2(\lambda+1)}\alpha^{\nu}}\, r ,$$

that is, the inequality in the lemma is also true for $\lambda+1$. The proof of Lemma 5.1 is completed by induction. □

**LEMMA 5.2.** *The inequality*

$$\rho^{(\lambda/q)} \ge \rho - r - 3r^{(\lambda/q)} \tag{5.25}$$

*is satisfied for each* $\lambda \in \{1,\dots,q\}$.

*Proof*. Let $i,j \in \{1,\dots,n\}$ $(i\ne j)$ be two indices for which $\rho_{ij}^{(\lambda/q)}$ is minimum and, consequently, equal to $\rho^{(\lambda/q)}$, that is

$$\rho^{(\lambda/q)} = \rho_{ij}^{(\lambda/q)} = |z_i^{(\lambda/q)} - z_j^{(\lambda/q)}| - r_j^{(\lambda/q)} .$$

Then

$$\begin{aligned}\rho \le \rho_{ij} &= |z_i - z_j| - r_j \\ &\le |z_i-\xi_i| + |\xi_i - z_i^{(\lambda/q)}| + \rho^{(\lambda/q)} + r_j^{(\lambda/q)} + |z_j^{(\lambda/q)} - \xi_i| + |\xi_j - z_j| - r_j .\end{aligned}$$

According to the assertion 1° of Theorem 5.2 we have $\xi_i \in Z_i^{(\lambda/q)}$ and $\xi_j \in Z_j^{(\lambda/q)}$. Hence

$$|\xi_i - z_i^{(\lambda/q)}| < r_i^{(\lambda/q)}, \qquad |\xi_j - z_j^{(\lambda/q)}| < r_j^{(\lambda/q)} .$$

Therefore, the previous inequality becomes

$$\rho \le \rho^{(\lambda/q)} + r + 3r^{(\lambda/q)}.$$

The lemma is proved. □

The following theorem, established in [131], gives the convergence speed of TS(1,q), defined by (5.16).

**THEOREM 5.3.** *Let* $\xi_i \in Z_i^{(0)}$ *for all* $i = 1,\dots,n$ *and*

$$4\sqrt{\frac{N-\mu}{\mu}}\,\theta^{(0)} \le 1. \tag{5.26}$$

*Then the iterative process TS(1,q) is convergent and*

$$\theta^{(m+\frac{\lambda}{q})} \le 5\left(\frac{N-\mu}{\mu}\right)^{\lambda}\left(\theta^{(m)}\right)^{2\lambda+1}$$

*for each* $\lambda \in \{1,\dots,q\}$ *and* $m = 0,1,\dots$ .

*Especially,*

$$\theta^{(m+1)} \le 5\left(\frac{N-\mu}{\mu}\right)^{q}\left(\theta^{(m)}\right)^{2q+1} \qquad (m = 0,1,\dots). \tag{5.27}$$

*Proof.* It is sufficient to prove that

$$\alpha^{(\lambda/q)} \le 5\alpha^{2\lambda+1} \qquad (\lambda \in \{1,\dots,q\})$$

under the condition $\alpha \le \frac{1}{4}$ . We will distinguish two cases: (I) $\lambda = 1$ and (II) $\lambda > 1$.

(I) $\lambda = 1$. From the inequalities (5.17) and (5.25) for $\lambda = 1$ it follows

$$r^{(1/q)} \le \frac{\alpha^2}{1 - 2\alpha - 2\alpha^2}\,r$$

and

$$\rho^{(1/q)} \ge \rho - r - 3r^{(1/q)} \ge \rho\,\frac{1 - 3\alpha - \alpha^3}{1 - 2\alpha - 2\alpha^2} .$$

Hence

$$\alpha^{(1/q)} = \sqrt{\frac{N-\mu}{\mu}}\cdot\frac{r^{(1/q)}}{\rho^{(1/q)}} \le \frac{\alpha^3}{1 - 3\alpha - \alpha^3} < 5\alpha^3 .$$

(II) $\lambda > 1$. From Lemmas 5.1 and 5.2 we get

$$r^{(\lambda/q)} \le \frac{(1-\alpha)\alpha^{2\lambda}}{1 - 3\alpha}\,r$$

and

$$\rho^{(\lambda/q)} \ge \rho - r - 3r^{(\lambda/q)} \ge \rho\,\frac{(1-\alpha)(1-3\alpha-3\alpha^{2\lambda+1})}{1 - 3\alpha} ,$$

so that

$$\alpha^{(\lambda/q)} = \sqrt{\frac{N-\mu}{\mu}} \cdot \frac{r^{(\lambda/q)}}{\rho^{(\lambda/q)}} \le \frac{\alpha^{2\lambda+1}}{1-3\alpha-3\alpha^{2\lambda+1}} \le \frac{\alpha^{2\lambda+1}}{1-3\alpha-\alpha^3} < 5\alpha^{2\lambda+1}.$$

Setting $\lambda=q$ in the last inequality we obtain (5.27), which means that the convergence order of iterative method TS(1,q) is equal to 2q+1. □

**REMARK 2.** Taking $q = 1$ in Theorem 5.3 we observe that the convergence condition (5.26) is improved in reference to the corresponding condition

$$6 \, ( \max_{1 \le i \le n} \mu_i)^{\frac{N-\mu}{\mu}} \theta^{(0)}) \le 1,$$

that has been established in [33] for TS(1,1). ®

On the basis of a great number of examples, tested by Algorithms TS(k,q) and SS(k,q), one can conclude that the efficiency of these algorithms does not considerably increase when the parameters k and q rise. Rapid convergence (using high k and q) asks for high computational cost. When a high k is used then the iterative formula requires higher polynomial derivatives and more terms (in the form of sums) of Bell's polynomial. The enlargement of q leads to the use of the same values of a polynomial P and its derivatives q times, so that the improvement of inclusive approximations may be inadequate. The practical execution of Algorithms TS(k,q) shows that the values of k and q should be taken to be 2 at most. Among such procedures the iterative interval processes TS(1,1) and SS(1,1) are described in Sections 4.1 and 4.2 and Newton's method TS(1,q) with q steps in this section. The iterative methods TS(2,1), SS(2,1) and SS(2,2) are considered in the next section.

## 5.3. HALLEY-LIKE ALGORITHMS

Let $\sigma_i = \frac{P^{(i)}(z)}{P(z)}$ $(i = 1,2)$. The functions $\Delta_{k,i}(z)$ and $F_{k,i}(z)$, introduced in Section 5.1, for $k = 2$ become

$$\Delta_{2,i}(z) = \frac{1}{2\mu_i}\left(\frac{1}{\mu_i} + 1\right)\left(\frac{P'(z)}{P(z)}\right)^2 - \frac{1}{2\mu_i} \cdot \frac{P''(z)}{P(z)}$$

$$= \frac{1}{2\mu_i}\left(\frac{1}{\mu_i} + 1\right)\sigma_1(z)^2 - \frac{1}{2\mu_i}\sigma_2(z),$$

$$F_{2,i}(z) = B_2\left(\frac{1}{\mu_i}\sum_{j\neq i}\frac{\mu_j}{z-\xi_j}, \frac{1}{\mu_i}\sum_{j\neq i}\frac{\mu_j}{(z-\xi_j)^2}\right)$$

$$= \frac{1}{2\mu_i^2}\left(\sum_{\substack{j=1\\ j\neq i}}^{n}\frac{\mu_j}{z-\xi_j}\right)^2 + \frac{1}{2\mu_i}\sum_{\substack{j=1\\ j\neq i}}^{n}\frac{\mu_j}{(z-\xi_j)^2}.$$

In addition, we have

$$\Delta_{1,i}(z) = \frac{1}{\mu_i}\cdot\frac{P'(z)}{P(z)} = \frac{1}{\mu_i}\sigma_1(z).$$

Now, the fixed point relation (5.5) for $k = 2$ yields

$$\xi_i = z - \frac{\Delta_{1,i}(z)}{\Delta_{2,i}(z) - B_2\left(\frac{1}{\mu_i}\sum_{j\neq i}\frac{\mu_j}{z-\xi_j}, \frac{1}{\mu_i}\sum_{j\neq i}\frac{\mu_j}{(z-\xi_j)^2}\right)},$$

that is,

$$\xi_i = z - \frac{1}{\frac{\sigma_1(z)}{2}\left(1+\frac{1}{\mu_i}\right) - \frac{\sigma_2(z)}{2\sigma_1(z)} - \frac{1}{2\sigma_1(z)}\left[\frac{1}{\mu_i}\left(\sum_{\substack{j=1\\ j\neq i}}^{n}\frac{\mu_j}{z-\xi_j}\right)^2 + \sum_{\substack{j=1\\ j\neq i}}^{n}\frac{\mu_j}{(z-\xi_j)^2}\right]}. \tag{5.28}$$

We note that (5.28) can be also derived using $\sigma_1(z)$, $\sigma_1'(z)$ and $\sigma_2(z)$, where, according to the factorization (5.2),

$$\sigma_1^{(\nu)}(z) = \frac{d^{\nu+1}}{dz^{\nu+1}}(\log P(z)) = \sum_{j=1}^{n}\frac{\mu_j}{(z-\xi_j)^{\nu+1}} \quad (\nu = 0,1;\ \sigma_1^{(0)}(z)\equiv\sigma_1(z))$$

and then, evaluating $\xi_i$ from the relation

$$\sigma_2(z) = \sigma_1(z)^2 + \sigma_1'(z).$$

We observe that the function

$$f(z) = \frac{1}{h(z)} = \frac{\sigma_1(z)}{2}(1 + \frac{1}{\mu}) - \frac{\sigma_2(z)}{2\sigma_1(z)} = \frac{P'(z)}{2P(z)}(1 + \frac{1}{\mu}) - \frac{P''(z)}{2P'(z)} \tag{5.29}$$

in the denominator of (5.28) appears in the Halley iterative formula

$$\hat{z} = z - h(z) \tag{5.30}$$

for the determination of a zero of a polynomial with the multiplicity $\mu$ (see [45]). The iterative method (5.30) has *cubic* convergence.

The additional term

$$\delta_i(z,\xi) = \frac{u(z)}{2} q_i(z,\xi), \tag{5.31}$$

where

$$u(z) = \frac{P(z)}{P'(z)},$$

$$q_i(z,\xi) = \frac{1}{\mu_i}\left(\sum_{\substack{j=1\\ j\neq i}}^{n} \frac{\mu_j}{z-\xi_j}\right)^2 + \sum_{\substack{j=1\\ j\neq i}}^{n} \frac{\mu_j}{(z-\xi_j)^2}, \tag{5.32}$$

and $\xi = (\xi_1,\ldots,\xi_n)$ is the vector of zeros, has the role of a correction term in the fixed point relation (5.28).

Using (5.29) and (5.31), the relation (5.28) can be rewritten as

$$\xi_i = z - \frac{1}{f(z) - \delta_i(z,\xi)} \quad (i = 1,\ldots,n). \tag{5.33}$$

We will now desribe some algorithms of Halley's type, presented by Petković [93].

ALGORITHM FOR SIMULTANEOUS IMPROVING K INCLUSION DISKS

Assume we have found k disjoint disks $Z_1^{(0)},\ldots,Z_k^{(0)}$ that contain the zeros $\xi_1,\ldots,\xi_k$ and these disks are located inside the disk $\{c\,;\,R\}$. All other zeros $\xi_{k+1},\ldots,\xi_n$ are supposed to lie in the region $W^{(0)} = \{z: |z-c| > R\}$ (the exterior of $\{c;R\}$). This kind of distribution appears in the case when the refinement of only a certain group of zeros, clustered around a center c, is of interest. As in §4.4, we can take $c = 0$ because of the possibility of a translation onto the origin; thus, $W^{(0)} = \{z: |z| > R\}$.

Let us introduce for $i = 1,\ldots,k$ $(k \geq 2)$

$$A_i^{(0)} = \sum_{\substack{j=1\\ j\neq i}}^{k} \frac{\mu_j}{z_i^{(0)} - Z_j^{(0)}},$$

$$C_i^{(0)} = \sum_{\substack{j=1\\ j\neq i}}^{k} \mu_j \left(\frac{1}{z_i^{(0)} - Z_j^{(0)}}\right)^2,$$

$$B_i^{(0)} = \sum_{j=k+1}^{n} \frac{\mu_j}{z_i^{(0)} - W^{(0)}} = \frac{N - N_1}{z_i^{(0)} - W^{(0)}},$$

$$D_i^{(0)} = \sum_{j=k+1}^{n} \mu_j \left(\frac{1}{z_i^{(0)} - W^{(0)}}\right)^2 = (N-N_1)\left(\frac{1}{z_i^{(0)} - W^{(0)}}\right)^2,$$

where $N_1 = \mu_1 + \cdots + \mu_k$ and $z_i^{(0)} = \text{mid } Z_i^{(0)}$.

**REMARK 3.** Since $\text{rad}\,((1/Z)^2) < \text{rad}\,(1/Z^2)$ (in accordance with (2.20')), to obtain a smaller radius in calculating $C_i$ and $D_i$, it is better to sum the squares of inverse disks rather than the inversion of disks' squares.®

Taking $z = z_i^{(0)}$ in (5.28) and using the inclusion isotonicity property, we obatin from (5.28) (that is, (5.33))

$$\xi_i \in z_i^{(0)} - \frac{1}{f(z_i^{(0)}) - \frac{1}{2}u(z_i^{(0)})\left[\frac{1}{\mu_i}(A_i^{(0)} + B_i^{(0)})^2 + C_i^{(0)} + D_i^{(0)}\right]}.$$

For the next outer approximation to $\xi_i$ we take the set

$$Z_1^{(1)} = z_i^{(0)} - \frac{1}{f(z_i^{(0)}) - \frac{1}{2}u(z_i^{(0)})\left[\frac{1}{\mu_i}(A_i^{(0)} + B_i^{(0)})^2 + C_i^{(0)} + D_i^{(0)}\right]}. \tag{5.34}$$

It is obvious that $\xi_i \in Z_i^{(1)}$ $(i = 1,\ldots,k)$. According to the properties of circular arithmetic, it follows that the regions $A_i^{(0)}$, $B_i^{(0)}$, $C_i^{(0)}$ and $D_i^{(0)}$ are circular regions. The region $Z_i^{(1)}$ will also be a closed interior of a circle if the disk in the denominator of (5.34) does not contain the origin. The conditions under which the inversion of the denominator gives a circular region will be included.

Let $Z_i^{(m)} = \{z_i^{(m)}; r_i^{(m)}\}$ be a disk with center $z_i^{(m)} = \text{mid } Z_i^{(m)}$ and radius $r_i^{(m)} = \text{rad } Z_i^{(m)}$ $(i = 1,\ldots,k)$, where $m = 0,1,\ldots$ is the iteration index, and let $A_i^{(m)}$, $B_i^{(m)}$, $C_i^{(m)}$ and $D_i^{(m)}$ be the circular regions defined in the same way as previously (the iteration index "0" is replaced by "m" and $W^{(m+1)} = W^{(m)}$ $(m=0,1,\ldots))$. If $P(z_i^{(m)}) \neq 0$, then the successive inclusion approximations to the zeros $\xi_1,\ldots,\xi_k$ are given by

$$Z_i^{(m+1)} = z_i^{(m)} - \frac{1}{f(z_i^{(m)}) - \frac{1}{2}u(z_i^{(m)})\left[\frac{1}{\mu_i}(A_i^{(m)}+B_i^{(m)})^2 + C_i^{(m)} + D_i^{(m)}\right]} \tag{5.35}$$

$$(i=1,\ldots,k;\ m=0,1,\ldots)\ .$$

**REMARK 4.** If $P(z_i^{(m)}) = 0$, then we have found the exact zero $\xi_i$, and we continue the iterative procedure for the remaining zeros taking $Z_i^{(m+1)} = \{z_i^{(m)}\ ;\ 0\}$ and do not iterate further for index i. ®

In the sequel the convergence analysis of the iterative method (5.35) will be presented. This analysis is similar to that given in [33] and [88]. We will prove that $z_i^{(m)} \to \xi_i$ $(i=1,\ldots,k)$ as $m\to\infty$, that is, the radii $r_i^{(m)}$ of the inclusion disks $Z_i^{(m)}$ tend to zero. The rate of convergence is proved to be at least *cubic*.

We introduce the following notations:

$$r^{(m)} = \max_{1\le i\le k} r_i^{(m)}, \qquad \mu = \min_{1\le i\le k} \mu_i, \qquad \alpha = \frac{1}{\mu}(4N^2 - 2\mu^2),$$

$$\rho^{(m)} = \min\left\{\min_{\substack{i,j\\ i\neq j}} \{|z_i^{(m)} - z_j^{(m)}| - r_j^{(m)}\},\quad \min_{1\le i\le k}\{R - |z_i^{(m)}|\}\right\}.$$

For brevity, the circular region in the denominator of (5.35) will be denoted by $Q_i^{(m)}$, that is,

$$Q_i^{(m)} = \frac{1}{\mu_i}(A_i^{(m)} + B_i^{(m)})^2 + C_i^{(m)} + D_i^{(m)}.$$

Clearly, $Q_i^{(m)}$ is circular extension of $q_i(z_i^{(m)},\xi)$ given by (5.32).

First, we will give some assertions necessary for the convergence analysis of the method (5.35). For the sake of simplicity, we omit the iteration index. In the sequel, we will always assume that $N > \mu$.

**LEMMA 5.3.** *For arbitrary real numbers $x\ge 1$, $y\ge 1$, the inequality*

$$5x^2(10y^2 - 2y - 1) - 9xy - 2y^2 > 0 \tag{5.36}$$

*is valid.*

The proof of the inequality (5.36) is elementary and will be omitted.

**LEMMA 5.4.** *Under the condition*

$$\rho > 5(N-\mu)r \tag{5.37}$$

*the disk* $f(z_i) - \frac{1}{2}u(z_i)Q_i$ *does not contain the origin.*

*Proof.* Using the inclusions derived in [33], we obtain

$$A_i \subset \left\{0\ ;\ \frac{N_1}{\rho}\right\}, \qquad B_i \subset \left\{0\ ;\ \frac{2(N-N_1)}{\rho}\right\},$$

$$C_i \subset \left\{0\ ;\ \frac{N_1}{\rho^2}\right\}, \qquad D_i \subset \left\{0\ ;\ \frac{4(N-N_1)}{\rho^2}\right\}.$$

Therefore, it follows

$$Q_i \subset \left\{0\ ;\ \frac{\frac{1}{\mu_i}(2N-N_1)^2 + 4N-3N_1}{\rho^2}\right\} \subseteq \left\{0\ ;\ \frac{\alpha}{\rho^2}\right\}. \tag{5.38}$$

The requirement $0 \notin f(z_i) - \frac{1}{2}u(z_i)Q_i$ from the lemma is equivalent to

$$|\text{mid}\,(f(z_i) - \tfrac{1}{2}u(z_i)Q_i)| > \text{rad}\,(f(z_i) - \tfrac{1}{2}u(z_i)Q_i),$$

that is, in regard to (5.29) and (5.38), to the inequality

$$\left|\frac{P'(z_i)}{2P(z_i)}(1+\frac{1}{\mu_i}) - \frac{P''(z_i)}{2P'(z_i)}\right| > \frac{\alpha}{2\rho^2}\left|\frac{P(z_i)}{P'(z_i)}\right|,$$

or

$$\left|\frac{P'(z_i)^2 - P(z_i)P''(z_i)}{P(z_i)^2} + \frac{1}{\mu_i}\left(\frac{P'(z_i)}{P(z_i)}\right)^2\right| > \frac{\alpha}{\rho^2}. \tag{5.39}$$

Using (5.37) and the relations

$$\frac{P'(z)}{P(z)} = \sum_{j=1}^{n} \mu_j(z-\xi_j)^{-1}, \qquad \frac{P'(z)^2 - P''(z)P(z)}{P(z)^2} = \sum_{j=1}^{n} \mu_j(z-\xi_j)^{-2},$$

the left-hand side of the inequality (5.39) becomes

$$\left|\frac{P'(z_i)^2 - P(z_i)P''(z_i)}{P(z_i)^2} + \frac{1}{\mu_i}\left(\frac{P'(z_i)}{P(z_i)}\right)^2\right|$$

$$= \left|\frac{2\mu_i}{(z_i-\xi_i)^2} + \sum_{j\neq i}\frac{\mu_j}{(z_i-\xi_j)^2} + \frac{2}{z_i-\xi_i}\sum_{j\neq i}\frac{\mu_j}{z_i-\xi_j} + \frac{1}{\mu_i}\left(\sum_{j\neq i}\frac{\mu_j}{z_i-\xi_j}\right)^2\right|$$

$$\geq \frac{2\mu_i}{|z_i-\xi_i|^2} - \sum_{j\neq i}\frac{\mu_j}{|z_i-\xi_j|^2} - \frac{2}{|z_i-\xi_i|}\sum_{j\neq i}\frac{\mu_j}{|z_i-\xi_j|} - \frac{1}{\mu_i}\left(\sum_{j\neq i}\frac{\mu_j}{|z_i-\xi_j|}\right)^2$$

$$\geq \frac{2\mu_i}{r^2} - \frac{N-\mu_i}{\rho^2} - \frac{2}{r}\cdot\frac{N-\mu_i}{\rho} - \frac{1}{\mu_i}\left(\frac{N-\mu_i}{\rho}\right)^2$$

$$> \frac{50\,\mu(N-\mu)^2}{\rho^2} - \frac{N-\mu}{\rho^2} - \frac{10(N-\mu)^2}{\rho^2} - \frac{1}{\mu}\cdot\frac{(N-\mu)^2}{\rho^2} = \frac{1}{\rho^2}\Phi(N,\mu),$$

where

$$\Phi(N,\mu) = (N-\mu)\left[(N-\mu)\left(50\mu - 10 - \frac{1}{\mu}\right) - 1\right].$$

To prove the inequality (5.39), it suffices to show that

$$\Phi(N,\mu) > \frac{1}{\mu}(4N^2 - 2\mu^2) = \alpha,$$

that is,

$$(N-\mu)\left[(N-\mu)(50\mu^2 - 10\mu - 1) - \mu\right] > 4N^2 - 2\mu^2.$$

Putting $N-\mu = x \geq 2$ and $\mu = y \geq 1$, we obtain the inequality (5.36) from Lemma 5.3. □

**LEMMA 5.5.** *If (5.37) holds, then*

$$|u(z_i)| < \frac{5}{5\mu-1}r, \tag{5.40}$$

$$|f(z_i)| > \frac{1}{r}\cdot\frac{9\mu-1}{10}. \tag{5.41}$$

*Proof.* First, we prove the inequality (5.40). Since

$$\frac{1}{|u(z_i)|} = \left|\sum_{j=1}^{n} \frac{\mu_j}{z_i - \xi_j}\right| > \frac{\mu_i}{|z_i - \xi_i|} - \sum_{j\neq i} \frac{\mu_j}{|z_i - \xi_j|}$$

$$> \frac{\mu_i}{r} - \frac{N-\mu_i}{\rho} > \frac{\mu}{r} - \frac{N-\mu}{\rho} = \frac{\mu\rho - (N-\mu)r}{\rho r},$$

using (5.37) we estimate

$$|u(z_i)| < \frac{\rho r}{\mu\rho - (N-\mu)r} = r\left[\frac{1}{\mu - (N-\mu)\frac{r}{\rho}}\right] < \frac{5r}{5\mu - 1}.$$

To prove (5.41) we transform $f(z_i)$ as follows

$$2|f(z_i)| = \left|\frac{P'(z_i)}{P(z_i)}\left(1 + \frac{1}{\mu_i}\right) - \frac{P''(z_i)}{P'(z_i)}\right|$$

$$= \left|\left(\frac{P'(z_i)^2 - P''(z_i)P(z_i)}{P(z_i)^2}\right) \Big/ \left(\frac{P'(z_i)}{P(z_i)}\right) + \frac{1}{\mu_i}\cdot\frac{P'(z_i)}{P(z_i)}\right|$$

$$= \left| \left( \sum_{j=1}^{n} \frac{\mu_j}{(z_i - \xi_j)^2} \right) \Big/ \left( \sum_{j=1}^{n} \frac{\mu_j}{z_i - \xi_j} \right) + \frac{1}{\mu_i} \sum_{j=1}^{n} \frac{\mu_j}{z_i - \xi_j} \right| .$$

After a short rearrangement, we obtain

$$2|f(z_i)| = \left| \frac{1}{z_i - \xi_i} \left( 1 + \frac{1 + \beta_i}{1 + \gamma_i} \right) + \frac{1}{\mu_i} \sum_{j \neq i} \frac{\mu_j}{z_i - \xi_j} \right| , \qquad (5.42)$$

where

$$\beta_i = \frac{(z_i - \xi_i)^2}{\mu_i} \sum_{j \neq i} \frac{\mu_j}{(z_i - \xi_j)^2} ,$$

$$\gamma_i = \frac{z_i - \xi_i}{\mu_i} \sum_{j \neq i} \frac{\mu_j}{z_i - \xi_j} .$$

Using the inequality $\frac{r}{\rho} < \frac{1}{5(N-\mu)}$ (the condition (5.37)), we find the following bounds for the complex numbers $\beta_i$ and $\gamma_i$:

$$|\beta_i| = \frac{|z_i - \xi_i|^2}{\mu_i} \left| \sum_{j \neq i} \frac{\mu_j}{(z_i - \xi_j)^2} \right| < \frac{r_i^2}{\mu_i} \sum_{j \neq i} \frac{\mu_j}{|z_i - \xi_j|^2}$$

$$< \frac{r^2}{\mu} \cdot \frac{N - \mu}{\rho^2} < \frac{1}{25\mu(N-\mu)} \le \frac{1}{50} ,$$

and

$$|\gamma_i| = \frac{|z_i - \xi_i|}{\mu_i} \left| \sum_{j \neq i} \frac{\mu_j}{z_i - \xi_j} \right| < \frac{r_i}{\mu_i} \sum_{j \neq i} \frac{\mu_j}{|z_i - \xi_j|} < \frac{r}{\rho}(N-\mu) < \frac{1}{5} .$$

Let $B = \{0 ; \frac{1}{50}\}$ and $\Gamma = \{0 ; \frac{1}{5}\}$ be the disks containing $\beta_i$ and $\gamma_i$, respectively. Then

$$\min_{\substack{\beta_i \in B \\ \gamma_i \in \Gamma}} \left| 1 + \frac{1 + \beta_i}{1 + \gamma_i} \right| = 1 + \frac{1 - \frac{1}{50}}{1 + \frac{1}{5}} > \frac{9}{5} ,$$

so that

$$\left| 1 + \frac{1 + \beta_i}{1 + \gamma_i} \right| > \frac{9}{5} .$$

Using the last inequality, from (5.42) there follows

$$2|f(z_i)| \ge \frac{1}{|z_i - \xi_i|} \left| 1 + \frac{1 + \beta_i}{1 + \gamma_i} \right| - \frac{1}{\mu_i} \sum_{j \neq i} \frac{\mu_j}{|z_i - \xi_j|}$$

$$> \frac{9}{5r} - \frac{1}{\mu_i} \cdot \frac{N - \mu_i}{\rho} \ge \frac{9}{5r} - \frac{N-\mu}{\mu\rho} > \frac{1}{r}\left(\frac{9}{5} - \frac{1}{5\mu}\right) = \frac{9\mu - 1}{5\mu r} ,$$

wherefrom

$$|f(z_i)| > \frac{1}{r}\cdot\frac{9\mu-1}{10\mu}\,. \qquad \square$$

**THEOREM 5.4.** *Let the sequences of intervals* $(Z_i^{(m)})$ $(i=1,\dots,k)$ *be defined by* (5.35). *Then, under the condition*

$$\rho^{(0)} > 5(N-\mu)r^{(0)}, \tag{5.43}$$

*for each* $i=1,\dots,k$ *and* $m=0,1,\dots$, *we have*

$1^o$ $\xi_i \in Z_i^{(m)}$;

$2^o$ $r^{(m+1)} < \dfrac{\alpha}{\mu}\cdot\dfrac{1.25\,r^{(m)^3}}{(\rho^{(0)}-5r^{(0)})^2}$.

*Proof.* First, we will prove the assertion $1^o$. Assume that $\xi_i \in Z_i^{(m)}$ $(i=1,\dots,k)$. Since the zeros $\xi_{k+1},\dots,\xi_n$ are located within the interior of $W^{(m)}$ $(=W^{(0)})$ and

$$\sum_{\substack{j=1\\ j\neq i}}^{k} \frac{\mu_j}{z_i^{(m)}-\xi_j} \in \sum_{\substack{j=1\\ j\neq i}}^{k} \frac{\mu_j}{z_i^{(m)}-Z_j^{(m)}} = A_i^{(m)},$$

$$\sum_{\substack{j=1\\ j\neq i}}^{k} \frac{\mu_j}{(z_i^{(m)}-\xi_j)^2} \in \sum_{\substack{j=1\\ j\neq i}}^{k} \mu_j\left(\frac{1}{z_i^{(m)}-Z_j^{(m)}}\right)^2 = C_i^{(m)},$$

from (5.33) and (5.35) we conclude that

$$\xi_i \in z_i^{(m)} - \frac{1}{f(z_i^{(m)}) - \frac{1}{2}u(z_i^{(m)})Q_i^{(m)}} = Z_i^{(m+1)}.$$

Since $\xi_i \in Z_i^{(0)}$, the assertion $1^o$ follows by mathematical induction.

Let us prove now the assertion $2^o$. Using the inclusion (5.38) we find

$$r_i^{(m+1)} = \operatorname{rad}\left(\frac{1}{f(z_i^{(m)}) - \frac{1}{2}u(z_i^{(m)})Q_i^{(m)}}\right)$$

$$< \operatorname{rad}\left(\frac{1}{\left\{f(z_i^{(m)})\,;\ \frac{\alpha}{2\rho^{(m)^2}}|u(z_i^{(m)})|\right\}}\right)$$

$$= \frac{\alpha|u(z_i^{(m)})|/2\rho^{(m)^2}}{|f(z_i^{(m)})|^2 - \alpha^2|u(z_i^{(m)})|^2 / 4\rho^{(m)^4}} .$$

According to the bounds (5.40) and (5.41), and the initial condition (5.43), for m = 0 we obtain

$$r_i^{(1)} \le r^{(1)} < \frac{\frac{\alpha}{2}\cdot\frac{5r^{(0)}}{(5\mu-1)\rho^{(0)^2}}}{\frac{1}{r^{(0)^2}}\left(\frac{9\mu-1}{10\mu}\right)^2 - \frac{\alpha^2}{4}\cdot\frac{25r^{(0)^2}}{(5\mu-1)^2\rho^{(0)^4}}}$$

$$= \frac{250\alpha\mu^2(5\mu-1)r^{(0)^3}/\rho^{(0)^2}}{(9\mu-1)^2(5\mu-1)^2 - 625\alpha^2\mu^2\left[\frac{r^{(0)}}{\rho^{(0)}}\right]^4}$$

$$< \frac{250(4N^2-2\mu^2)\mu(5\mu-1)r^{(0)^3}/\rho^{(0)^2}}{(9\mu-1)^2(5\mu-1)^2 - \left[\frac{\alpha\mu}{(N-\mu)^2}\right]^2} .$$

Since

$$\frac{\alpha\mu}{(N-\mu)^2} \le 2\mu^2 + 8\mu + 4,$$

we then have

$$r^{(1)} < \frac{250(4N^2-2\mu^2)\mu(5\mu-1)}{(9\mu-1)^2(5\mu-1)^2 - (2\mu^2+8\mu+4)^2}\cdot\frac{r^{(0)^3}}{\rho^{(0)^2}} . \qquad (5.44)$$

It is easy to show that

$$\frac{(5\mu-1)\mu^3}{(9\mu-1)^2(5\mu-1)^2 - (2\mu^2+8\mu+4)^2} \le \frac{1}{207}$$

for every $\mu \ge 1$. According to this, from (5.44) it follows that

$$r^{(1)} < \frac{250(4N^2-2\mu^2)}{207\mu^2}\cdot\frac{r^{(0)^3}}{\rho^{(0)^2}} < \frac{1.25(4N^2-2\mu^2)}{\mu^2}\cdot\frac{r^{(0)^3}}{\rho^{(0)^2}} ,$$

that is,

$$r^{(1)} < 1.25\,\frac{\alpha}{\mu}\cdot\frac{r^{(0)^3}}{\rho^{(0)^2}} . \qquad (5.45)$$

Hence

$$r^{(1)} < 1.25\,\frac{\alpha}{\mu}\cdot\frac{r^{(0)^3}}{[\rho^{(0)} - 5r^{(0)}]^2} \qquad (5.46)$$

and

$$r^{(1)} < 1.25\,\frac{\alpha}{\mu}\cdot\frac{r^{(0)}}{\left[\dfrac{\rho^{(0)}}{r^{(0)}}\right]^2} < \frac{1.25\alpha r^{(0)}}{25\mu(N-\mu)^2} = \frac{r^{(0)}}{\omega(N,\mu)} \ ,$$

where

$$\omega(N,\mu) = \frac{10(N-\mu)^2}{2N^2/\mu^2 - 1} \ .$$

In view of the fact that

$$\min_{\substack{N\geq 3\\ 1\leq\mu<N}} \omega(N,\mu) = \omega(3,1) = \frac{40}{17} > 2,$$

we have

$$r^{(1)} < \frac{r^{(0)}}{2}. \tag{5.47}$$

The disks $Z_1^{(1)},\ldots,Z_k^{(1)}$ will be nonoverlapping if

$$\rho^{(0)} > |z_i^{(1)} - z_i^{(0)}| + 3r^{(1)}$$

holds (see [33]). Since $|z_i^{(1)} - z_i^{(0)}| \leq r^{(0)} + r^{(1)}$, by (5.43) and (5.47) we obtain

$$\rho^{(0)} > 3r^{(0)} > r^{(0)} + 4r^{(1)} > |z_i^{(1)} - z_i^{(0)}| + 3r^{(1)}.$$

Starting from the inequality

$$\rho^{(1)} \geq \rho^{(0)} - r^{(0)} - 3r^{(1)}, \tag{5.48}$$

which follows from a geometric construction (see [33], [88], [131]), and using (5.43) and (5.47), we find

$$\rho^{(1)} > 5(N-\mu)r^{(0)} - r^{(0)} - \frac{3r^{(0)}}{2} > 2[5(N-\mu)-\tfrac{5}{2}]r^{(1)};$$

hence

$$\rho^{(1)} > 5(N-\mu)r^{(1)}. \tag{5.49}$$

We will now prove the assertion 2° by induction. We assume the following inequalities are true for any $m \geq 1$:

$$r^{(m)} < \frac{1.25\alpha}{\mu}\cdot\frac{{r^{(m-1)}}^3}{{\rho^{(m-1)}}^2}, \tag{5.50}$$

$$r^{(m)} < \frac{r^{(m-1)}}{2}, \tag{5.51}$$

$$\rho^{(m)} > 5(N-\mu)r^{(m)}, \tag{5.52}$$

$$\rho^{(m)} \geq \rho^{(m-1)} - r^{(m-1)} - 3r^{(m)}. \tag{5.53}$$

The last four relations have already been proved for $m = 1$. We will now prove these relations are valid for index $m+1$ $(m \geq 0)$.

By entirely analogous reasoning as for $m = 0$ and by (5.52), we find

$$r^{(m+1)} < \frac{1.25\alpha}{\mu} \cdot \frac{r^{(m)^3}}{\rho^{(m)^2}} < \frac{r^{(m)}}{2}.$$

Using the consideration similar to that used for $m = 0$, we obtain

$$\rho^{(m+1)} \geq \rho^{(m)} - r^{(m)} - 3r^{(m+1)} > 5(N-\mu)r^{(m+1)}.$$

Similarly, applying (5.51) and (5.53) we get

$$\rho^{(m)} > \rho^{(0)} - 5r^{(0)},$$

so that

$$r^{(m+1)} < \frac{1.25\alpha}{\mu(\rho^{(0)} - 5r^{(0)})^2} r^{(m)^3}.$$

Finally, we must prove that the interval process (5.35) is always defined under the initial condition (5.43). We have already proved that (5.43) implies the inequalities

$$\rho^{(m)} > 5(N-\mu)r^{(m)} \quad (m = 1,2,\dots).$$

Therefore, Lemma 5.4 is applicable for each $m = 0,1,\dots$ so that

$$0 \notin f(z_i^{(m)}) - \tfrac{1}{2}u(z_i^{(m)})Q_i^{(m)}.$$

Thus, the inversion of denominator in (5.35) is a disk in each iterative step. □

## ALGORITHM FOR SIMULTANEOUS IMPROVING ALL INCLUSION DISKS

Assume now that we have found an array of n initial nonoverlapping disks $Z_i^{(0)} = \{c_i^{(0)}; r_i^{(0)}\}$ $(i = 1,\dots,n)$ satisfying $z_i^{(0)} \neq \xi_i \in Z_i^{(0)}$. Starting from (5.28) and using the inclusion isotonicity property, we can establish the following algorithm for successive inclusion approximations:

$$z_i^{(m+1)} = z_i^{(m)} - \frac{1}{f(z_i^{(m)}) - \frac{1}{2}u(z_i^{(m)})\left[\frac{1}{\mu_i}A_i^{(m)^2} + C_i^{(m)}\right]} \tag{5.54}$$

$$(i = 1,\ldots,n;\ m = 0,1,\ldots),$$

where

$$A_i^{(m)} = \sum_{\substack{j=1\\ j\neq i}}^{n} \frac{\mu_j}{z_i^{(m)} - Z_j^{(m)}}, \qquad C_i^{(m)} = \sum_{\substack{j=1\\ j\neq i}}^{n} \mu_j \left(\frac{1}{z_i^{(m)} - Z_j^{(m)}}\right)^2 .$$

Let us introduce the following:

$$r^{(m)} = \max_{1\le i\le n} r_i^{(m)}, \qquad \rho^{(m)} = \min_{\substack{i,j\\ i\neq j}} \{|z_i^{(m)} - z_j^{(m)}| - r_j^{(m)}\},$$

$$\mu = \min_{1\le i\le n} \mu_i, \qquad \gamma = N/\mu,$$

$$Q_i^{(m)} = \frac{1}{\mu_i}A_i^{(m)^2} + C_i^{(m)}. \tag{5.55}$$

Before turning to the convergence theorem of the algorithm (5.54), we will derive some necessary inclusions. The notation $z_i$, $Z_i$, $r_i$, $r$, $\rho$, $A_i$, $C_i$ and $Q_i$ will be used instead of $z_i^{(m)}$, $Z_i^{(m)}$, $r_i^{(m)}$, $r^{(m)}$, $\rho^{(m)}$, $A_i^{(m)}$, $C_i^{(m)}$ and $Q_i^{(m)}$, respectively.

Assume that the condition

$$\rho > 3(N-\mu)r \tag{5.56}$$

is valid. Using the inclusions given in [88], we find

$$A_i = \sum_{\substack{j=1\\ j\neq i}}^{n} \frac{\mu_j}{z_i - Z_j} \subset \sum_{\substack{j=1\\ j\neq i}}^{n} \mu_j \left\{\frac{1}{z_i - z_j};\ \frac{r}{\rho^2}\right\} = \left\{\sum_{\substack{j=1\\ j\neq i}}^{n} \frac{\mu_j}{z_i - z_j};\ \frac{(N-\mu_i)r}{\rho^2}\right\},$$

and hence

$$A_i^2 \subset \left\{\left(\sum_{\substack{j=1\\ j\neq i}}^{n} \frac{\mu_j}{z_i - z_j}\right)^2;\ 2\left|\sum_{\substack{j=1\\ j\neq i}}^{n} \frac{\mu_j}{z_i - z_j}\right| \cdot \frac{(N-\mu_i)r}{\rho^2} + \frac{(N-\mu_i)^2 r^2}{\rho^4}\right\}$$

$$\subset \left\{\left(\sum_{\substack{j=1\\ j\neq i}}^{n} \frac{\mu_j}{z_i - z_j}\right)^2;\ \frac{3(N-\mu)^2 r}{\rho^3}\right\}.$$

Further, we have

$$C_i = \sum_{\substack{j=1\\ j\neq i}}^{n} \mu_j \left(\frac{1}{z_i - z_j}\right)^2 \subset \sum_{\substack{j=1\\ j\neq i}}^{n} \mu_j \left\{ \frac{1}{(z_i-z_j)^2} ; \frac{2r}{\rho^3} \right\}$$

$$= \left\{ \sum_{\substack{j=1\\ j\neq i}}^{n} \frac{\mu_j}{(z_i-z_j)^2} ; \frac{2(N-\mu_i)r}{\rho^3} \right\} \subseteq \left\{ \sum_{\substack{j=1\\ j\neq i}}^{n} \frac{\mu_j}{(z_i-z_j)^2} ; \frac{2(N-\mu)r}{\rho^3} \right\}.$$

This yields

$$Q_i = \frac{1}{\mu_i} A_i^2 + C_i \subset \left\{ \frac{1}{\mu_i}\left(\sum_{j\neq i} \frac{\mu_j}{z_i-z_j}\right)^2 + \sum_{j\neq i} \frac{\mu_j}{(z_i-z_j)^2} ; \left(\frac{3N}{\mu} - 1\right)\frac{(N-\mu)r}{\rho^3} \right\},$$

that is,

$$Q_i \subset \{\tilde{q}_i ; \eta\}, \tag{5.57}$$

where

$$\tilde{q}_i = \frac{1}{\mu_i}\left(\sum_{j\neq i} \frac{\mu_j}{z_i-z_j}\right)^2 + \sum_{j\neq i} \frac{\mu_j}{(z_i-z_j)^2}$$

and

$$\eta = \left(\frac{3N}{\mu} - 1\right) \frac{(N-\mu)r}{\rho^3} .$$

The complex quantity $\tilde{q}_i$ is an approximation to $q_i(z_i,\xi)$ defined by (5.32), that is obtained by substituting the exact zeros $\xi_1,\ldots,\xi_n$ by their approximations $z_1,\ldots,z_n$.

**LEMMA 5.6.** *If the condition (5.56) is valid, then*

$$|u(z_i)| < \frac{3}{3\mu-1} r , \tag{5.58}$$

$$|f(z_i)| > \frac{1}{r}\cdot\frac{5\mu-1}{6\mu} , \tag{5.59}$$

$$|f(z_i) - \tfrac{1}{2}u(z_i)\tilde{q}_i| > \frac{5\mu - \frac{7}{4}}{6\mu r} . \tag{5.60}$$

*P r o o f.* The inequalities (5.58) and (5.59) can be proved in a way similar to (5.40) and (5.41). To prove (5.60) we first estimate

$$|\tilde{q}_i| < \frac{1}{\mu_i}\left(\sum_{j\neq i} \frac{\mu_j}{|z_i-z_j|}\right)^2 + \sum_{j\neq i} \frac{\mu_j}{|z_i-z_j|^2} < \frac{1}{\mu}\cdot\frac{(N-\mu)^2}{\rho^2} + \frac{N-\mu}{\rho^2} = \frac{N(N-\mu)}{\mu\rho^2} .$$

Now, using (5.56), we find

$$|f(z_i) - \frac{1}{2}u(z_i)\tilde{q}_i| > |f(z_i)| - \frac{1}{2}|u(z_i)||\tilde{q}_i| > \frac{1}{r}\left[\frac{5\mu-1}{6\mu} - \frac{3N(N-\mu)}{2\mu(3\mu-1)}\cdot\frac{r^2}{\rho^2}\right]$$

$$> \frac{1}{6\mu r}\left[5\mu - 1 - \frac{N}{(3\mu-1)(N-\mu)}\right] > \frac{5\mu - \frac{7}{4}}{6\mu r}$$

because of $\frac{N}{(N-\mu)(3\mu-1)} \le \frac{3}{4}$ . □

**LEMMA 5.7.** *If (5.56) holds, then* $0 \notin f(z_i) - \frac{1}{2}u(z_i)Q_i$, *where* $Q_i$ *is the disk defined by (5.55).*

The proof of this assertion is similar to the proof of Lemma 5.4 and will be omitted.

The following theorem gives conditions that guarantee $\xi_i \in Z_i^{(m)}$ and $\text{rad } Z_i^{(m)} \to 0$ when $m \to \infty$ .

**THEOREM 5.5.** *Under the condition*

$$\rho^{(0)} > 3(N-\mu)r^{(0)}, \tag{5.61}$$

*the iterative process (5.54) converges with*

$1^o$ $\quad \xi_i \in Z_i^{(m)}$

*and*

$2^o$ $\quad r^{(m+1)} < \dfrac{3(3\gamma-1)(\gamma-1)r^{(m)^4}}{\left(\rho^{(0)} - \frac{7}{3}r^{(0)}\right)^3}$

*for each* $i = 1,\dots,n$ *and* $m = 0,1,\dots$ .

*P r o o f.* From the inclusion (5.57) there follows

$$f(z_i^{(0)}) - \frac{1}{2}u(z_i^{(0)})Q_i^{(0)} \subset f(z_i^{(0)}) - \frac{1}{2}\left\{u(z_i^{(0)})\tilde{q}_i^{(0)} \; ; \; |u(z_i^{(0)})|\eta^{(0)}\right\} .$$

According to this and the bounds given in Lemma 5.6, after an extensive but elementary estimation procedure, we get

$$r_i^{(1)} \le r^{(1)} < \frac{\frac{1}{2}|u(z_i^{(0)})|\eta^{(0)}}{|f(z_i^{(0)}) - \frac{1}{2}u(z_i^{(0)})\tilde{q}_i^{(0)}|^2 - \frac{1}{4}|u(z_i^{(0)})|^2\eta^{(0)^2}}$$

$$< \frac{3(3\gamma-1)(\gamma-1)r^{(0)^4}}{\rho^{(0)^3}} .$$

Hence

$$r^{(1)} < \frac{3(3\gamma-1)(\gamma-1)r^{(0)^4}}{(\rho^{(0)} - \frac{7}{3}r^{(0)})^3} \tag{5.62}$$

and

$$r^{(1)} < \frac{2}{9}r^{(0)} < \frac{r^{(0)}}{4} \ . \tag{5.63}$$

Using (5.48), (5.61) and (5.63), we find

$$\rho^{(1)} > 3(N-\mu)r^{(1)} \ . \tag{5.64}$$

The last three relations, including also (5.48), can be regarded as the first part of a proof by induction with $m = 1$. As in Theorem 5.4 these relations can be supposed true for any $m \geq 1$ and proved to be true for $m+1$.

From the relations

$$r^{(m+1)} < \frac{2}{9}r^{(m)} < \frac{r^{(m)}}{4} \tag{5.65}$$

and

$$r^{(m+1)} < \frac{3(3\gamma-1)(\gamma-1)r^{(m)^4}}{(\rho^{(0)} - \frac{7}{3}r^{(0)})^3},$$

we conclude that the sequence $(r^{(m)})$ tends to zero with a rate of convergence of at least *four*.

The proof of the assertion 1° of Theorem 5.5 is the same as in the previous theorem. Similarly, since (5.62) implies the inequality

$$\rho^{(m)} > 3(N-\mu)r^{(m)}$$

for each $m = 1,2,\ldots$ , according to Lemma 5.7 it follows that the interval process (5.54) is defined in each iteration. □

**REMARK 5.** If we write $Z_i^{(m+1)}$, defined by (5.35) or (5.54), as

$$Z_i^{(m+1)} = z_i^{(m)} - h(z_i^{(m)}) \Big/ \left(1 - \frac{u(z_i^{(m)})h(z_i^{(m)})Q_i^{(m)}}{2}\right),$$

we observe the similarity of the iterative formulas (5.35) and (5.54) with Halley's formula (5.30). ®

The order of convergence of the basic (*total-step*) method (5.54) can be increased if, in the calculation of the inclusion disk $Z_i^{(m+1)}$, we use the already known disks $Z_1^{(m+1)},\ldots,Z_{i-1}^{(m+1)}$ as in the preceding chapters. In this way, we obtain the accelerated (*single-step*) interval method

$$z_i^{(m+1)} = z_i^{(m)} - \frac{1}{f(z_i^{(m)}) - \frac{1}{2}u(z_i^{(m)})\left[\frac{1}{\mu_i}\tilde{A}_i^{(m)\,2} + \tilde{C}_i^{(m)}\right]} \tag{5.66}$$

$$(i = 1,\dots,n;\ m = 0,1,\dots) ,$$

where

$$\tilde{A}_i^{(m)} = \sum_{j<i} \frac{\mu_j}{z_i^{(m)} - z_j^{(m+1)}} + \sum_{j>i} \frac{\mu_j}{z_i^{(m)} - z_j^{(m)}} ,$$

$$\tilde{C}_i^{(m)} = \sum_{j<i} \mu_j\left(\frac{1}{z_i^{(m)} - z_j^{(m+1)}}\right)^2 + \sum_{j>i} \mu_j\left(\frac{1}{z_i^{(m)} - z_j^{(m)}}\right)^2 .$$

The convergence properties of the single-step method (5.66) are treated in the following theorem:

**THEOREM 5.6.** *Under the condition (5.61) the iterative method (5.66) is convergent with the R-order of convergence*

$$O_R((5.66),\xi) \geq \rho(X_n),$$

*where $\rho(X_n)$ is the spectral radius of the matrix $X_n$ given by*

$$X_2 = \begin{bmatrix} 3 & 1 \\ 3 & 4 \end{bmatrix}, \quad X_n = \begin{bmatrix} 3 & 1 & & & & & \\ & 3 & 1 & & & \text{O} & \\ & & \cdot & \cdot & & & \\ & & & \cdot & \cdot & & \\ & \text{O} & & & \cdot & & \\ & & & & & 3 & 1 \\ 3 & 1 & 0 & \cdot & \cdot & 0 & 3 \end{bmatrix} \quad (n \geq 3). \tag{5.67}$$

*P r o o f.* For reasons of simplicity, we will omit the iteration index and use the abbreviations as well as some of the inclusions and inequalities employed for the total-step method. In addition, we will write $\hat{Z}_i$, $\hat{z}_i$, $\hat{r}_i$, $\hat{r}$, $\tilde{A}_i$, $\tilde{C}_i$ instead of $Z_i^{(m+1)}$, $z_i^{(m+1)}$, $r_i^{(m+1)}$, $r^{(m+1)}$, $\tilde{A}_i^{(m)}$, $\tilde{C}_i^{(m)}$, respectively.

First, applying the properties of circular arithmetic we obtain the following inclusions:

$$\sum_{j=1}^{i-1} \frac{\mu_j}{z_i-\hat{z}_j} \subset \left\{ \sum_{j=1}^{i-1} \frac{\mu_j}{z_i-\hat{z}_j} \; ; \; \frac{1}{\rho(\rho-2\hat{r})} \sum_{j=1}^{i-1} \mu_j \hat{r}_j \right\} ,$$

$$\sum_{j=i+1}^{n} \frac{\mu_j}{z_i-z_j} \subset \left\{ \sum_{j=i+1}^{n} \frac{\mu_j}{z_i-z_j} \; ; \; \frac{1}{\rho^2} \sum_{j=i+1}^{n} \mu_j r_j \right\} .$$

Hence

$$\tilde{A}_i^2 \subset \left\{ \sum_{j<i} \frac{\mu_j}{z_i-\hat{z}_j} + \sum_{j>i} \frac{\mu_j}{z_i-z_j} \; ; \; a_1 \left[ \frac{1}{\rho(\rho-2\hat{r})} \sum_{j<i} \mu_j \hat{r}_j + \frac{1}{\rho^2} \sum_{j>i} \mu_j r_j \right] \right\}$$

and

$$\tilde{C}_i \subset \left\{ \sum_{j<i} \frac{\mu_j}{(z_i-\hat{z}_j)^2} + \sum_{j>i} \frac{\mu_j}{(z_i-z_j)^2} \; ; \; \frac{2}{(\rho-2\hat{r})^2} \sum_{j<i} \mu_j \hat{r}_j + \frac{2}{\rho^2} \sum_{j>i} \mu_j r_j \right\},$$

where

$$a_1 = \left[ \frac{2}{\rho-\hat{r}} + \frac{\hat{r}}{\rho(\rho-2\hat{r})} \right] \sum_{j<i} \mu_j + \left( \frac{2}{\rho} + \frac{r}{\rho^2} \right) \sum_{j>i} \mu_j .$$

The above inclusions yield

$$\tilde{Q}_i = \frac{1}{\mu_i} \tilde{A}_i^2 + \tilde{C}_i \subset \{ w_i \; ; \; \eta_i \} ,$$

where

$$w_i = \frac{1}{\mu_i} \left( \sum_{j<i} \frac{\mu_j}{z_i-\hat{z}_j} + \sum_{j>i} \frac{\mu_j}{z_i-z_j} \right)^2 + \sum_{j<i} \frac{\mu_j}{(z_i-\hat{z}_j)^2} + \sum_{j>i} \frac{\mu_j}{(z_i-z_j)^2} ,$$

$$\eta_i = \frac{a_2}{\rho(\rho-2\hat{r})} \sum_{j<i} \mu_j \hat{r}_j + \frac{a_3}{\rho^2} \sum_{j>i} \mu_j r_j ,$$

$$a_2 = \frac{a_1}{\mu_i} + \frac{2}{\rho-2\hat{r}} , \quad a_3 = \frac{a_1}{\mu_i} + \frac{2}{\rho} .$$

Using the estimations

$$\frac{r}{\rho} < \frac{1}{3(N-\mu)} \le \frac{1}{6} ,$$

$$a_1 < \frac{\frac{13}{6}(N-\mu_i)}{\rho} ,$$

$$a_3 < a_2 < \frac{\frac{13}{6}N + \frac{3}{200}\mu_i}{\rho\mu_i} ,$$

we obtain

$$\eta_i < \frac{a_2}{\rho}\left(\frac{1}{\rho-2\hat{r}}\sum_{j<i}\mu_j\hat{r}_j + \frac{1}{\rho}\sum_{j>i}\mu_j r_j\right)$$

$$< \frac{\frac{13}{6}N + \frac{3}{200}\mu_i}{\mu_i\rho^2}\left(\frac{1}{\rho-2\hat{r}}\sum_{j<i}\mu_j\hat{r}_j + \frac{1}{\rho}\sum_{j>i}\mu_j r_j\right)$$

$$< \frac{\frac{13}{6}N + \frac{3}{200}\mu_i}{\mu_i\rho^2}\cdot\frac{r}{\rho}(N-\mu_i).$$

Let

$$c_i = f(z_i) - \frac{1}{2}u(z_i)w_i, \qquad d_i = \frac{1}{2}|u(z_i)|\eta_i.$$

We estimate

$$d_i < \frac{1}{2}\cdot\frac{3r_i}{3\mu_i-1}\cdot\frac{\frac{13}{6}N + \frac{3}{200}\mu_i}{\mu_i\rho^2}\left(\frac{1}{\rho-2\hat{r}}\sum_{j<i}\mu_j\hat{r}_j + \frac{1}{\rho}\sum_{j>i}\mu_j r_j\right),$$

wherefrom

$$d_i < \frac{\frac{39}{12}N + \frac{1}{40}\mu_i}{\mu_i(3\mu_i-1)}\cdot\frac{r_i}{\rho^2}\left(\frac{1}{\rho-2\hat{r}}\sum_{j<i}\mu_j\hat{r}_j + \frac{1}{\rho}\sum_{j>i}\mu_j r_j\right) \tag{5.68}$$

and

$$d_i < \frac{(\frac{39}{12}N + \frac{1}{40}\mu_i)r_i}{\mu_i(3\mu_i-1)\rho^2}\cdot\frac{r}{\rho}(N-\mu_i) < \frac{1}{16}\cdot\frac{1}{\mu_i r_i}. \tag{5.69}$$

Furthermore, since

$$|w_i| < \frac{1}{\mu_i}\left(\frac{1}{\rho-\hat{r}}\sum_{j<i}\mu_j + \frac{1}{\rho}\sum_{j>i}\mu_j\right)^2 + \frac{1}{(\rho-\hat{r})^2}\sum_{j<i}\mu_j + \frac{1}{\rho^2}\sum_{j>i}\mu_j$$

$$< \frac{(N-\mu_i)^2}{\mu_i(\rho-\hat{r})^2} + \frac{N-\mu_i}{(\rho-\hat{r})^2} < \frac{N}{8\mu_i(N-\mu_i)r_i^2},$$

it follows

$$|c_i| = |f(z_i) - \frac{1}{2}u(z_i)w_i| \geq |f(z_i)| - \frac{1}{2}|u(z_i)||w_i|$$

$$> \frac{5\mu_i-1}{6\mu_i r_i} - \frac{1}{2}\cdot\frac{3r_i}{3\mu_i-1}\cdot\frac{N}{8\mu_i(N-\mu_i)r_i^2},$$

that is,

$$|c_i| > \frac{5\mu_i - \frac{7}{4}}{6\mu_i r_i} \ . \tag{5.70}$$

By the bounds (5.69) and (5.70), we find

$$|c_i|^2 - d_i^2 > \left(\frac{5\mu_i - \frac{7}{4}}{6\mu_i r_i}\right)^2 - \left(\frac{1}{16\mu_i r_i}\right)^2 > \frac{7}{25 r_i^2} \ , \tag{5.71}$$

which means that the disk $\{c_i ; d_i\}$ has its inverse disk and the iterative process (5.66) is defined because

$$f(z_i) - \frac{1}{2}u(z_i)\tilde{Q}_i \subset \{c_i ; d_i\} \ .$$

Using the inclusion

$$\hat{Z}_i \subset z_i - \{c_i ; d_i\}^{-1} \ ,$$

the bounds (5.68) and (5.71) and the inequalities

$$\rho > \rho^{(0)} - \frac{15}{7}r^{(0)} \ ,$$

$$\rho - 2\hat{r} > \rho^{(0)} - \frac{13}{5}r^{(0)} \ ,$$

obtained by (5.53), (5.61) and (5.65), we furnish

$$\hat{r}_i = \operatorname{rad} \hat{Z}_i < \frac{d_i}{|c_i|^2 - d_i^2}$$

$$< \frac{25}{7} r_i^2 \, \frac{\frac{39}{12}N + \frac{1}{40}\mu_i}{\mu_i(3\mu_i - 1)} \cdot \frac{r_i}{\rho^2}\left(\frac{1}{\rho - 2\hat{r}} \sum_{j<i} \mu_j \hat{r}_j + \frac{1}{\rho} \sum_{j>i} \mu_j r_j\right)$$

$$< a\mu_i^2 r_i^3 \left(\sum_{j<i} \mu_j \hat{r}_j + \sum_{j>i} \mu_j r_j\right) \ ,$$

where

$$a = \frac{12(N - \mu_i)}{\mu_i^3 [\rho^{(0)} - \frac{15}{7}r^{(0)}][\rho^{(0)} - \frac{13}{5}r^{(0)}]} \ .$$

Thus, we have derived the relations

$$r_i^{(m+1)} < a\mu_i^2 (r_i^{(m)})^3 \left(\sum_{j=1}^{i-1} \mu_j r_j^{(m+1)} + \sum_{j=i+1}^{n} \mu_j r_j^{(m)}\right) \quad (i=1,\dots,n). \tag{5.72}$$

Substituting

$$r_i^{(m)} = \frac{v_i^{(m)}}{\mu_i[(n-1)a]^{1/3}} \qquad (i = 1,\dots,n) \tag{5.73}$$

in (5.72), we get

$$v_i^{(m+1)} < \frac{(v_i^{(m)})^3}{n-1}\left(\sum_{j=1}^{i-1} v_j^{(m+1)} + \sum_{j=i+1}^{n} v_j^{(m)}\right) \quad (i=1,\dots,n). \tag{5.74}$$

By (5.61) we obtain from (5.73)

$$(v_i^{(0)})^3 = \mu_i^3 (r_i^{(0)})^3 (n-1)a = \left(\frac{r^{(0)}}{\rho^{(0)}}\right)^3 \frac{12(n-1)(N-\mu_i)}{\left[1 - \frac{15}{7}\frac{r^{(0)}}{\rho^{(0)}}\right]^2 \left[1 - \frac{13}{5}\frac{r^{(0)}}{\rho^{(0)}}\right]}$$

$$< \frac{12}{27(N-\mu)\left[1 - \frac{15/7}{3(N-\mu)}\right]^2 \left[1 - \frac{13/5}{3(N-\mu)}\right]} < 1,$$

which proves that

$$v = \max_{1 \le i \le n} v_i^{(0)} < 1 . \tag{5.75}$$

According to (5.74) and (5.75) we conclude from (5.72) that the sequences $(r_i^{(m)})$ $(i=1,\dots,n)$ converge to zero. Furthermore, from (5.74) and (5.75) we obtain

$$v_i^{(m+1)} \le v^{t_i^{(m+1)}} \quad (i=1,\dots,n;\ m=0,1,\dots). \tag{5.76}$$

The components $t_1^{(m)},\dots,t_n^{(m)}$ of the vector $\mathbf{t}^{(m)} = [t_1^{(m)} \cdots t_m^{(m)}]^T$ can be successively calculated by

$$\mathbf{t}^{(m+1)} = X_n \mathbf{t}^{(m)} \quad (m = 0,1,\dots),$$

where $\mathbf{t}^{(0)} = [1 \cdots 1]^T$ and $X_n$ is the matrix defined by (5.67).

Finally, starting from (5.76) and using the same consideration as in [4], we can prove that the lower bound of $O_R((5.66),\xi)$ is given by

$$O_R((5.66),\xi) \ge \rho(X_n),$$

where $\rho(X_n)$ is the spectral radius of the matrix $X_n$. □

The lower bound of the R-order of convergence of the iterative process (5.66) can be computed easily by applying the *power method* for finding the dominant eigenvalue of $X_n$ (that is, the spectral radius $\rho(X_n)$). The values of $\rho(X_n)$ have already been displayed in Table 5.1 (for k=2 and q=1).

To reduce the number of numerical operations, the Gauss-Seidel procedure can be applied to the basic iterative formula (5.54) $s$ $(>1)$ times using the same values of polynomial P and its derivatives P' and P" at $s$ successive steps (Halley's method with *s steps* ; see the recent paper [131]). However, if the number of repetitions $s$ is large ( say, $s \geq 3$ ), then Halley's method with $s$ steps is not efficient enough. Namely, applying the single-step method (5.66) the desired accuracy will be attained very quickly (by few iterations) so that the advantage of the method with $s$ repetitions - the reduced number of numerical operations - will not be dominant because the convergence of this method is slow compared to Algorithm (5.66). The above given discussion was verified in a great number of numerical examples.

In this regard, we will shortly present Halley's method with just two steps. Starting from the initial disks $Z_1^{(0)},\ldots,Z_n^{(0)}$ containing the exact zeros $\xi_1,\ldots,\xi_n$, we obtain the following algorithm:

$$
\begin{aligned}
W_i^{(m)} &= z_i^{(m)} - \frac{1}{f(z_i^{(m)}) - \frac{1}{2}u(z_i^{(m)})Q_1^{(m)}}, \\
Z_i^{(m)} &= z_i^{(m)} - \frac{1}{f(z_i^{(m)}) - \frac{1}{2}u(z_i^{(m)})Q_2^{(m)}}
\end{aligned}
\tag{5.77}
$$

$$(i=1,\ldots,n;\ m=0,1,\ldots),$$

where

$$
Q_1^{(m)} = \frac{1}{\mu_i}\left(\sum_{j=1}^{i-1}\frac{\mu_j}{z_i^{(m)}-W_j^{(m)}} + \sum_{j=i+1}^{n}\frac{\mu_j}{z_i^{(m)}-Z_j^{(m)}}\right)^2
+ \sum_{j=1}^{i-1}\mu_j\left(\frac{1}{z_i^{(m)}-W_j^{(m)}}\right)^2 + \sum_{j=i+1}^{n}\mu_j\left(\frac{1}{z_i^{(m)}-Z_j^{(m)}}\right)^2,
$$

and

$$
Q_2^{(m)} = \frac{1}{\mu_i}\left(\sum_{j=1}^{i-1}\frac{\mu_j}{z_i^{(m)}-Z_j^{(m+1)}} + \sum_{j=i+1}^{n}\frac{\mu_j}{z_i^{(m)}-W_j^{(m)}}\right)^2
+ \sum_{j=1}^{i-1}\mu_j\left(\frac{1}{z_i^{(m)}-Z_j^{(m+1)}}\right)^2 + \sum_{j=i+1}^{n}\mu_j\left(\frac{1}{z_i^{(m)}-W_j^{(m)}}\right)^2 .
$$

We observe that the above algorithm consists of two iterations when the method (5.66) is applied and $f(z_i^{(m)})$ and $u(z_i^{(m)})$ remain unchanged.

The convergence order of the iterative procedure (5.77) can be determined by the result presented in [131]. Let $Y_n$ be the $n \times n$ matrix expressed by

$$Y_2 = \begin{bmatrix} 6 & 4 \\ 6 & 7 \end{bmatrix}, \quad Y_3 = \begin{bmatrix} 3 & 3 & 1 \\ 3 & 4 & 3 \\ 3 & 3 & 4 \end{bmatrix}, \quad Y_n = \begin{bmatrix} 3 & 3 & 1 & & & & \\ 0 & 3 & 3 & 1 & & O & \\ & & & \cdot & & & \\ & O & & & \cdot & & \\ & & & & \cdot & 3 & 1 \\ 3 & 1 & 0 & & & 3 & 3 \\ 3 & 3 & 1 & 0 & \ldots & 0 & 3 \end{bmatrix} \quad (n \geq 4).$$

The lower bound of the R-order of convergence is given by

$$O_R((5.77), \xi) \geq \rho(Y_n).$$

The values of the spectral radius $\rho(Y_n)$ are shown in Table 5.2.

| $n$ | 2 | 3 | 4 | 5 | 6 | 7 | 8 | 9 | 10 | 20 |
|---|---|---|---|---|---|---|---|---|---|---|
| $\rho(Y_n)$ | 11.424 | 9.000 | 8.347 | 8.000 | 7.798 | 7.664 | 7.568 | 7.496 | 7.441 | 7.221 |

Table 5.2

The realization of Algorithm (5.77) on a computer is simple because it occupies the same storage space for the disks $W_i^{(m)}$ and $Z_i^{(m)}$, the vector $(z_1^{(m)}, \ldots, z_n^{(m)})$ and the values $P(z_i^{(m)})$, $P'(z_i^{(m)})$ and $P''(z_i^{(m)})$ $(i = 1, \ldots, n)$, already calculated in the previous iterative step. Algorithm (5.77) is suitable when the requested accuracy of improved approximations does not necessarily have to be high but a reduction of numerical operations is desirable.

HALLEY'S COMBINED METHODS

In a similar way as in §3.1 and §4.4, the interval Halley-like method (5.54) can be combined with some of the point iterative methods presented in this book. Let $Z_1^{(0)}, \ldots, Z_n^{(0)}$ be the initial disks containing the zeros $\xi_1, \ldots, \xi_n$. Then the improved circular approximations to these zeros are given by

$$z_i^{(M,1)} = z_i^{(M)} - \frac{1}{f(z_i^{(M)}) - \frac{1}{2}u(z_i^{(M)})\left[\frac{1}{\mu_i}A_i^{(M)^2} + C_i^{(M)}\right]} \quad (i = 1, \ldots, n),$$

where

$$A_i^{(M)} = \sum_{\substack{j=1 \\ j\neq i}}^{n} \frac{\mu_j}{z_i^{(M)} - z_j^{(0)}}, \quad C_i^{(M)} = \sum_{\substack{j=1 \\ j\neq i}}^{n} \mu_j \left(\frac{1}{z_i^{(M)} - z_j^{(0)}}\right)^2,$$

and $z_i^{(M)}$ is the point approximation to $\xi_i$, obtained by applying an iterative method in ordinary complex arithmetic. We establish the following four combined methods:

$(K_W)$: ((3.3),(5.54)),

$(K_N)$: ((4.126),(5.54)),

$(K_M)$: ((4.119),(5.54)),

$(K_{MN})$: ((4.122),(5.54)).

We note that the first combination $(K_W)$ cannot be employed for multiple zeros. In such case we may apply Newton's method (4.126) instead of Weierstrass' method (3.3) (the combined method $(K_N)$).

Let $k_p$ be the convergence order of the point method which is applied M times. Then we have the following estimation of the new circular approximations produced by Halley's combined method

$$r^{(M,1)} = O\left((r^{(0)})^{3k_p^M + 1}\right),$$

where $r^{(0)} = \max_{1\le i\le n} r_i^{(0)}$ and $r^{(M,1)} = \max_{1\le i\le n} \operatorname{rad} Z_i^{(M,1)}$.

## ANALYSIS OF NUMERICAL STABILITY

In practice, the application of any iterative formula requires that rounding errors are included in each evaluation. We will now analyze the iterative methods (5.35) and (5.54) in the presence of rounding errors. Accordingly, the single errors relative to the evaluation of P, P' and P" will be taken into consideration in finding f(z) and u(z).

We will present a qualitative analysis of numerical stability of the iterative methods (5.35) and (5.54) omitting the details concerning the initial convergence conditions. Let $\Delta P$, $\Delta P'$ and $\Delta P''$ be upper bounds for the absolute value of the rounding errors that appear in the evaluation of P, P' and P", respectively.

Regarding the discussion presented in § 4.3, the stopping criterion that determines the maximal number of iterations, is based on the comparison of $\Delta P_i^{(m)}$ to $|P(z_i^{(m)})|$ for each i; namely, the iterative process is terminated when at least one of the inequalities $\Delta P_i^{(m)} < |P(z_i^{(m)})|$ $(i = 1,\ldots,k\ (\leq n))$ is not valid. The last inequalities are also very useful as conditions under which the inversions of certain disks, occurring in the iterative formulas (5.35) and (5.54), are defined. To use these conditions, we will rewrite (5.35) and (5.54) in the form

$$z_i^{(m+1)} = z_i^{(m)} - \frac{2\sigma_1(z_i^{(m)})}{(1+\frac{1}{\mu_i})[\sigma_1(z_i^{(m)})]^2 - \sigma_2(z_i^{(m)}) - \{w_i^{(m)}\ ;\ \eta_i^{(m)}\}} \tag{5.78}$$

$$(i = 1,\ldots,k\ (\leq n);\ m = 0,1,\ldots).$$

The disk $\{w_i^{(m)}\ ;\ \eta_i^{(m)}\}$ denotes the circular region in the denominator of the formula (5.35) (instead of $Q_i^{(m)}$) and (5.54) (instead of $\frac{1}{\mu_i}A_i^{(m)^2}+C_i^{(m)}$).

According to the previous discussion, we have to replace the complex value of the polynomial P by the disk $\{P\ ;\ \Delta P\}$ and its derivatives by the disks $\{P'\ ;\ \Delta P'\}$ and $\{P''\ ;\ \Delta P''\}$ with centers P, P', P" and radii ΔP, ΔP', ΔP". Circular extensions of the scalar quantities $\sigma_1(z)$ and $\sigma_2(z)$ (which appear in (5.78)), obtained by the above substitutions, are the disks

$$S_1 = \frac{\{P'\ ;\ \Delta P'\}}{\{P\ ;\ \Delta P\}} = \left\{\frac{P'\bar{P}}{|P|^2-\Delta P^2}\ ;\ \frac{|P'|\Delta P + |P|\Delta P' + \Delta P\Delta P'}{|P|^2-\Delta P^2}\right\} = \{c^{(1)}\ ;\ \delta^{(1)}\},$$

$$S_2 = \frac{\{P''\ ;\ \Delta P''\}}{\{P\ ;\ \Delta P\}} = \left\{\frac{P''\bar{P}}{|P|^2-\Delta P^2}\ ;\ \frac{|P''|\Delta P + |P|\Delta P'' + \Delta P\Delta P''}{|P|^2-\Delta P^2}\right\} = \{c^{(2)}\ ;\ \delta^{(2)}\}.$$

The inversion of the disk $\{P\ ;\ \Delta P\}$ (which occurs in the evaluation of $S_1$ and $S_2$) will produce a disk if $\Delta P < |P|$ holds, which coincides with the stopping criterion.

To simplify our analysis, we will omit the iteration index and write $z_i$, $r_i$, $\hat{r}_i$, $r$, $\hat{r}$, $Z_i$, $\hat{Z}_i$, $w_i$, $\eta_i$, $c_i^{(1)}$, $c_i^{(2)}$, $\delta_i^{(1)}$, $\delta_i^{(2)}$ instead of $z_i^{(m)}$, $r_i^{(m)}$, $r_i^{(m+1)}$, $r^{(m)}$, $r^{(m+1)}$, $Z_i^{(m)}$, $Z_i^{(m+1)}$, $w_i^{(m)}$, $\eta_i^{(m)}$, mid $S_1(z_i^{(m)})$, mid $S_2(z_i^{(m)})$, rad $S_1(z_i^{(m)})$, rad $S_2(z_i^{(m)})$, respectively. The iterative formula (5.78) in the presence of rounding error becomes

$$\hat{Z}_i = z_i - \frac{2\{c_i^{(1)}; \delta_i^{(1)}\}}{(1+\frac{1}{\mu_i})\{c_i^{(1)}; \delta_i^{(1)}\}^2 - \{c_i^{(2)}; \delta_i^{(2)}\} - \{w_i; \eta_i\}} \tag{5.79}$$

$$(i = 1,\ldots,k\ (\le n))\ .$$

Let $\delta = \max\{\delta_1^{(1)},\ldots,\delta_k^{(1)},\delta_1^{(2)},\ldots,\delta_k^{(2)}\}$. Then

$$\{c_i^{(j)}; \delta_i^{(j)}\} \subseteq \{c_i^{(j)}; \delta\} \qquad (j = 1,2;\ i = 1,\ldots,k)$$

and from (5.79) we obtain

$$\hat{Z}_i \subseteq z_i - \frac{2\{c_i^{(1)}; \delta\}}{(1+\frac{1}{\mu_i})\{c_i^{(1)}; \delta\}^2 - \{c_i^{(2)}; \delta\} - \{w_i; \eta_i\}}$$

$$= z_i - \frac{2\{c_i^{(1)}; \delta\}}{\{v_i; d_i\}} = z_i - \frac{2\{c_i^{(1)}; \delta\}\cdot\{\bar{v}_i; d_i\}}{|v_i|^2 - d_i^2},$$

that is,

$$\hat{Z}_i \subseteq z_i - \frac{2\{c_i^{(1)}\bar{v}_i;\ |c_i^{(1)}|d_i + |v_i|\delta + d_i\delta\}}{|v_i|^2 - d_i^2}, \tag{5.80}$$

where

$$v_i = (1+\frac{1}{\mu_i})(c_i^{(1)})^2 - c_i^{(2)} - w_i\,,$$

$$d_i = (1+\frac{1}{\mu_i})(2|c_i^{(1)}|\delta + \delta^2) + \delta + \eta_i\,. \tag{5.81}$$

From (5.80) it follows

$$\hat{r}_i = \operatorname{rad} \hat{Z}_i \le \frac{2|c_i^{(1)}| + 2|v_i|\delta + 2d_i\delta}{|v_i|^2 - d_i^2}\,. \tag{5.82}$$

It is easy to show that $|c_i^{(j)}| = O(1/r)$ $(j = 1,2)$ (see, e.g., [89] ). According to this and (5.81) we find $|v_i| = O(1/r^2)$. Furthermore, from the iterative formulas (5.35) and (5.54) we observe that

$$Q_i^{(m)} = O(1), \qquad \frac{1}{\mu_i}A_i^{(m)2} + C_i^{(m)} = O(r)\,.$$

Therefore, $\eta_i = O(1)$ for the method (5.35) and $\eta_i = O(r)$ for (5.54). Using this facts and (5.81) we then have

$$d_i = \frac{\alpha_i \delta + \beta_i^{(\lambda)} r^\lambda}{r} \qquad (\alpha_i, \beta_i^{(\lambda)} \in \mathbb{R}^+, \quad \lambda \in \{1,2\}),$$

where $\lambda = 1$ for the method (5.35) and $\lambda = 2$ for (5.54). In view of this, from (5.82) we find

$$\hat{r} \le r^2(\gamma_1 \delta + \gamma_2^{(\lambda)} r^\lambda + \gamma_3 r \delta^2 + \gamma_4^{(\lambda)} \delta r^{\lambda+1}), \tag{5.83}$$

where $\gamma_1, \gamma_2^{(\lambda)}, \gamma_3, \gamma_4^{(\lambda)} \in \mathbb{R}^+$ and $\lambda \in \{1,2\}$.

According to (5.83) we can deduce the following conclusions relative to the convergence rate of the iterative methods (5.35) and (5.54) in the presence of rounding errors:

(*i*) The method (5.54) *preserves* the convergence rate (equal to *four*) only if $\delta = O(r^2)$. In this case the convergence order of (5.35) is, obviously, *three* ;

(*ii*) If the value of the incorporating roundoff error $\delta$ is of the same order of magnitude as the disk size, that is $\delta = O(r)$, the convergence order of the iterative processes (5.35) and (5.54) is *three* . Thus, the method (5.35) *preserves* its convergence speed while it *reduces* to *three* for (5.54). This kind of (slightly unexpected) behavior can be explaned taking into account the dominant influence of the term $|v_i|$ (which is proportional to $1/r^2$) in the numerator of (5.82);

(*iii*) The convergence order of both methods is only *quadratic* if $r = o(\delta)$ (the incorporating roundoff error exceeds significantly the radius of an inclusion disk). Of course, this discussion does not appear reasonable if the condition $\Delta P_i^{(m)} < |P(z_i^{(m)})|$ is not satisfied. Moreover, in that case (namely, $\delta > r$ ), a more general conclusion may be drawn for the fact that the values of the radii are close to the accuracy limit of the (single or double precision) arithmetic of the used computer (the case when $r \sim 10^{-q}$, where q is the number of significant decimal digits of the mantissa). So, any iterative process realized in circular arithmetic cannot guarantee the inclusion of zeros in the desribed case. The following simple examples explain the above statement. If $Z_1 = \{c_1 ; r_1\}$ and $Z_2 = \{c_2 ; r_2\}$ are disks, then the exact complex-valued set $\{z_1 z_2 : z_1 \in Z_1, z_2 \in Z_2\}$ need not be completely contained in the disk $Z = Z_1 \cdot Z_2 = \{c_1 c_2 ; |c_1| r_2 + |c_2| r_1 + r_1 r_2\}$

(definition (2.11) of multiplication, introduced in [37]) because the center mid Z and the radius rad Z are calculated with certain errors (for instance, the center is dislocated or the radius is less than the exact value). A similar situation may occur in the inversion of a disk, namely

$$\left\{\frac{1}{z} : z \in Z = \{c; r\}, \quad |c| > r\right\} \not\subseteq Z^{-1} = \left\{\frac{\bar{c}}{|c|^2 - r^2}; \frac{r}{|c|^2 - r^2}\right\};$$

or even a worse one, although $0 \notin Z$, it is possible that $0 \in Z^{-1}$ and any evaluation is broken.

Some of the above mentioned difficulties can be avoided if, instead of the circular arithmetic, the rounded complex rectangular arithmetic (constructed by the rounded real interval arithmetic, see [109] ) is employed. Although the operations of rectangular arithmetic are complicated and, consequently, any theoretical consideration is not simple, this arithmetic possesses a useful property that takes into account rounding errors. For example, to obtain the improved rectangular approximations (instead of circular ones), the basic formula (5.54) may be written in the following form

$$R_i^{(m+1)} = \left\{ z_i^{(m)} - \frac{1}{f(z_i^{(m)}) - \frac{1}{2}u(z_i^{(m)})\left[\frac{1}{\mu_i}G_i^{(m)2} + H_i^{(m)}\right]} \right\} \cap R_i^{(m)}$$

$$(i = 1,\ldots,n; \ m = 0,1,\ldots),$$

where

$$G_i^{(m)} = \sum_{\substack{j=1\\ j\neq i}}^{n} \frac{\mu_j}{z_i^{(m)} - R_j^{(m)}}, \qquad H_i^{(m)} = \sum_{\substack{j=1\\ j\neq i}}^{n} \mu_j \left(\frac{1}{z_i^{(m)} - R_j^{(m)}}\right)^2$$

and $R_i^{(m)}$ is the rectangle with the center $z_i^{(m)}$. The use of intersection of two rectangles in the above formula provides the *monotonicity* of interval sequences (that is, $R_i^{(0)} \supseteq R_i^{(1)} \supseteq R_i^{(2)} \supseteq \cdots$ ) and even the convergence of interval sequences $(R_i^{(m)})$ in certain cases when the initial rectangles $R_1^{(0)},\ldots,R_n^{(0)}$ are not small enough.

**EXAMPLE 1.** The iterative methods (5.35), (5.54) and (5.66), considered in this section, were illustrated numerically by means of the polynomial ([33])

$$P(z) = z^9 + (-2+3i)z^8 + (48-6i)z^7 + (-94+152i)z^6 + (522-298i)z^5 + (-950+1974i)z^4$$
$$+ (-1400-3650i)z^3 + (3750+1200i)z^2 + (-1875+1250i)z - 625i. \tag{5.84}$$

The exact zeros of P are $\xi_1 = 1$, $\xi_2 = -i$, $\xi_3 = -5i$ and $\xi_4 = 5i$ with respective multiplicities $\mu_1 = 2$, $\mu_2 = 3$, $\mu_3 = 2$, $\mu_4 = 2$.

For Algorithm (5.35) the initial disks were taken to be

$$Z_1^{(0)} = \{1.1 + 0.3i \ ; \ 0.4\}, \qquad Z_2^{(0)} = \{0.3 - 0.8i \ ; \ 0.4\},$$

$$W^{(0)} = \{ z: \ |z| \geq 4 \}.$$

After the second iteration we obtained

$$Z_1^{(2)} = \{1.000001 - 4.9 \times 10^{-7} i \ ; \ 8.31 \times 10^{-6}\},$$

$$Z_2^{(2)} = \{1.2 \times 10^{-9} - 0.999999998i \ ; \ 1.8 \times 10^{-8}\}.$$

Applying Algorithm (5.54) the initial disks $Z_1^{(0)}$ and $Z_2^{(0)}$ were selected to be the same as before, while $Z_3^{(0)}$ and $Z_4^{(0)}$ were given by

$$Z_3^{(0)} = \{0.2 - 4.7i \ ; \ 0.4\}, \qquad Z_4^{(0)} = \{0.2 + 4.7i \ ; \ 0.4\}.$$

For the second iteration we obtained the following inclusion disks:

$$Z_1^{(2)} = \{0.99999995 + 9.2 \times 10^{-9} i \ ; \ 2. \times 10^{-7}\},$$

$$Z_2^{(2)} = \{-1.6 \times 10^{-9} - 1.0000000008i \ ; \ 4.9 \times 10^{-9}\},$$

$$Z_3^{(2)} = \{-9.4 \times 10^{-12} - 5.000000000005i \ ; \ 4.7 \times 10^{-11}\},$$

$$Z_4^{(2)} = \{-4.3 \times 10^{-14} + 4.9999999999998i \ ; \ 7.4 \times 10^{-12}\}.$$

The better results were obtained applying the single-step method (5.66). The second iteration furnished

$$Z_1^{(2)} = \{1.0000000075 + 1.7 \times 10^{-9} i \ ; \ 2.8 \times 10^{-8}\},$$

$$Z_2^{(2)} = \{3.9 \times 10^{-15} - 1.0000000000000019 i \ ; \ 1.3 \times 10^{-14}\},$$

$$Z_3^{(2)} = \{1.4 \times 10^{-18} - 5.0000000000000000035 i \ ; \ 6.1 \times 10^{-18}\},$$

$$Z_4^{(2)} = \{-4.5 \times 10^{-24} + 4.99999999999999999999999992 i \ ; \ 3.4 \times 10^{-23}\}.$$

Many numerical examples show that the quotient $\rho^{(0)}/r^{(0)}$, which appears in the initial convergence conditions, may take a much smaller value than $5(N-\mu)$ (the inequality (5.43)) or $3(N-\mu)$ (the inequality (5.61)). For example, in the case of Algorithm (5.35) and the polynomial (5.84) we have $\rho^{(0)}/r^{(0)} \cong 2.4$, which is considerably less that $5(N-\mu) = 35$. In fact, the estimations and inequalities, used to prove the convergence theorems, are not sharp enough and so they require too wide initial conditions (5.43) and (5.61). In practice, the initial disks may be chosen under weaker conditions. Sometimes, the interval method can converge although the initial disks are not disjoint. In those cases it is necessary that the inequalities $|z_i^{(0)} - z_j^{(0)}| > \max(r_i^{(0)}, r_j^{(0)})$ $(i \neq j)$ hold so that the inversion of disks $z_i - Z_j$ $(i,j = 1,\ldots,n;\ i \neq j)$ can also produce disks.

The following example illustrates the above discussion.

**EXAMPLE 2.** We considered the polynomial

$$P(z) = z^{11} + (-7+2i)z^{10} + (11-14i)z^{9} + (19+24i)z^{8} + (-70+24i)z^{7}$$

$$+ (42-116i)z^{6} + (198+108i)z^{5} + (-234+280i)z^{4} + (-491-360i)z^{3}$$

$$+ (45-702i)z^{2} + (351-270i)z + 135,$$

whose zeros $\xi_1 = -1$, $\xi_2 = 3$, $\xi_3 = -i$, $\xi_4 = 1-2i$, $\xi_5 = 1+2i$ have the multiplicities $\mu_1 = 4$, $\mu_2 = 3$, $\mu_3 = 2$, $\mu_4 = 1$, $\mu_5 = 1$. The following initial disks were taken:

$$Z_1^{(0)} = \{-0.8 + 0.2i\ ;\ 1\},\quad Z_2^{(0)} = \{2.7 + 0.2i\ ;\ 1\},\quad Z_3^{(0)} = \{0.2 - 1.2i\ ;\ 1\},$$

$$Z_4^{(0)} = \{1.2 - 2.1i\ ;\ 1\},\quad Z_5^{(0)} = \{1.2 + 2.1i\ ;\ 1\}.$$

For these disks the value $\rho^{(0)}/r^{(0)} \cong 0.35$ is much smaller than $3(N-\mu) = 30$, which follows from (5.61). Moreover, the disks $Z_1^{(0)}$ and $Z_3^{(0)}$ as well as $Z_3^{(0)}$ and $Z_4^{(0)}$ are overlapping. Although the initial disks are obviously badly separated, the interval method (5.66) demonstrated good behavior. The radii of inclusion disks, obtained after the first iteration, were in the range $[4.9 \times 10^{-4}, 1.6 \times 10^{-2}]$, while after the second iteration the following disks were computed:

$$Z_1^{(2)} = \{-0.999999999944 - 2.2 \times 10^{-11} i\ ;\ 2.05 \times 10^{-10}\},$$

$$Z_2^{(2)} = \{2.99999999999975 + 2.7 \times 10^{-12} i\ ;\ 1. \times 10^{-11}\},$$

$$Z_3^{(2)} = \{2. \times 10^{-11} - 1.00000000016 i\ ;\ 2.4 \times 10^{-9}\},$$

$$Z_4^{(2)} = \{1.0000000000000013 - 1.9999999999999973 i\ ;\ 1.7 \times 10^{-14}\},$$

$$Z_5^{(2)} = \{1.000000000000000000000005 + 2.000000000000000000000004 i\ ;\ 1.2 \times 10^{-21}\}.$$

## 5.4. BELL'S ALGORITHMS FOR A SINGLE COMPLEX ZERO

Let

$$P(z) = z^N + a_{N-1}z^{N-1} + \cdots + a_1 z + a_0 \qquad (a_i \in \mathbb{C})$$

be a monic polynomial of degree $N \geq 3$ with simple or multiple complex zeros $\xi_1,\ldots,\xi_n$ $(n \leq N)$ of the multiplicity $\mu_1,\ldots,\mu_n$. Assume that we have found the inclusion disk $\{z: |z-a| \leq R\}$, denoted shortly by $\{a; R\}$, with center $a$ and radius $R$ containing only one zero, say $\xi_i$, of $P$. All other zeros are supposed to lie in the region $W = \{w: |w-a| > R\}$, that is, in the exterior of the disk $\{a; R\}$. Without loss of generality, we will adopt that the required zero is denoted with $\xi_1$. Moreover, we will write $\xi$ and $\mu$ instead of $\xi_1$ and $\mu_1$ for brevity.

Using the inclusion isotonicity property we have for $z \in \{a; R\}$

$$\left(\frac{1}{z-\xi_j}\right)^{\nu} \in \left(\frac{1}{z-W}\right)^{\nu} \qquad (j = 2,\ldots,n;\ \nu \text{ is a natural number}). \tag{5.85}$$

Since $z \notin W$, that is $|z-a| < R$, the inversion of the region

$$z - W = \{w: |w-(z-a)| > R\}$$

is a closed interior of a circle, given by

$$V = (z-W)^{-1} = \left\{ w: \left| w + \frac{\bar{z}-\bar{a}}{R^2 - |z-a|^2} \right| \leq \frac{R}{R^2 - |z-a|^2} \right\} = \{h; d\},$$

where

$$h = \operatorname{mid} V = \frac{\bar{a}-\bar{z}}{R^2 - |z-a|^2}, \qquad d = \operatorname{rad} V = \frac{R}{R^2 - |z-a|^2}$$

(see § 2.2). According to (2.12), we have

$$V^{\nu} = \{h; d\}^{\nu} = \{h^{\nu}; (|h|+d)^{\nu} - |h|^{\nu}\}.$$

Let $Z^{(m)} = \{z^{(m)}; r^{(m)}\}$ be a disk with center $z^{(m)} = \operatorname{mid} Z^{(m)}$ and radius $r^{(m)} = \operatorname{rad} Z^{(m)}$ $(m = 0,1,\ldots)$. For an initial inclusion disk $Z^{(0)}$ we take $Z^{(0)} = \{a; R\}$, that is, $z^{(0)} = a$ and $r^{(0)} = R$. We will use the following notations :

$$V^{(m)} = \{h^{(m)}; d^{(m)}\},$$

$$h^{(m)} = \frac{\bar{a}-\bar{z}^{(m)}}{R^2 - |z^{(m)}-a|^2}, \qquad d^{(m)} = \frac{R}{R^2 - |z^{(m)}-a|^2}.$$

In view of (5.85), circular extension of

$$s_{\nu,1}(z^{(m)}) = s_\nu(z^{(m)}) = \frac{1}{\mu}\sum_{j=2}^{n}\mu_j(z^{(m)}-\xi_j)^{-\nu}$$

is the disk

$$S_{\nu,1}^{(m)} = S_\nu^{(m)} = \frac{1}{\mu}\sum_{j=2}^{n}\mu_j\left(\frac{1}{z^{(m)}-W}\right)^\nu = \frac{N-\mu}{\mu}(V^{(m)})^\nu .$$

Therefore, the iterative formula (5.12) that defines the family of interval methods, has the following form in the case of a multiple complex zero

$$Z^{(m+1)} = z^{(m)} - \frac{\Delta_{k-1}(z^{(m)})}{\Delta_k(z^{(m)}) - B_k\left(\frac{N-\mu}{\mu}V^{(m)},\ldots,\frac{N-\mu}{\mu}(V^{(m)})^k\right)} \qquad (5.86)$$

$$(m = 0,1,\ldots),$$

where $\Delta_k$ is given by (5.6) and $\mu_i = \mu$.

The values of the radius

$$\text{rad}\, B_k\left(\frac{N-\mu}{\mu}V^{(m)},\ldots,\frac{N-\mu}{\mu}(V^{(m)})^k\right)$$

belong to a reasonably narrow interval, depending on the difference $|z^{(m)} - a|$, and they do not have any significant influence on the convergence rate of the iterative method (5.86). However, the *main* role of the interval $B_k$ is to *provide* the inclusion of the zero $\xi$ within the circular region $Z^{(m+1)}$, obtained by (5.86).

Following the convergence analysis presented in [127], it can be easily shown that the sequence $(r^{(m)})$, $r^{(m)} = \text{rad}\, Z^{(m)}$, under some suitable initial conditions tends to zero with the convergence order equal to k+1.

For small computational cost, the following iterative formulas, obtained as special cases from (5.86) setting k = 1 and k = 2, have the most practical importance:

$$Z^{(m+1)} = z^{(m)} - \frac{\mu}{\frac{P'(z^{(m)})}{P(z^{(m)})} - (N-\mu)\{h^{(m)}; d^{(m)}\}} \quad (m = 0,1,\ldots) \qquad (5.87)$$

*(Newton-like algorithm for a multiple zero);*

$$z^{(m+1)} = z^{(m)} - \frac{1}{(1+\frac{1}{\mu})\frac{P'(z^{(m)})}{2P(z^{(m)})} - \frac{P''(z^{(m)})}{2P'(z^{(m)})} - \frac{P(z^{(m)})}{2P'(z^{(m)})}\cdot\frac{N(N-\mu)}{\mu}\{h^{(m)};d^{(m)}\}^2}$$

$$(m = 0,1,\ldots) \qquad (5.88)$$

*(Halley-like algorithm for a multiple zero).*

In this section we will analyze the Newton-like method (5.87) in detail (see [92]). Assume we have found an initial disk $Z^{(0)} = \{a ; R\}$ so that the condition

$$\left|\frac{P(a)}{P'(a)}\right| < \frac{R}{2(\mu+1)(N-\mu)} \qquad (5.89)$$

is satisfied, and introduce

$$\rho^{(m)} = R - |z^{(m)} - a|, \quad (m = 1,2,\ldots).$$

We begin by considering the first iteration $(m = 0)$. The inversion of the region $Q = \{z: |z-q| > R\}$, when $0 \notin Q$ (i.e. $|q| < R$), is given by

$$Q^{-1} = \left\{\frac{-\bar{q}}{R^2 - |q|^2} ; \frac{R}{R^2 - |q|^2}\right\}. \qquad (5.90)$$

According to this we obtain

$$\frac{N-\mu}{z^{(0)} - W} = (N-\mu)\left\{0 ; \frac{1}{R}\right\} = \left\{0 ; \frac{N-\mu}{R}\right\} \quad (z^{(0)} = a).$$

Applying the inequality (5.89) we find

$$\left|\frac{P'(z^{(0)})}{P(z^{(0)})}\right| = \left|\frac{P'(a)}{P(a)}\right| > \frac{2(\mu+1)(N-\mu)}{R} > \frac{N-\mu}{R} = \operatorname{rad} \frac{N-\mu}{z^{(0)} - W},$$

which means that the disk in the denominator of formula (5.87) does not contain the origin when $m = 0$. Therefore, its inversion is a disk so that $Z^{(1)}$ is also a disk.

Under the condition (5.89) we find the following upper bounds:

$$r^{(1)} = \operatorname{rad} Z^{(1)} = \operatorname{rad} \frac{\mu}{\left\{\frac{P'(a)}{P(a)} ; \frac{N-\mu}{R}\right\}}$$

$$= \frac{\frac{\mu(N-\mu)}{R}}{\left|\frac{P'(a)}{P(a)}\right|^2 - \frac{(N-\mu)^2}{R^2}} < \frac{\frac{\mu(N-\mu)}{R}}{\left[\frac{2(\mu+1)(N-\mu)}{R}\right]^2 - \frac{(N-\mu)^2}{R^2}},$$

that is,

$$r^{(1)} < \frac{R}{15(N-\mu)} \tag{5.91}$$

because of

$$\frac{\mu}{4(\mu+1)^2 - 1} \le \frac{1}{15} .$$

Furthermore, we have

$$|z^{(1)} - z^{(0)}| = |z^{(1)} - a| = \frac{\mu\left|\frac{P'(a)}{P(a)}\right|}{\left|\frac{P'(a)}{P(a)}\right|^2 - \frac{(N-\mu)^2}{R^2}} < \frac{\frac{2\mu(\mu+1)(N-\mu)}{R}}{\frac{4(\mu+1)^2(N-\mu)^2}{R^2} - \frac{(N-\mu)^2}{R^2}}$$

$$= \frac{2\mu(\mu+1)R}{(N-\mu)[4(\mu+1)^2 - 1]} .$$

Since

$$\frac{2\mu(\mu+1)}{4(\mu+1)^2 - 1} \le \frac{4}{15} ,$$

we estimate

$$|z^{(1)} - a| < \frac{4R}{15(N-\mu)} . \tag{5.92}$$

Now, we will prove that the condition (5.89) implies the inequality

$$\rho^{(1)} > 6(N-\mu)r^{(1)} . \tag{5.93}$$

Using the inequality (5.92) we find

$$\rho^{(1)} = R - |z^{(1)} - a| > R - \frac{4R}{15(N-\mu)} = R\left[1 - \frac{4}{15(N-\mu)}\right] ,$$

so that it suffices to show (taking into account the inequality (5.91))

$$R\left[1 - \frac{4}{15(N-\mu)}\right] > 6(N-\mu)\frac{R}{15(N-\mu)} ,$$

or

$$1 - \frac{4}{15(N-\mu)} > \frac{2}{5} .$$

The last inequality is obvious by virtue of

$$\min_{1 \le \mu < N}\left(1 - \frac{4}{15(N-\mu)}\right) = \frac{11}{15} .$$

The analysis of the first iterative step shows that (*i*) a new disk-approximation $Z^{(1)}$ includes the zero $\xi$; (*ii*) this disk is contracted because of

$$r^{(1)} < \frac{R}{15(N-\mu)} \leq \frac{R}{15} .$$

Besides, the initial condition (5.89) induces the condition (5.93) which will be suitably used in the following.

Now, we can analyze the iterative process (5.87) beginning with $m \geq 1$ and starting from the inclusive disk $Z^{(1)}$ with the assumption that the inequality (5.93) (induced by the initial condition (5.89)) holds. Using (5.90) the iterative formula (5.87) can be rewritten in the form

$$Z^{(m+1)} = Z^{(m)} - \frac{\mu}{\{c^{(m)} ; \eta^{(m)}\}} \quad (m = 1,2,\ldots), \tag{5.94}$$

where

$$c^{(m)} = \frac{P'(z^{(m)})}{P(z^{(m)})} - \frac{(N-\mu)(\bar{z}^{(m)} - \bar{a})}{R^2 - |z^{(m)}-a|^2}$$

and

$$\eta^{(m)} = \frac{(N-\mu)R}{R^2 - |z^{(m)}-a|^2} .$$

For simplicity, in further analysis we will omit the iteration index always when the possibility of any confusion does not exist.

**LEMMA 5.8.** *If the inequality*

$$\rho > 6(N-\mu)r \tag{5.95}$$

*holds, then* $0 \notin \{c ; \eta\}$ *and*

$$\frac{\mu|c|}{|c|^2 - \eta^2} < \frac{8}{5}r . \tag{5.96}$$

*Proof.* First, by using (5.95) we find the bound

$$|c| = \left| \frac{P'(z)}{P(z)} - \frac{(N-\mu)(\bar{z}-\bar{a})}{R^2 - |z-a|^2} \right| > \left| \frac{\mu}{|z-\xi|} - \sum_{j=2}^{n} \frac{\mu_j}{|z-\xi_j|} - \frac{(N-\mu)|z-a|}{R^2 - |z-a|^2} \right|$$

$$> \frac{\mu}{r} - \frac{N-\mu}{\rho} - \frac{(N-\mu)(R-\rho)}{R^2 - (R-\rho)^2} > \frac{1}{r}\left[\mu - \frac{1}{6}\left(1 + \frac{R-\rho}{2R-\rho}\right)\right] ,$$

that is,

$$|c| > \frac{1}{r}(\mu - \frac{1}{3}) \geq \frac{2}{3r} . \tag{5.97}$$

Furthermore, from

$$\eta = \frac{(N-\mu)R}{R^2 - |z-a|^2} = \frac{(N-\mu)R}{R^2 - (R-\rho)^2},$$

it follows

$$\eta < \frac{N-\mu}{\rho}. \qquad (5.98)$$

In view of (5.95) we have $\frac{2}{3r} > \frac{N-\mu}{\rho}$, so that

$$|c| > \frac{2}{3r} > \frac{N-\mu}{\rho} > \eta,$$

which means that $0 \notin \{c\,;\eta\}$.

Now, we will prove (5.96). Using the inequalities (5.95), (5.97) and (5.98), we find

$$\frac{\mu|c|}{|c|^2 - \eta^2} < \frac{\frac{\mu}{r}(\mu-\frac{1}{3})}{\frac{1}{r^2}(\mu-\frac{1}{3})^2 - (\frac{N-\mu}{\rho})^2} < \frac{\mu(\mu-\frac{1}{3})r}{(\mu-\frac{1}{3})^2 - \frac{1}{36}} = \frac{(\mu^2-\frac{\mu}{3})r}{\mu^2 - \frac{2}{3}\mu + \frac{1}{12}} \le \frac{8}{5}r,$$

because of

$$\frac{\mu^2 - \mu/3}{\mu^2 - \frac{2}{3}\mu + \frac{1}{12}} \le \frac{8}{5}. \qquad \square$$

**THEOREM 5.7.** *Let the sequence of circular intervals* $(Z^{(m)})$ $(m = 0,1,\ldots)$ *be defined by the iterative formula (5.87), assuming that the initial disk* $Z^{(0)} = \{a\,;R\}$ *is chosen so that the condition (5.89) is satisfied. Then, in each iterative step, the following is true:*

$1^o$ $\xi \in Z^{(m)}$;

$2^o$ $r^{(m+1)} < \frac{6(N-\mu)}{R} {r^{(m)}}^2$.

*P r o o f.* The proof of the assertion $1^o$ follows from the construction of the method (5.86), based on the inclusion isotonicity and the fixed point relation (5.5), and the fact that $z^{(m)} \in \{a\,;R\}$ for each $m = 0,1,\ldots$, which is obvious because of $R - |z^{(m)} - a| = \rho^{(m)} > 6(N-\mu)r^{(m)} > 0$. To prove the convergence rate of the iterative method (5.87) is quadratic (the assertion $2^o$) we recall that the condition (5.89) causes the inequality (5.93) which will be used in the convergence analysis.

From equation (5.94) we obtain

$$r^{(2)} = \operatorname{rad} Z^{(2)} = \frac{\mu\eta^{(1)}}{|c^{(1)}|^2 - {\eta^{(1)}}^2}.$$

Using the estimations

$$\frac{\mu}{(\mu-\frac{1}{3})^2-\frac{1}{36}} < 3,$$

$$|c^{(1)}| > \frac{1}{r^{(1)}}(\mu-\frac{1}{3}),$$

and

$$\eta^{(1)} < \frac{N-\mu}{\rho^{(1)}},$$

we find by Lemma 5.8 and (5.93)

$$r^{(2)} < \frac{\mu(N-\mu)r^{(1)^2}}{\rho^{(1)}\left[(\mu-\frac{1}{3})^2-(N-\mu)^2\left(\frac{r^{(1)}}{\rho^{(1)}}\right)^2\right]} < \frac{\mu(N-\mu)r^{(1)^2}}{\rho^{(1)}[(\mu-\frac{1}{3})^2-\frac{1}{36}]}$$

$$< \frac{3(N-\mu)r^{(1)^2}}{\rho^{(1)}}.$$

Since

$$\rho^{(1)} = R - |z^{(1)}-a| > R\left[1-\frac{4}{15(N-\mu)}\right] \geq \frac{11}{15}R > \frac{R}{2},$$

we then have

$$r^{(2)} < \frac{6(N-\mu)r^{(1)^2}}{R}$$

and

$$r^{(2)} < \frac{3(N-\mu)r^{(1)}}{\rho^{(1)}/r^{(1)}} < \frac{r^{(1)}}{2}$$

because of (5.93).

By virtue of (5.96) we find

$$\rho^{(2)} = R - |z^{(2)}-a| = R - \left|z^{(1)} - a - \frac{\mu\bar{c}^{(1)}}{|c^{(1)}|^2-\eta^{(1)2}}\right|$$

$$> R - |z^{(1)}-a| - \frac{\mu|c^{(1)}|}{|c^{(1)}|^2-\eta^{(1)2}} = \rho^{(1)} - \frac{\mu|c^{(1)}|}{|c^{(1)}|^2-\eta^{(1)2}},$$

that is,

$$\rho^{(2)} > \rho^{(1)} - \frac{8}{5}r^{(1)}.$$

Using the inequalities (5.93) and $r^{(2)} < r^{(1)}/2$, one obtains

$$\rho^{(2)} > \rho^{(1)} - \frac{8}{5}r^{(1)} > 6(N-\mu)r^{(1)} - \frac{8}{5}r^{(1)} = [6(N-\mu) - \frac{8}{5}]r^{(1)}$$

$$> 2[6(N-\mu) - \frac{8}{5}]r^{(2)} > 6(N-\mu)r^{(2)}.$$

The assertion $2^{\circ}$ will be proved by induction. Assume that for $m \geq 2$ the following is true:

$$r^{(m)} < \frac{3(N-\mu)}{\rho^{(m-1)}} r^{(m-1)^2}, \tag{5.99}$$

$$r^{(m)} < \frac{r^{(m-1)}}{2}, \tag{5.100}$$

$$\rho^{(m)} > 6(N-\mu)r^{(m)}, \tag{5.101}$$

$$\rho^{(m)} > \rho^{(m-1)} - \frac{8}{5}r^{(m-1)}. \tag{5.102}$$

These relations has already proved for $m = 2$. We will prove that they are valid for the index $m+1$.

By the above consideration for $m = 2$ and (5.101), we obtain

$$r^{(m+1)} < \frac{3(N-\mu)}{\rho^{(m)}} r^{(m)^2} < \frac{r^{(m)}}{2}. \tag{5.103}$$

In a similar way as for $m = 1$, it is easy to show that

$$\rho^{(m+1)} > 6(N-\mu)r^{(m+1)} \quad \text{and} \quad \rho^{(m+1)} > \rho^{(m)} - \frac{8}{5}r^{(m)}.$$

By the successive application of (5.100) and (5.102), it follows

$$\begin{aligned}
\rho^{(m)} &> \rho^{(m-1)} - \frac{8}{5}r^{(m-1)} > \rho^{(m-2)} - \frac{8}{5}r^{(m-2)} - \frac{8}{5}r^{(m-1)} \\
&> \rho^{(m-2)} - \frac{8}{5}r^{(m-2)} - \frac{8}{5}\frac{r^{(m-1)}}{2} = \rho^{(m-2)} - \frac{8}{5}r^{(m-2)}(1+\frac{1}{2}) \\
&> \rho^{(m-3)} - \frac{8}{5}r^{(m-3)} - \frac{8}{5}r^{(m-2)}(1+\frac{1}{2}) \\
&> \rho^{(m-3)} - \frac{8}{5}r^{(m-3)} - \frac{8}{5}\frac{r^{(m-3)}}{2}(1+\frac{1}{2}) > \rho^{(m-3)} - \frac{8}{5}r^{(m-3)}[1+\frac{1}{2}(1+\frac{1}{2})] \\
&\;\;\vdots \\
&> \rho^{(1)} - \frac{8}{5}r^{(1)}(1 + \frac{1}{2} + \frac{1}{2^2} + \cdots) = \rho^{(1)} - \frac{16}{5}r^{(1)}.
\end{aligned}$$

Since, in accordance with (5.92),

$$\rho^{(1)} = R - |z^{(1)} - a| > R - \frac{4R}{15(N-\mu)} \geq \frac{11}{15}R$$

and

$$r^{(1)} < \frac{R}{15(N-\mu)} \leq \frac{R}{15},$$

we obtain

$$\rho^{(m)} > \rho^{(1)} - \frac{16}{5} r^{(1)} > \frac{11}{15} R - \frac{16}{5} \cdot \frac{R}{15} > \frac{R}{2} .$$

Therefore, from the last inequality and (5.103) we find

$$r^{(m+1)} < \frac{6(N-\mu)}{R} r^{(m)^2} .$$

Finally, we prove that, under the condition (5.89), the iterative process (5.87) is defined in each iterative step. Namely, since (5.89) implies the inequality (5.93) from which there follows the inequality (5.101) for each $m = 1,2,\ldots$, Lemma 5.8 is applicable for each m, which means that $0 \in \{c^{(m)} ; \eta^{(m)}\}$. □

The Halley-like method, given by (5.88), has *cubic* convergence. Precisely, the following theorem is valid.

**THEOREM 5.8.** *Let the iterative formula (5.88) produce the sequence of disks $(Z^{(m)})$ $(m=0,1,\ldots)$, assuming that the initial disk $Z^{(0)} = \{a ; R\}$ is chosen so that the conditions*

$$\left| \frac{P(a)}{P'(a)} \right| < \frac{4R}{5N(N-\mu)} = \frac{1}{A}$$

*and*

$$\left| \frac{P''(a)}{P'(a)} \right| < \frac{(\mu+1)RA - 4(N-1)}{\mu R}$$

*hold. Then, in each iterative step, we have*

$1^o$ $\xi \in Z^{(m)}$;

$2^o$ $r^{(m+1)} < \dfrac{9N(N-\mu)}{\mu^2 R^2} r^{(m)^3}$ .

For the proof of Theorem 5.8 see [92].

In order to illustrate numerically the algorithms (5.87) and (5.88) for $\mu > 1$, we have constructed the initial disk $Z^{(0)} = \{a ; R\}$ using the results given in [51, p. 454]:

(i) the disk $|z-a| \le |P(a)|^{1/N}$ contains at least one zero of P;

(ii) the disk $|z-a| \le N|P(a)/P'(a)|$ $(P'(a) \ne 0)$ (Laguerre's disk) contains at least one zero of P.

The number of zeros inside the above disks can be determined by inclusion tests (see § 2.1).

**EXAMPLE 3.** The polynomial

$$P(z) = z^9 + (-2+3i)z^8 + (48-6i)z^7 + (-94+152i)z^6 + (522-298i)z^5 + (-950+1974i)z^4$$
$$+ (-1400-3650i)z^3 + (3750+1200i)z^2 + (-1875+1250i)z - 625i$$

(the example taken from [33], see also Example 1, § 5.3) has the factorization

$$P(z) = (z-5i)^2(z+5i)^2(z-1)^2(z+i)^3.$$

If we choose $a = 4.5\,i$, then we calculate for the given polynomial

$P(4.5i) = -33784.5 + 72261.3\,i,$

$P'(4.5i) = -207125.2 - 88415.3\,i,$

$P(4.5i)/P'(4.5i) = 0.012 - 0.354\,i,$

$|P(4.5i)|^{1/9} = 3.505,$

$9|P(4.5i)/P'(4.5i)| = 3.19.$

Therefore,

$$Z^{(0)} = \{4.5i\ ;\ 3.505\} \quad \text{in Case (i)},$$

$$Z^{(0)} = \{4.5i\ ;\ 3.19\} \quad \text{in Case (ii)}.$$

By the inclusion test due to Marden [67], we establish that these disks contain only one zero of P.

To estimate the multiplicity of the requested zero contained in the region $Z^{(0)}$, we use Lagouanelle's limiting formula (2.2). Let I(s) denote the integer which is the closest to a real positive number s. In practice, the actual computation of (2.2) is

$$\mu = I\left(\left|\frac{P'(z)^2}{P'(z)^2 - P''(z)P(z)}\right|\right), \tag{5.104}$$

where the center $z = a$ of the starting disk $Z^{(0)}$ is taken as an approximation to the zero $\xi$. To obtain a precise estimation of multiplicity, it is suitable to use already calculated values P(a) and P'(a) for computation an improved approximation relative to the center a by the Newton iterative formula

$$z = a - \frac{P(a)}{P'(a)}. \tag{5.105}$$

For $a = 4.5i$ we obtain from (5.105)

$$z = 4.5i - (0.012 - 0.354i) = -0.012 + 4.854i.$$

Now, using (5.104) we find $\mu = I(1.7) = 2.$

We applied the algorithms (5.87) and (5.88) in both Cases (i) and (ii). For the initial disk $Z^{(0)} = \{4.5i \; ; \; 3.504\}$ the Newton-like algorithm (5.87) produced the following disks that contain the exact zero $\xi = 5i$:

$$Z^{(1)} = \{4.7 \times 10^{-2} + 5.92i \; ; \; 1.006\},$$

$$Z^{(2)} = \{3.0 \times 10^{-2} + 5.17i \; ; \; 0.436\},$$

$$Z^{(3)} = \{6.4 \times 10^{-3} + 5.015i \; ; \; 2.65 \times 10^{-2}\},$$

$$Z^{(4)} = \{1.3 \times 10^{-4} + 5.00011i \; ; \; 2.62 \times 10^{-4}\},$$

$$Z^{(5)} = \{2.1 \times 10^{-8} + 4.999999995i \; ; \; 3.14 \times 10^{-8}\}.$$

For the same initial disk, the Halley-like algorithm (5.88) gave

$$Z^{(1)} = \{-7.2 \times 10^{-3} + 4.98i \; ; \; 0.102\},$$

$$Z^{(2)} = \{-8.6 \times 10^{-7} + 4.9999993i \; ; \; 4.1 \times 10^{-6}\}.$$

We note that the convergence of the Newton-like method is slow at the beginning of iterative procedure and it is quadratic after the third iterative step. Moreover, the approximation $z^{(1)} = \text{mid}\, Z^{(1)}$ is worse compared to $a = \text{mid}\, Z^{(0)}$ (namely, $|z^{(1)} - 5i| > |a - 5i|$). It must be pointed out, however, that the order of convergence is not always indicative of the initial speed of convergence. The Halley-like method converges very fast; as an illustration, this method gave $r^{(3)} = \text{rad}\, Z^{(3)} = 5.6 \times 10^{-19}$, while the iterative formula (5.87) only $r^{(6)} = 9.54 \times 10^{-16}$.

In the second case ($Z^{(0)} = \{4.5i \; ; \; 3.19\}$) both methods showed similar behavior as before. The radii of inclusive disks, obtained by the iterative formula (5.87) and (5.88) are given in Table 5.3 .

| | $r^{(1)}$ | $r^{(2)}$ | $r^{(3)}$ | $r^{(4)}$ | $r^{(5)}$ | $r^{(6)}$ |
|---|---|---|---|---|---|---|
| Newton-like method | 1.39 | 0.83 | $3.59\times10^{-2}$ | $4.34\times10^{-4}$ | $8.44\times10^{-8}$ | $3.12\times10^{-15}$ |
| Halley-like method | 0.126 | $7.15\times10^{-7}$ | $1.93\times10^{-21}$ | | | |

Table 5.3

**EXAMPLE 4.** We considered the polynomial P whose factorization is

$$P(z) = (z+1)^3(z-3)^2(z-5)^2(z^2-2z+5)^2 \; .$$

Taking $a = -1.5$, according to Case (i) we have constructed the inclusion disk $Z^{(0)}$ with the radius $R = |P(-1.5)|^{1/11} = 2.335$ and established that this disk contains only one zero of P (by the inclusion test given in [67]). The multiplicity of this zero, evaluated by the same procedure as in the previous example, was $\mu = 3$.

Applying the Newton-like algorithm (5.87), as result we obtained

$Z^{(1)} = \{-1.017 \;;\; 0.214\}$,

$Z^{(2)} = \{-1.000095 \;;\; 3.33 \times 10^{-4}\}$,

$Z^{(3)} = \{-1.0000000033 \;;\; 1.09 \times 10^{-8}\}$.

The Halley-like algorithm (5.88) produced the disks

$Z^{(1)} = \{-1.0052 \;;\; 8.28 \times 10^{-2}\}$

and

$Z^{(2)} = \{-1.000000024 \;;\; 1.98 \times 10^{-7}\}$.

## 5.5. BELL'S POLYNOMIALS AND PARALLEL ITERATIONS IN COMPLEX ARITHMETIC

Let $\Delta_{k,i}(z)$, $s_{\lambda,i}(z)$ and $F_{k,i}(z)$ be given by (5.6), (5.8) and (5.10), respectively, as in § 5.1. Then, the fixed point relation (5.5) can be written in the form

$$\xi_i = z - \frac{\Delta_{k-1,i}(z)}{\Delta_{k,i}(z) - B_k(s_{1,i}(z),\ldots,s_{k,i}(z))} . \tag{5.106}$$

Assume that reasonably good approximations $z_1,\ldots,z_n$ of the zeros $\xi_1,\ldots,\xi_n$ have been found. Taking $z = z_i$ and $\xi_j := z_j$ in (5.8), we obtain the following approximation $\hat{s}_{\lambda,i}(z_i)$ of $s_{\lambda,i}(z_i)$,

$$\hat{s}_{\lambda,i}(z_i) = \frac{1}{\mu_i} \sum_{\substack{j=1\\ j\neq i}}^{n} \frac{\mu_j}{(z_i - z_j)^{\lambda}} .$$

Therefore, the right-hand side of (5.106) for $z = z_i$ will present a new approximation $\hat{z}_i$ to the zero $\xi_i$, that is,

$$\hat{z}_i = z_i - \frac{\Delta_{k,i}(z_i)}{\Delta_{k,i}(z_i) - B_k(\hat{s}_{1,i}(z_i),\ldots,\hat{s}_{k,i}(z_i))} \quad (i=1,\ldots,n). \tag{5.107}$$

Formula (5.107) defines a family of parallel iterations in ordinary complex arithmetic, based on Bell's polynomials. The convergence order of the generalized SIP (5.107) is k + 2, the same one as in the case of interval iterations (5.12).

In view of the computational cost, the iterative methods which follow from (5.107) for $k = 1$ and $k = 2$ have a great importance. The total-step method obtained from (5.107) for $k = 1$ and its modifications have already been considered in § 4.4. For this reason, we will now present only the point iterative methods of Halley's type (the case $k=2$, see [96]). For simplicity, we will omit the iteration index.

Let $z \in \mathbb{C}$ and let

$$h_1(z) = \frac{P'(z)}{P(z)}, \qquad \delta(z) = \frac{P''(z)}{P'(z)}$$

$$g(z) = \frac{1}{H(z)} = \frac{1}{2}\left[(1+\frac{1}{\mu})h_1(z) - \delta(z)\right]$$

be the notations used earlier in § 4.4 and § 5.3. Here, $\mu$ is the multiplicity of a desired zero. By means of

$$N(z) = \frac{\mu}{h_1(z)}$$

and

$$H(z) = 2\left[(1+\frac{1}{\mu})h_1(z) - \delta(z)\right]^{-1}$$

we define Newton's and Halley's corrections, as in § 4.4. Besides, let

$$\Sigma_i(\mathbf{a},\mathbf{b}) = \frac{1}{\mu_i}\left[\sum_{j=1}^{i-1}\mu_j(z-a_j)^{-1} + \sum_{j=i+1}^{n}\mu_j(z-b_j)^{-1}\right]^2$$

$$+ \sum_{j=1}^{i-1}\mu_j(z-a_j)^{-2} + \sum_{j=i+1}^{n}\mu_j(z-b_j)^{-2},$$

where $\mathbf{a} = (a_1,\ldots,a_n)$ and $\mathbf{b} = (b_1,\ldots,b_n)$ are some vectors. In particular, according to the above, we have, for instance,

$$\Sigma_i(\mathbf{a},\mathbf{a}) = \frac{1}{\mu_i}\left(\sum_{\substack{j=1\\ j\neq i}}^{n}\mu_j(z-a_j)^{-1}\right)^2 + \sum_{\substack{j=1\\ j\neq i}}^{n}\mu_j(z-a_j)^{-2}.$$

Assume that sufficiently close approximations $z_1,\ldots,z_n$ to the zeros $\xi_1,\ldots,\xi_n$ have been found. We introduce the following vectors:

$\mathbf{z} = (z_1,\ldots,z_n)$ (the former approximations),

$\hat{\mathbf{z}} = (\hat{z}_1,\ldots,\hat{z}_n)$ (the new approximations),

$\mathbf{z}_N = (z_{N,1},\ldots,z_{N,n})$, $z_{N,i} = z_i - N(z_i)$ (Newton's approximations),

$\mathbf{z}_H = (z_{H,1},\dots,z_{H,n})$, $z_{H,i} = z_i - H(z_i)$ (Halley's approximations).

In calculating the approximations $z_{N,i}$ and $z_{H,i}$, and in all formulas where the function $g(z)$ appears, one has to take $\mu = \mu_i$.

The iterative methods of Halley's type for determining all zeros simultaneously are listed below using the same notations as in § 4.1 and § 4.4.

(TS): $$\hat{z}_i = z_i - \left[g(z_i) - \frac{1}{2h_1(z_i)}\Sigma_i(\mathbf{z},\mathbf{z})\right]^{-1} \qquad (i = 1,\dots,n),$$

(SS): $$\hat{z}_i = z_i - \left[g(z_i) - \frac{1}{2h_1(z_i)}\Sigma_i(\hat{\mathbf{z}},\mathbf{z})\right]^{-1} \qquad (i = 1,\dots,n),$$

(TSN): $$\hat{z}_i = z_i - \left[g(z_i) - \frac{1}{2h_1(z_i)}\Sigma_i(\mathbf{z}_N,\mathbf{z}_N)\right]^{-1} \qquad (i = 1,\dots,n),$$

(SSN): $$\hat{z}_i = z_i - \left[g(z_i) - \frac{1}{2h_1(z_i)}\Sigma_i(\hat{\mathbf{z}},\mathbf{z}_N)\right]^{-1} \qquad (i = 1,\dots,n),$$

(TSH): $$\hat{z}_i = z_i - \left[g(z_i) - \frac{1}{2h_1(z_i)}\Sigma_i(\mathbf{z}_H,\mathbf{z}_H)\right]^{-1} \qquad (i = 1,\dots,n),$$

(SSH): $$\hat{z}_i = z_i - \left[g(z_i) - \frac{1}{2h_1(z_i)}\Sigma_i(\hat{\mathbf{z}},\mathbf{z}_H)\right]^{-1} \qquad (i = 1,\dots,n).$$

The relations of the form (2.29) are identical for the Halley-like algorithms presented above and the corresponding square root algorithms (see § 4.4). For this reason, the R-order of convergence of the single-step Halley-like algorithms (SS), (SSN) and (SSH) is given by Theorem 4.11 for $k = 2$. The order of convergence of the total-step methods (TS), (TSN) and (TSH) is *four, five* and *six*, respectively.

**REMARK 6.** The Halley-like iterative methods for determining simple complex zeros have been considered in detail by Wang and Wu [126]. These authors have also presented a brief description of algorithms for multiple zeros, realized in a parallel fashion (total-step methods). ®

**EXAMPLE 5.** The algorithms (TS), (SS), (TSN), (SSN), (TSH) and (SSH) were applied for the improvement of zeros of the polynomial

$$P(z) = z^9 - 7z^8 + 20z^7 - 28z^6 - 18z^5 + 110z^4 - 92z^3 - 44z^2 + 345z + 225$$

(see Example 11 in §4.4). The exact zeros of this polynomial are $\xi_1 = 1+2i$, $\xi_2 = 1-2i$,

$\xi_3 = -1$ and $\xi_4 = 3$, with the multiplicities $\mu_1 = 2$, $\mu_2 = 2$, $\mu_3 = 3$ and $\mu_4 = 2$. As the initial approximations to these zeros the following complex numbers were taken:

$$z_1^{(0)} = 1.7 + 2.7i, \quad z_2^{(0)} = 1.7 - 2.7i, \quad z_3^{(0)} = -0.3 - 0.7i, \quad z_4^{(0)} = 2.4 - 0.6i.$$

Numerical results, obtained in the second iteration, are given in Table 5.4.

| method | i | $\mathrm{Re}\{z_i^{(2)}\}$ | $\mathrm{Im}\{z_i^{(2)}\}$ |
|---|---|---|---|
| (TS) | 1 | 0.999999703872727 | 1.999999577023530 |
| | 2 | 1.000004966234449 | -1.999858354626263 |
| | 3 | -1.000001724263487 | $1.28 \times 10^{-6}$ |
| | 4 | 3.000175153200852 | $4.58 \times 10^{-5}$ |
| (SS) | 1 | 0.999999603833368 | 2.000000538829041 |
| | 2 | 0.999997513035036 | -2.000168291520113 |
| | 3 | -1.000001434643141 | $8.31 \times 10^{-7}$ |
| | 4 | 3.000000000400398 | $4.03 \times 10^{-9}$ |
| (TSN) | 1 | 1.000005463270708 | 1.999990357789566 |
| | 2 | 1.000000009930465 | -2.000000025656453 |
| | 3 | -0.999999370541218 | $-2.44 \times 10^{-7}$ |
| | 4 | 2.999980969476169 | $5.24 \times 10^{-6}$ |
| (SSN) | 1 | 0.999998904155992 | 1.999998927299469 |
| | 2 | 0.999999988521851 | -1.999999982758255 |
| | 3 | -1.000000001576391 | $5.07 \times 10^{-9}$ |
| | 4 | 3.000000000001254 | $-3.26 \times 10^{-12}$ |
| (TSH) | 1 | 1.000000002444691 | 2.000000000565806 |
| | 2 | 1.000000002639924 | -2.000000001014728 |
| | 3 | -0.999999999964674 | $-2.81 \times 10^{-12}$ |
| | 4 | 3.000000003876174 | $-3.25 \times 10^{-10}$ |
| (SSH) | 1 | 1.000000000020514 | 2.000000000101261 |
| | 2 | 1.000000000012086 | -1.999999999988034 |
| | 3 | -1.000000000000157 | $-2.86 \times 10^{-13}$ |
| | 4 | 3.000000000000029 | $-2.88 \times 10^{-14}$ |

Table 5.4

CHAPTER 6

# COMPUTATIONAL EFFICIENCY OF SIMULTANEOUS METHODS

Most of the iterative methods for the simultaneous determination of polynomial complex zeros, presented in Chapters 3, 4 and 5, will be analyzed and mutually compared regarding their computational cost and convergence speed. We will compare these methods in view of computational efficiency for various values of the polynomial degree and several types of digital computers. For simplicity, only the case of simple complex zeros will be treated. An estimation of computational efficiency of the considered class of methods provides their ranking which is of interest in designing a package of algorithms for the simultaneous approximation of polynomial zeros, where automatic procedure selection is desired.

## 6.1. MEASURE OF COMPUTATIONAL EFFICIENCY OF SIMULTANEOUS METHODS

In practice, it is important to know certain characteristics of any zero-finding algorithm relative to the number of numerical operations in calculating the zeros with the wanted accuracy, convergence speed, processor time of a computer, taking possesion of a storage space at a computer, the number of central processors available to the user, etc. An estimation of the efficiency of iterative methods for improving, simultaneously, approximations to the polynomial zeros taking into consideration the above-mentioned points, given by the *coefficient of efficiency*, has been introduced in [70]. This coefficient takes into account (1) the R-order of convergence (in the sense of the definition introduced by Ortega and Rheinboldt [78], see §2.3) and (2) the number of basic arithmetic operations per iteration, taken with certain *weight* depending on processor time. This definition of efficiency enables simultaneous methods to be compared with various structures (e.g., those with derivatives or without them, in serial or parallel fashion, realized in circular complex arithmetic or ordinary complex arithmetic).

Let $(z^{(m)})$ be an iterative sequence generated by any iterative function (shorter IF) solving a nonlinear (algebraic or transcendental) equation $f(z) = 0$. A measure of the information used by an IF and a measure of the efficiency of the IF are required. Taking the informational usage $d$ of an IF as the number of new pieces of information required per iteration, Traub [122, p. 11] introduced the following definition:

The informational efficiency E is the convergence order of IF $r$, divided by the informational usage $d$; that is

$$^{*}E = \frac{r}{d}\ .$$

The informational usage $d$ is the total number of new function evaluations (the values of f and its derivatives) per iteration.

Ostrowski [79, p. 20] gave an alternative definition of efficiency, the *efficiency index*,

$$E = r^{1/d}\ .$$

The concept of informational efficiency does not take into account the cost of evaluating the values of function f and its derivatives. The *computational efficiency* of an iterative function $\phi$ relative to f, which does take these costs into consideration, was introduced by Traub [122, pp. 260-264]:

$$E(\phi, f) = r^{1/\Theta},$$

where $\Theta = \sum \theta_j$ and $\theta_j$ is the cost of evaluating $f^{(j)}$. If the informational usage of $\phi$ is $d$, and $\theta_j$ is independent of j, then $E(\phi, f)$ is independent of f and reduces to

$$E = E(\phi) = r^{1/d}.$$

Now, let P be a monic algebraic polynomial of degree n with real or complex zeros $\xi_1, \ldots, \xi_n$; that is

$$P(z) = z^n + a_{n-1}z^{n-1} + \cdots + a_1 z + a_0 \doteq \prod_{i=1}^{n}(z - \xi_i) \qquad (a_i \in \mathbb{C}).$$

Applying any simultaneous method for the determination of all zeros of the polynomial P, one forms n sequences $(z_1^{(m)}),\dots,(z_n^{(m)})$ $(m = 1,2,\dots)$ starting with reasonably good initial approximations $z_1^{(0)},\dots,z_n^{(0)}$. The previous definitions of efficiency refer to only *one* sequence and they cannot be applied directly for simultaneous methods where n mutually dependent sequences are produced. Furthermore, in generating these sequences by some iterative formula, apart from the evaluations of polynomial and (eventually) its derivatives, the necessity for evaluation of some arithmetic expressions (sums, products, and so on), depending on the approximations of polynomial zeros (the terms of sequences $(z_i^{(m)})$) appears. In order to define the cost of iteration, a heterogenous (intermixed) structure imposes the necessity of introducing the total number of basic arithmetic operations (addition , subtraction, multiplication, division) for all zeros per iteration.

In practical application of simultaneous methods, the Gauss - Seidel approach (a serial fashion) is often used. Not only is the accelerated convergence attained without additional evaluations, but this procedure is also favorable relative to the occupation of storage space at a digital computer. The analysis of the convergence order of these (single-step) methods is provided by the concept of the R-order of convergence. Since the R-order depends on the polynomial degree n, we will denote it with r(n) in the following.

It is obvious that any simultaneous method is more efficient if its R-order of convergence is greater and the total number of basic arithmetic operations per iteration is smaller. Defining a coefficient of efficiency, it is also necessary to take into consideration the processor time needed for execution of the mentioned operations. Because of that, we will correspond to each operation the *weight* that is (1) proportional to the number of elementary steps (period clocks) necessary in the execution of this operation in the arithmetic units of the computer and (2) normalized in reference to the addition. These weights will be denoted by $w_A$, $w_S$, $w_M$, and $w_D$ for addition, subtraction, multiplication, and division, respectively, setting $w_A = 1$ because of the normalization.

Let us analyze now the number of necessary arithmetic operations per iteration. We presume that the Horner scheme is used for the evaluation of the given polynomial and its derivatives (if they appear) and assume in our analysis that the computer used to implement algorithms would execute only *real arithmetic operations*.

In evaluating the polynomial values at n point approximations to the zeros, taking into account the weights of the basic operations, the corresponding cost of evaluation of the polynomial P can be defined as follows:

$$G(n) = w_A \bar{A}(n) + w_M \bar{M}(n).$$

Here $\bar{A}(n)$ and $\bar{M}(n)$ denote the number of additions and multiplications, respectively. It is well known that

$$\bar{A}(n) = 4n^2, \quad \bar{M}(n) = 4n^2$$

for a complex polynomial of degree n using the Horner scheme. The other more economical schemes are known, but we will use the Horner scheme for simplicity.

The number of all operations which are necessary in realization of one iteration, including $\bar{A}(n)$ and $\bar{M}(n)$, will be denoted by

$A(n)$ (additions),

$S(n)$ (subtractions),

$M(n)$ (multiplications),

$D(n)$ (divisions).

The quantities $A(n)$ and $M(n)$ include the operations necessary for evaluation of the derivatives of the polynomial (by the Horner scheme), when they appear.

The total cost of the evaluation (for all zeros) per iteration is equal to

$$T(n) = w_A A(n) + w_S S(n) + w_M M(n) + w_D D(n). \tag{6.1}$$

It is convenient to introduce the *normalized cost* of evaluation

$$\theta(n) = \frac{T(n)}{G(n)};$$

that is,

$$\Theta(n) = \frac{w_A A(n) + w_S S(n) + w_M M(n) + w_D D(n)}{G(n)} . \qquad (6.2)$$

In practice, it is most frequently $w_A \cong w_S$, so that the total number of additions and subtractions will be denoted by AS(n). Besides, taking $w_S \cong w_A = 1$ because of the normalization, we obtain $G(n) = 4n^2(1 + w_M)$ and the formulas (6.1) and (6.2) now become

$$T(n) = AS(n) + w_M M(n) + w_D D(n), \qquad (6.3)$$

$$\Theta(n) = \frac{AS(n) + w_M M(n) + w_D D(n)}{4n^2(1 + w_M)} . \qquad (6.4)$$

The weights $w_M$ and $w_D$ in (6.3) and (6.4) are *normalized* in relation to $w_A$.

T(n) given by (6.3) is the total cost of the evaluation per iteration expressed by the normalized "time" of addition. If $t_A$ is the real CPU*) time (expressed, for instance, in μsec) needed for one oparation of addition, then $T(n)\cdot t_A$ theoretically gives an estimation of the real time (in μsec) for performing one iteration for all zeros.

Investigating the problem of a genuine estimation of computational efficiency of an iterative method for solving equations, we have started from the fact that any method is more efficient the smaller its computational amount of work for a given accuracy ε. Assuming that (complex) zeros of tested polynomials are normalized to lie in the unit disk and starting with the same initial approximations $z_1^{(0)},\dots,z_n^{(0)}$ to the zeros $\xi_1, \dots,\xi_n$, a stopping criterion can be given by

$$\max_{1 \le i \le n} |z_i^{(m)} - \xi_i| < \varepsilon = 10^{-q} ,$$

where m is the iteration index and q is the number of significant decimal digits at the approximations $z_1^{(m)},\dots,z_n^{(m)}$. If $|z_i^{(0)} - \xi| \sim 10^{-1}$ and r is the order of convergence of applied simultaneous iterative method, then the (theoretical) number of iterative steps, necessary for obtaining the accuracy ε, can be determined approximately as $m \cong \frac{\log q}{\log r}$ (following from $10^{-q} = 10^{-r^m}$).

A computational efficiency E is, obviously, proportional to the reciprocal value of total computational cost Θ considering a complete zero-finding procedure consisting of m iterative steps; hence

$$E = \frac{1}{m\Theta} = \frac{1}{\log q}\,\frac{\log r}{\Theta} .$$

---

*) CPU time - central processor unit time

Estimating iterative methods for some fixed accuracy $\varepsilon = 10^{-q}$, it is sufficient to compare the values of $\frac{\log r}{\Theta}$ . Furthermore, to normalize values of E close to 1 we can calculate $r^{1/\Theta}$ instead of $\frac{\log r}{\Theta}$. Finally, according to the previous analysis and the results by Traub ([122, Appendix C]) we may define the computational efficiency of a simultaneous iterative process (shorter SIP) for finding polynomial zeros:

*Definition.* If $r(n)$ is the R-order of convergence of the simultaneous iterative process SIP and $\Theta(n)$ is the normalized cost of evaluation, then

$$E(SIP,n) = r(n)^{1/\Theta(n)} \tag{6.5}$$

will be called the *coefficient of the efficiency* of SIP.

An alternative definition of the coefficient of efficiency can be introduced by (see. p. 222)

$$^{*}E(SIP,n) = \frac{r(n)}{\Theta(n)} . \tag{6.6}$$

We have measured the CPU times (for three digital computers), necessary for obtaining the required accuracy applying most SIP. Although any measuring of a 'proper' CPU time is not simple and depends on many parameters, we have concluded that the average "rating list" of the applied simultaneous methods (formed according to the CPU times) considerably coincides with the corresponding rating list that is obtained by calculating the computational efficiency by (6.5). In particular, (6.5) is convenient for estimation of computational efficiency of combined methods (see § 6.4). Therefore, in the following we will use the definition (6.5) of computational efficiency.

Formula (6.6) demonstrates good agreement with the "CPU" rating list too, especially in the cases of the most efficient and the least efficient methods. We recall that (6.6) showed even slightly better agreement with the CPU rating list, compared to (6.5), for real polynomials having only real zeros (see [70]). Such behavior of the computational efficiency defined by (6.5) and (6.6) occured due to different computational costs of the methods of the same type employing various kinds of arithmetics (e.g. real arithmetic, complex arithmetic or circular interval arithmetic). A typical example is Weierstrass' method whose computational efficiency considerably depends on the applied arithmetic (cf. the results from [70] and Sections 6.2 and 6.3).

In view of (6.5) it is clear that the efficiency measure depends on the architecture of the arithmetic unit and on the software applied in the calculation of some library functions such as square root, sine, cosine and arc tg (required in the implementation of zero-finding algorithms). The optimization of number of arithmetic operations is also of importance. For this reason, the number of operations given for the algorithms which are considered in Chapters 3, 4 and 5 can insignificantly differ from the number obtained by some other procedure for counting over operations (for example, applying some more economical schemes for the evaluation of polynomials and the k-th root of a real or complex number, for multiplication of two complex numbers (cf. [32]), optimizing some parts of the program in the implementation on a computer, etc.). Therefore, the listed number of operations has rather to be regarded as approximate. For the previous facts, slight variations of the values of the coefficients of efficiency, calculated by (6.5) (or by the corresponding formula given in §6.4 for combined methods), are possible. In spite of that, the averaged values of the coefficients of efficiency, obtained by applying various computing machines, provide reliable ranking of the estimated simultaneous methods. *)

In the following sections we will estimate three kinds of simultaneous methods: (1) iterative methods in "point" complex arithmetic, denoted with the prefix "P", (2) iterative methods in complex circular arithmetic (with the prefix "I") and (3) combined methods (with the prefix "K"). The single-step methods, obtained from the basic (total-step) methods by the Gauss-Seidel procedure, have the additional suffix "a" (indicating an accelerated method). For example, (P5) denotes the iterative method in "point" complex arithmetic numbered as "5"; (P5a) is the single-step method constructed from the total-step method (P5). For easier citation, the listed methods are named by connecting the corresponding methods with the original formulas introduced by the titled authors (as Weierstrass, Newton, Halley, Maehly, Börsch-Supan). The SIP (I8) and (I8a) are named Bell's disk iterations because they are constructed by Bell's polynomials (see §5.2), although these methods were introduced and studied by Wang and Zheng (see [127],[130]).

---

*) All computations necessary for Tables 6.4, 6.5, 6.7, 6.8, 6.12 and 6.13 were done at the Rechenzentrum at the University of Oldenburg.

## 6.2. ITERATIVE METHODS IN COMPLEX ARITHMETIC

First we will give a review of the most frequently used simultaneous methods for polynomial complex zeros, realized in (point) complex arithmetic. Let us introduce some notations:

1° The approximations $z_1^{(m)},\dots,z_n^{(m)}$ of the zeros at the m-th iteration will be briefly denoted with $z_1,\dots,z_n$, and the new approximations $z_1^{(m+1)},\dots,z_n^{(m+1)}$, obtained by some simultaneous method, by $\hat{z}_1,\dots,\hat{z}_n$, respectively;

2° $N_i = P(z_i)/P'(z_i)$ (Newton's correction),

$$W_i = P(z_i)\Big/\prod_{\substack{j=1\\ j\neq i}}^{n}(z_i - z_j) \quad \text{(Weierstrass' correction)},$$

$$H_i = \left[\frac{P'(z_i)}{P(z_i)} - \frac{P''(z_i)}{2P'(z_i)}\right]^{-1} \quad \text{(Halley's correction)};$$

3° $$S_{k,i}(a,b) = \sum_{j=1}^{i-1}(z_i-a_j)^{-k} + \sum_{j=i+1}^{n}(z_i-b_j)^{-k},$$

where $\mathbf{a} = (a_1,\dots,a_n)$ and $\mathbf{b} = (b_1,\dots,b_n)$ are some vectors the components of which are complex numbers or disks;

4° $\mathbf{z} = (z_1,\dots,z_n)$ (the former approximations),

$\hat{\mathbf{z}} = (\hat{z}_1,\dots,\hat{z}_n)$ (the new approximations),

$\mathbf{z}_N = (z_{N,1},\dots,z_{N,n})$, $z_{N,i} = z_i - N_i$ (the Newton approximations),

$\mathbf{z}_W = (z_{W,1},\dots,z_{W,n})$, $z_{W,i} = z_i - W_i$ (the Weierstrass approximations),

$\mathbf{z}_H = (z_{H,1},\dots,z_{H,n})$, $z_{H,i} = z_i - H_i$ (the Halley approximations).

For the total-step and single-step methods the abbreviations TS and SS will be used.

In this section we will estimate the following iterative methods:

(*TS Weierstrass' method*, § 3.1)

$$\hat{z}_i = z_i - W_i = z_i - P(z_i)\Big/\prod_{\substack{j=1\\ j\neq i}}^{n}(z_i - z_j) \quad (r(n) = 2), \tag{P1}$$

(*SS Weierstrass' method,* § 3.1)

$$\hat{z}_i = z_i - \frac{P(z_i)}{\prod_{j<i} (z_i-\hat{z}_j) \prod_{j>i} (z_i-z_j)}, \tag{P1a}$$

(*TS Börsch-Supan's method,* § 3.2)

$$\hat{z}_i = z_i - \frac{W_i}{1 - \sum_{\substack{j=1 \\ j\neq i}}^{n} \frac{W_j}{z_j - z_i}} \qquad (r(n) = 3), \tag{P2}$$

(*TS Börsch-Supan's method with Weierstrass' correction,* § 3.2)

$$\hat{z}_i = z_i - \frac{W_i}{1 - \sum_{\substack{j=1 \\ j\neq i}}^{n} \frac{W_j}{z_j - z_{W,i}}} \qquad (r(n) = 4), \tag{P3}$$

(*TS Maehly's method,* § 4.1)

$$\hat{z}_i = z_i - [N_i^{-1} - S_{1,i}(z,z)]^{-1} \qquad (r(n) = 3), \tag{P4}$$

(*SS Maehly's method,* § 4.1)

$$\hat{z}_i = z_i - [N_i^{-1} - S_{1,i}(\hat{z},z)]^{-1}, \tag{P4a}$$

(*TS Maehly's method with Newton's correction,* § 4.1)

$$\hat{z}_i = z_i - [N_i^{-1} - S_{1,i}(z_N, z_N)]^{-1} \qquad (r(n) = 4), \tag{P5}$$

(*SS Maehly's method with Newton's correction,* § 4.1)

$$\hat{z}_i = z_i - [N_i^{-1} - S_{1,i}(\hat{z},z_N)]^{-1}, \tag{P5a}$$

(*TS Halley-like method,* § 5.5)

$$\hat{z}_i = z_i - \left\{ H_i^{-1} - \frac{N_i}{2}[S_{1,i}^2(z,z) + S_{2,i}(z,z)] \right\}^{-1} \tag{P6}$$

$$(r(n) = 4),$$

(*SS Halley-like method*, § 5.5)

$$\hat{z}_i = z_i - \left\{H_i^{-1} - \frac{N_i}{2}[S_{1,i}^2(\hat{z},z) + S_{2,i}(\hat{z},z)]\right\}^{-1}, \tag{P6a}$$

(*TS Halley-like method with Newton's corrections*, § 5.5)

$$\hat{z}_i = z_i - \left\{H_i^{-1} - \frac{N_i}{2}[S_{1,i}^2(z_N,z_N) + S_{2,i}(z_N,z_N)]\right\}^{-1} \tag{P7}$$

$(r(n) = 5)$,

(*SS Halley-like method with Newton's corrections*, § 5.5)

$$\hat{z}_i = z_i - \left\{H_i^{-1} - \frac{N_i}{2}[S_{1,i}^2(\hat{z},z_N) + S_{2,i}(\hat{z},z_N)]\right\}^{-1}, \tag{P7a}$$

(*TS Halley-like method with Halley's corrections*, § 5.5)

$$\hat{z}_i = z_i - \left\{H_i^{-1} - \frac{N_i}{2}[S_{1,i}^2(z_H,z_H) + S_{2,i}(z_H,z_H)]\right\}^{-1} \tag{P8}$$

$(r(n) = 6)$,

(*SS Halley-like method with Halley's corrections*, § 5.5)

$$\hat{z}_i = z_i - \left\{H_i^{-1} - \frac{N_i}{2}[S_{1,i}^2(\hat{z},z_H) + S_{2,i}(\hat{z},z_H)]\right\}^{-1}. \tag{P8a}$$

The convergence order $r(n)$ of the above-mentioned total-step methods is given after the corresponding formula. The lower bounds of the R-order of convergence $r(n)$ for the listed single-step methods are displayed in Table 6.1 for $n = 3(1)10$, 15 and 20. These bounds are calculated by

$$r(n) = p + t_n(p,q),$$

where $t_n(p,q)$ $(q,p+q)$ is the unique positive root of the equation (see Theorem 2.4)

$$t^n - tq^{n-1} - pq^{n-1} = 0,$$

and $p$ and $q$ are the integers appearing in the relation (2.29). The values of $p$ and $q$ are shown in Table 6.1 for each of the considered single-step methods. Table 6.1 also contains the lower bounds of the R-order of convergence of the single-step interval methods, whose efficiency is studied in § 6.3.

| | p=1,q=1 | p=2,q=1 | p=1,q=2 | p=3,q=1 | p=4,q=1 | p=3,q=2 | p=3,q=3 |
|---|---|---|---|---|---|---|---|
| | P1a<br>I1a | P4a<br>I3a | P5a | P6a<br>I4a<br>I6a | I5a<br>I7a | P7a | P8a |
| r(3) | 2.325 | 3.521 | 4.649 | 4.672 | 5.796 | 5.862 | 6.974 |
| r(4) | 2.221 | 3.353 | 4.441 | 4.453 | 5.534 | 5.585 | 6.662 |
| r(5) | 2.167 | 3.267 | 4.335 | 4.341 | 5.401 | 5.443 | 6.502 |
| r(6) | 2.135 | 3.215 | 4.269 | 4.274 | 5.321 | 5.357 | 6.404 |
| r(7) | 2.113 | 3.180 | 4.226 | 4.229 | 5.268 | 5.299 | 6.338 |
| r(8) | 2.097 | 3.154 | 4.194 | 4.196 | 5.230 | 5.257 | 6.291 |
| r(9) | 2.085 | 3.135 | 4.170 | 4.172 | 5.201 | 5.225 | 6.255 |
| r(10) | 2.076 | 3.121 | 4.152 | 4.153 | 5.179 | 5.200 | 6.227 |
| r(15) | 2.049 | 3.078 | 4.097 | 4.098 | 5.115 | 5.130 | 6.147 |
| r(20) | 2.036 | 3.057 | 4.072 | 4.073 | 5.085 | 5.096 | 6.109 |

Table 6.1 The values of the R-order of single-step methods

The review of the number of basic arithmetic operations AS(n), M(n) and D(n) for the methods (P1)-(P8a) is given in Table 6.2 as a function of the polynomial degree n.

| Point SIP | AS(n) | M(n) | D(n) |
|---|---|---|---|
| (P1),(P1a) | $8n^2 + n$ | $8n^2 + 2n$ | $2n$ |
| (P2) | $15n^2 - 6n$ | $14n^2 + 2n$ | $2n^2 + 2n$ |
| (P3) | $15n^2 - 4n$ | $14n^2 + 2n$ | $2n^2 + 2n$ |
| (P4),(P4a) | $13n^2 - 3n$ | $10n^2 + 2n$ | $2n^2 + 2n$ |
| (P5) | $13n^2$ | $10n^2 + 4n$ | $2n^2 + 4n$ |
| (P5a) | $13n^2 - 3$ | $10n^2 + 4n - 2$ | $2n^2 + 4n - 2$ |
| (P6),(P6a) | $19n^2 - 3n$ | $18n^2 + 12n$ | $2n^2 + 4n$ |
| (P7) | $19n^2$ | $18n^2 + 14n$ | $2n^2 + 6n$ |
| (P7a) | $19n^2 - 3$ | $18n^2 + 14n - 2$ | $2n^2 + 6n - 2$ |
| (P8) | $20n^2$ | $18n^2 + 12n$ | $2n^2 + 8n$ |
| (P8a) | $20n^2 - 3$ | $18n^2 + 12n - 2$ | $2n^2 + 8n - 2$ |

Table 6.2 The number of basic arithmetic operations

To compare the iterative methods (P1) - (P8a), we have computed the coefficients of the efficiency for these methods using (6.5). We have used the characteristics (period clocks) of the arithmetic units of the computing machines HONEYWELL DPS 6/92, VAX 11/780, IBM 4341 as well as of the supercomputer CRAY X-MP/2 (on the basis of data given in [64]). Typical values of the operation weights, obtained after normalization in regard to $w_A$, are given in Table 6.3.

| | $w_A$ | $w_S$ | $w_M$ | $w_D$ |
|---|---|---|---|---|
| HONEYWELL DPS 6/92 | 1.00 | 1.00 | 3.00 | 5.62 |
| VAX 11/780 | 1.00 | 1.00 | 1.50 | 5.25 |
| IBM 4341 | 1.00 | 1.00 | 1.50 | 12.37 |
| CRAY X-MP/2 | 1.00 | 1.00 | 1.17 | 2.33 |

Table 6.3 The operation weights normalized in relation to $w_A$

**REMARK 1.** The normalized cost of the evaluation $\Theta(n)$ can be expressed in the form

$$\Theta(n) = a + \frac{b}{n} + \frac{c}{n^2},$$

where a ( > 0), b ( > 0) and c are real constants. From Table 6.2 we observe that c = 0 for all total-step methods and the single-step methods without correction. ®

Using the values of the R-order of convergence r(n) (Table 6.1), the number of basic arithemtic operations (Table 6.2) and the values of operation weights $w_M$ and $w_D$ (Table 6.3), the coefficients of the efficiency E(SIP,n) were computed by (6.5) for n = 3(1)10, 15 and 20 and displayed in Table 6.4 for the above-mentioned computing machines and SIP (P1) - (P8a).

We observe from Table 6.4 that the order of magnitude of E(SIP,n) for the considered computers is preserved (with slight exceptions) when n varies. This fact enables us to form a rating of the methods (P1) - (P8a) related to their efficiency. Let $(E((p_1),n),\dots,E((p_{14}),n))$ ($p_j \in \{$P1, P1a, ..., P8a$\}$, j = 1,...,14) be the ordered 14-tuplet whose components satisfy

| | | n | | | | | | | | | |
|---|---|---|---|---|---|---|---|---|---|---|---|
| | Point SIP | 3 | 4 | 5 | 6 | 7 | 8 | 9 | 10 | 15 | 20 |
| HONEYWELL DPS 6/92 | (P1) | 1.338 | 1.354 | 1.365 | 1.372 | 1.378 | 1.382 | 1.385 | 1.388 | 1.396 | 1.401 |
| | (P1a) | 1.425 | 1.418 | 1.415 | 1.414 | 1.413 | 1.413 | 1.413 | 1.413 | 1.413 | 1.413 |
| | (P2) | 1.277 | 1.281 | 1.283 | 1.285 | 1.286 | 1.287 | 1.288 | 1.288 | 1.290 | 1.291 |
| | (P3) | 1.357 | 1.363 | 1.367 | 1.370 | 1.372 | 1.373 | 1.375 | 1.376 | 1.378 | 1.380 |
| | (P4) | 1.347 | 1.355 | 1.361 | 1.364 | 1.367 | 1.369 | 1.370 | 1.371 | 1.375 | 1.377 |
| | (P4a) | 1.407 | 1.398 | 1.393 | 1.391 | 1.389 | 1.388 | 1.388 | 1.387 | 1.385 | 1.385 |
| | (P5) | 1.401 | 1.423 | 1.437 | 1.447 | 1.455 | 1.461 | 1.465 | 1.469 | 1.480 | 1.486 |
| | (P5a) | 1.473 | 1.473 | 1.476 | 1.478 | 1.481 | 1.483 | 1.485 | 1.486 | 1.491 | 1.495 |
| | (P6) | 1.241 | 1.254 | 1.262 | 1.268 | 1.272 | 1.275 | 1.278 | 1.280 | 1.287 | 1.290 |
| | (P6a) | 1.271 | 1.276 | 1.279 | 1.282 | 1.284 | 1.286 | 1.288 | 1.289 | 1.293 | 1.295 |
| | (P7) | 1.265 | 1.284 | 1.296 | 1.305 | 1.311 | 1.316 | 1.320 | 1.324 | 1.334 | 1.340 |
| | (P7a) | 1.302 | 1.310 | 1.317 | 1.322 | 1.326 | 1.329 | 1.331 | 1.334 | 1.341 | 1.345 |
| | (P8) | 1.291 | 1.312 | 1.327 | 1.337 | 1.345 | 1.351 | 1.356 | 1.360 | 1.372 | 1.379 |
| | (P8a) | 1.327 | 1.338 | 1.347 | 1.353 | 1.359 | 1.363 | 1.366 | 1.369 | 1.378 | 1.383 |
| VAX 11/780 | (P1) | 1.322 | 1.341 | 1.353 | 1.362 | 1.369 | 1.374 | 1.378 | 1.381 | 1.392 | 1.397 |
| | (P1a) | 1.405 | 1.402 | 1.402 | 1.403 | 1.403 | 1.404 | 1.405 | 1.406 | 1.408 | 1.409 |
| | (P2) | 1.251 | 1.255 | 1.257 | 1.259 | 1.260 | 1.261 | 1.261 | 1.262 | 1.263 | 1.264 |
| | (P3) | 1.322 | 1.328 | 1.332 | 1.334 | 1.336 | 1.337 | 1.338 | 1.339 | 1.342 | 1.343 |
| | (P4) | 1.299 | 1.306 | 1.311 | 1.314 | 1.316 | 1.318 | 1.319 | 1.320 | 1.323 | 1.325 |
| | (P4a) | 1.349 | 1.342 | 1.339 | 1.337 | 1.335 | 1.334 | 1.334 | 1.333 | 1.332 | 1.332 |
| | (P5) | 1.339 | 1.358 | 1.371 | 1.380 | 1.387 | 1.392 | 1.397 | 1.400 | 1.411 | 1.416 |
| | (P5a) | 1.400 | 1.401 | 1.404 | 1.407 | 1.409 | 1.411 | 1.413 | 1.415 | 1.420 | 1.423 |
| | (P6) | 1.224 | 1.236 | 1.243 | 1.248 | 1.252 | 1.255 | 1.258 | 1.259 | 1.265 | 1.268 |
| | (P6a) | 1.252 | 1.256 | 1.259 | 1.262 | 1.264 | 1.265 | 1.266 | 1.267 | 1.271 | 1.272 |
| | (P7) | 1.243 | 1.260 | 1.272 | 1.280 | 1.286 | 1.291 | 1.295 | 1.298 | 1.308 | 1.313 |
| | (P7a) | 1.278 | 1.285 | 1.291 | 1.296 | 1.299 | 1.302 | 1.305 | 1.307 | 1.314 | 1.317 |
| | (P8) | 1.260 | 1.280 | 1.294 | 1.304 | 1.312 | 1.317 | 1.322 | 1.326 | 1.338 | 1.345 |
| | (P8a) | 1.293 | 1.304 | 1.313 | 1.319 | 1.324 | 1.329 | 1.332 | 1.335 | 1.344 | 1.349 |
| IBM 4341 | (P1) | 1.264 | 1.290 | 1.309 | 1.323 | 1.333 | 1.342 | 1.348 | 1.354 | 1.372 | 1.382 |
| | (P1a) | 1.330 | 1.341 | 1.350 | 1.358 | 1.364 | 1.369 | 1.373 | 1.376 | 1.387 | 1.393 |
| | (P2) | 1.175 | 1.181 | 1.184 | 1.186 | 1.188 | 1.189 | 1.190 | 1.191 | 1.193 | 1.194 |
| | (P3) | 1.224 | 1.231 | 1.236 | 1.239 | 1.241 | 1.243 | 1.245 | 1.246 | 1.249 | 1.251 |
| | (P4) | 1.197 | 1.205 | 1.210 | 1.213 | 1.216 | 1.217 | 1.219 | 1.220 | 1.224 | 1.226 |
| | (P4a) | 1.229 | 1.228 | 1.228 | 1.228 | 1.228 | 1.228 | 1.229 | 1.229 | 1.230 | 1.230 |
| | (P5) | 1.215 | 1.231 | 1.243 | 1.251 | 1.257 | 1.262 | 1.265 | 1.269 | 1.278 | 1.284 |
| | (P5a) | 1.254 | 1.259 | 1.264 | 1.268 | 1.271 | 1.274 | 1.276 | 1.278 | 1.285 | 1.288 |
| | (P6) | 1.162 | 1.173 | 1.180 | 1.185 | 1.189 | 1.192 | 1.195 | 1.197 | 1.203 | 1.206 |
| | (P6a) | 1.182 | 1.188 | 1.192 | 1.195 | 1.198 | 1.200 | 1.201 | 1.203 | 1.207 | 1.209 |
| | (P7) | 1.170 | 1.186 | 1.196 | 1.204 | 1.210 | 1.215 | 1.219 | 1.222 | 1.232 | 1.238 |
| | (P7a) | 1.195 | 1.204 | 1.211 | 1.216 | 1.220 | 1.224 | 1.227 | 1.229 | 1.237 | 1.241 |
| | (P8) | 1.176 | 1.194 | 1.207 | 1.217 | 1.225 | 1.231 | 1.235 | 1.240 | 1.253 | 1.260 |
| | (P8a) | 1.198 | 1.211 | 1.221 | 1.228 | 1.234 | 1.239 | 1.243 | 1.246 | 1.257 | 1.263 |
| CRAY X-MP/2 | (P1) | 1.350 | 1.364 | 1.373 | 1.380 | 1.384 | 1.388 | 1.391 | 1.393 | 1.400 | 1.403 |
| | (P1a) | 1.442 | 1.430 | 1.425 | 1.422 | 1.420 | 1.419 | 1.418 | 1.418 | 1.416 | 1.416 |
| | (P2) | 1.300 | 1.301 | 1.301 | 1.301 | 1.302 | 1.302 | 1.302 | 1.302 | 1.302 | 1.302 |
| | (P3) | 1.384 | 1.387 | 1.389 | 1.390 | 1.391 | 1.392 | 1.392 | 1.393 | 1.394 | 1.394 |
| | (P4) | 1.364 | 1.369 | 1.372 | 1.374 | 1.375 | 1.376 | 1.377 | 1.378 | 1.380 | 1.381 |
| | (P4a) | 1.428 | 1.413 | 1.406 | 1.402 | 1.399 | 1.396 | 1.395 | 1.394 | 1.390 | 1.388 |
| | (P5) | 1.424 | 1.442 | 1.454 | 1.462 | 1.468 | 1.472 | 1.476 | 1.479 | 1.488 | 1.492 |
| | (P5a) | 1.500 | 1.494 | 1.493 | 1.493 | 1.494 | 1.495 | 1.496 | 1.496 | 1.499 | 1.500 |
| | (P6) | 1.263 | 1.273 | 1.280 | 1.284 | 1.287 | 1.290 | 1.292 | 1.294 | 1.298 | 1.301 |
| | (P6a) | 1.297 | 1.297 | 1.298 | 1.300 | 1.301 | 1.301 | 1.302 | 1.303 | 1.304 | 1.305 |
| | (P7) | 1.290 | 1.306 | 1.317 | 1.324 | 1.329 | 1.334 | 1.337 | 1.340 | 1.348 | 1.353 |
| | (P7a) | 1.331 | 1.335 | 1.339 | 1.342 | 1.345 | 1.347 | 1.349 | 1.350 | 1.355 | 1.358 |
| | (P8) | 1.316 | 1.335 | 1.347 | 1.355 | 1.362 | 1.367 | 1.370 | 1.374 | 1.384 | 1.389 |
| | (P8a) | 1.355 | 1.362 | 1.368 | 1.373 | 1.376 | 1.379 | 1.382 | 1.384 | 1.390 | 1.394 |

Table 6.4 The values of E(SIP,n) (rounded to the third decimal digit)

| | Point SIP | n = 3 | 4 | 5 | 6 | 7 | 8 | 9 | 10 | 15 | 20 |
|---|---|---|---|---|---|---|---|---|---|---|---|
| HONEYWELL 6/92 | 1. | 5a | 5a | 5a | 5a | 5a | 5a | 5a | 5a | 5a | 5a |
| | 2. | 1a | 5 | 5 | 5 | 5 | 5 | 5 | 5 | 5 | 5 |
| | 3. | 4a | 1a | 1a | 1a | 1a | 1a | 1a | 1a | 1a | 1a |
| | 4. | 5 | 4a | 4a | 4a | 4a | 4a | 4a | 1 | 1 | 1 |
| | 5. | 3 | 3 | 3 | 1 | 1 | 1 | 1 | 4a | 4a | 4a |
| | 6. | 4 | 4 | 1 | 3 | 3 | 3 | 3 | 3 | 3 | 8a |
| | 7. | 1 | 1 | 4 | 4 | 4 | 4 | 4 | 4 | 8a | 3 |
| | 8. | 8a | 8a | 8a | 8a | 8a | 8a | 8a | 8a | 4 | 8 |
| | 9. | 7a | 8 | 8 | 8 | 8 | 8 | 8 | 8 | 8 | 4 |
| | 10. | 8 | 7a | 7a | 7a | 7a | 7a | 7a | 7a | 7a | 7a |
| | 11. | 2 | 7 | 7 | 7 | 7 | 7 | 7 | 7 | 7 | 7 |
| | 12. | 6a | 2 | 2 | 2 | 2 | 2 | 2 | 6a | 6a | 6a |
| | 13. | 7 | 6a | 6a | 6a | 6a | 6a | 6a | 2 | 2 | 2 |
| | 14. | 6 | 6 | 6 | 6 | 6 | 6 | 6 | 6 | 6 | 6 |
| VAX 11/780 | 1. | 1a | 1a | 5a | 5a | 5a | 5a | 5a | 5a | 5a | 5a |
| | 2. | 5a | 5a | 1a | 1a | 1a | 1a | 1a | 1a | 5 | 5 |
| | 3. | 4a | 5 | 5 | 5 | 5 | 5 | 5 | 5 | 1a | 1a |
| | 4. | 5 | 4a | 1 | 1 | 1 | 1 | 1 | 1 | 1 | 1 |
| | 5. | 1 | 1 | 4a | 4a | 3 | 3 | 3 | 3 | 8a | 8a |
| | 6. | 3 | 3 | 3 | 3 | 4a | 4a | 4a | 8a | 3 | 8 |
| | 7. | 4 | 4 | 8a | 8a | 8a | 8a | 8a | 4a | 8 | 3 |
| | 8. | 8a | 8a | 4 | 4 | 4 | 4 | 8 | 8 | 4a | 4a |
| | 9. | 7a | 7a | 8 | 8 | 8 | 8 | 4 | 4 | 4 | 4 |
| | 10. | 8 | 8 | 7a | 7a | 7a | 7a | 7a | 7a | 7a | 7a |
| | 11. | 6a | 7 | 7 | 7 | 7 | 7 | 7 | 7 | 7 | 7 |
| | 12. | 2 | 6a | 6a | 6a | 6a | 6a | 6a | 6a | 6a | 6a |
| | 13. | 7 | 2 | 2 | 2 | 2 | 2 | 2 | 2 | 6 | 6 |
| | 14. | 6 | 6 | 6 | 6 | 6 | 6 | 6 | 6 | 2 | 2 |
| IBM 4341 | 1. | 1a | 1a | 1a | 1a | 1a | 1a | 1a | 1a | 1a | 1a |
| | 2. | 1 | 1 | 1 | 1 | 1 | 1 | 1 | 1 | 1 | 1 |
| | 3. | 5a | 5a | 5a | 5a | 5a | 5a | 5a | 5a | 5a | 5a |
| | 4. | 4a | 5 | 5 | 5 | 5 | 5 | 5 | 5 | 5 | 5 |
| | 5. | 3 | 3 | 3 | 3 | 3 | 3 | 3 | 8a | 8a | 8a |
| | 6. | 5 | 4a | 4a | 8a | 8a | 8a | 8a | 3 | 8 | 8 |
| | 7. | 8a | 8a | 8a | 4a | 4a | 8 | 8 | 8 | 3 | 3 |
| | 8. | 4 | 4 | 7a | 8 | 8 | 4a | 4a | 7a | 7a | 7a |
| | 9. | 7a | 7a | 4 | 7a | 7a | 7a | 7a | 4a | 7 | 7 |
| | 10. | 6a | 8 | 8 | 4 | 4 | 4 | 4 | 7 | 4a | 4a |
| | 11. | 8 | 6a | 7 | 7 | 7 | 7 | 7 | 4 | 4 | 4 |
| | 12. | 2 | 7 | 6a | 6a | 6a | 6a | 6a | 6a | 6a | 6a |
| | 13. | 7 | 2 | 2 | 2 | 6 | 6 | 6 | 6 | 6 | 6 |
| | 14. | 6 | 6 | 6 | 6 | 2 | 2 | 2 | 2 | 2 | 2 |
| CRAY X-MP/2 | 1. | 5a | 5a | 5a | 5a | 5a | 5a | 5a | 5a | 5a | 5a |
| | 2. | 1a | 5 | 5 | 5 | 5 | 5 | 5 | 5 | 5 | 5 |
| | 3. | 4a | 1a | 1a | 1a | 1a | 1a | 1a | 1a | 1a | 1a |
| | 4. | 5 | 4a | 4a | 4a | 4a | 4a | 4a | 4a | 1 | 1 |
| | 5. | 3 | 3 | 3 | 3 | 3 | 3 | 3 | 1 | 3 | 3 |
| | 6. | 4 | 4 | 1 | 1 | 1 | 1 | 1 | 3 | 4a | 8a |
| | 7. | 8a | 1 | 4 | 4 | 8a | 8a | 8a | 8a | 8a | 8 |
| | 8. | 1 | 8a | 8a | 8a | 4 | 4 | 4 | 4 | 8 | 4a |
| | 9. | 7a | 7a | 8 | 8 | 8 | 8 | 8 | 8 | 4 | 4 |
| | 10. | 8 | 8 | 7a | 7a | 7a | 7a | 7a | 7a | 7a | 7a |
| | 11. | 2 | 7 | 7 | 7 | 7 | 7 | 7 | 7 | 7 | 7 |
| | 12. | 6a | 2 | 2 | 2 | 2 | 2 | 6a | 6a | 6a | 6a |
| | 13. | 7 | 6a | 6a | 6a | 6a | 6a | 2 | 2 | 2 | 2 |
| | 14. | 6 | 6 | 6 | 6 | 6 | 6 | 6 | 6 | 6 | 6 |

Table 6.5 The rating vectors $R_p$

$E((p_1),n) > E((p_2),n) > \cdots > E((p_{14}),n)$, and let $R_P = (p_1,\ldots,p_{14})$ be the *rating vector* of the iterative methods $(p_1),\ldots,(p_{14})$. The vector $R_P$ determines the position of each SIP in reference to its efficiency for a given computer. The rating of the considered methods, related to the given computers, is shown in Table 6.5.

From Table 6.5 we can draw the following conclusions (in the sense of definition (6.5)):

$(C_1)$ For three considered computers and for any polynomial degree n, SIP (P5a) is the most efficient. The exeption appears at the IBM computer, where Weierstrass' methods (P1a) and (P1) are superior. The explanation lies in the fact that the operation weight $w_D$ is great for the IBM ($w_D = 12.37$) so that the remaining methods are less efficient compared to (P1a) and (P1) which require the smallest number of divisions (only 2n for all n zeros). Further, we observe that the total-step method (P5), which is the basic one for (P5a), is also one of the most powerful. The same is true for the Weierstrass methods (P1a) and (P1).

$(C_2)$ SIP (P2), (P6) and (P6a) are the least efficient for all considered computers.

$(C_3)$ Weierstrass' methods (P1) and (P1a) belong to the top of the rating list, especially in the case of the IBM computer, which has already been discussed. The single-step method (P1a) is also very efficient when implemented at the VAX computer, especially for small n. Its efficiency decreases when n grows. SIP (4a) behaves similarly in regard to n and it applies to all four computers. Börsch-Supan's method with Weierstrass' corrections (P3) possesses considerable efficiency for all four computing machines.

$(C_4)$ For all considered computers, SIP (P5a) and (P5) with Newton's corrections have considerable advantage in relation to SIP (P4a) and (P4) (without corrections). The reason is the improved convergence rate of (P5a) and (P5) (compared to the basic methods (P4a) and (P4)), which is attained by a slight increase of the number of additional operations. The same conclusion is valid for the methods (P8a) and (P8), compared to (P7a) and (P7), for (P7a) and (P7), compared to (P6a) and (P6), as well as for (P3) in relation to (P2).

The rating lists of the methods (P1) - (P8a) (given in Table 6.5)) and the above conclusions have been verified by the experimental results, which were obtained by measuring the CPU times corresponding to HONEYWELL DPS 6/92, VAX 11/780 and IBM 4341. In particular, this means that the definition (6.5) of the coefficient of efficiency of SIP is really applicable and describes a real situation in the practical realization of an iterative process on a computer (for more details, see [70]).

**REMARK 2.** The same approach to the computational efficiency, described in this Chapter, was applied in [70] for the real polynomials with real zeros only. The performed analysis showed that Weierstrass' methods (P1a) and (P1) are the most efficient in the case of real zeros. The decrease of the efficiency of SIP (P1a) and (P1) for complex zeros occurs because of the considerable increase of the number of operations due to multiplying of complex numbers. ®

## 6.3. ITERATIVE METHODS IN CIRCULAR ARITHMETIC

In this section we will use some of the notations introduced in the previous section and the following additional ones:

$1^\circ$ The circular approximations $Z_1^{(m)},\dots,Z_n^{(m)}$ to the zeros in the m-th iteration will be shortly denoted with $Z_1,\dots,Z_n$, and the new disk-approximations $Z_1^{(m+1)},\dots,Z_n^{(m+1)}$ by $\hat{Z}_1,\dots,\hat{Z}_n$. According to this, we introduce the vectors

$$\mathbf{Z} = (Z_1,\dots,Z_n), \qquad \hat{\mathbf{Z}} = (\hat{Z}_1,\dots,\hat{Z}_n);$$

$2^\circ$ $$h_k(z) = \frac{(-1)^{k-1}}{(k-1)!}\cdot\frac{d^{k-1}}{dz^{k-1}}\left(\frac{P'(z)}{P(z)}\right) \qquad (k = 1,2,\dots);$$

$3^\circ$ $$\sigma_k(z) = \frac{P^{(k)}(z)}{k!P(z)} \qquad (k = 1,2,\dots),$$

$$\Delta_2(z) = \sigma_1(z)^2 - \sigma_2(z),$$

$$\Delta_3(z) = \sigma_1(z)[\Delta_2(z) - \sigma_2(z)] + \sigma_3(z) \ .$$

We now give a review of the simultaneous methods in circular arithmetic, considered previously in Chapters 3, 4 and 5.

(*TS Weierstrass' method,* § 3.1)

$$\hat{Z}_i = z_i - \frac{P(z_i)}{\prod_{\substack{j=1 \\ j\neq i}}^{n} (z_i - Z_j)} \qquad (r(n) = 2), \tag{I1}$$

(*SS Weierstrass' method,* § 3.1)

$$\hat{Z}_i = z_i - \frac{P(z_i)}{\prod_{j<i} (z_i - \hat{Z}_j) \prod_{j>i} (z_i - Z_j)} , \tag{I1a}$$

(*TS Börsch-Supan's method,* § 3.2)

$$\hat{Z}_i = z_i - \frac{W_i}{1 - \sum_{\substack{j=1 \\ j\neq i}}^{n} \frac{W_j}{z_j - Z_i}} \qquad (r(n) = 3), \tag{I2}$$

(*TS Maehly's method,* § 4.1)

$$\hat{Z}_i = z_i - [N_i^{-1} - S_{1,i}(Z, Z)]^{-1} \qquad (r(n) = 3), \tag{I3}$$

(*SS Maehly's method,* § 4.2)

$$\hat{Z}_i = z_i - [N_i^{-1} - S_{1,i}(\hat{Z}, Z)]^{-1} , \tag{I3a}$$

(*TS Square root iterations,* § 4.1)

$$\hat{Z}_i = z_i - \left\{ [h_2(z_i) - S_{2,i}(Z, Z)]_*^{1/2} \right\}^{-1} \qquad (r(n) = 4), \tag{I4}$$

(*SS Square root iterations,* § 4.2)

$$\hat{Z}_i = z_i - \left\{ [h_2(z_i) - S_{2,i}(\hat{Z}, Z)]_*^{1/2} \right\}^{-1} , \tag{I4a}$$

(*TS Third root iterations,* § 4.1)

$$\hat{Z}_i = z_i - \left\{ [h_3(z_i) - S_{3,i}(Z, Z)]_*^{1/3} \right\}^{-1} \qquad (r(n) = 5), \tag{I5}$$

(*SS Third root iterations*, § 4.2)

$$\hat{z}_i = z_i - \left\{ [h_3(z_i) - S_{3,i}(\hat{Z},Z)]_*^{1/3} \right\}^{-1}, \tag{I5a}$$

(*TS Halley-like method*, § 5.3)

$$\hat{z}_i = z_i - \left\{ H_i^{-1} - \frac{N_i}{2}[S_{1,i}^2(Z,Z) + S_{2,i}(Z,Z)] \right\}^{-1} \quad (r(n) = 4), \tag{I6}$$

(*SS Halley-like method*, § 5.3)

$$\hat{z}_i = z_i - \left\{ H_i^{-1} - \frac{N_i}{2}[S_{1,i}^2(\hat{Z},Z) + S_{2,i}(\hat{Z},Z)] \right\}^{-1}, \tag{I6a}$$

(*TS Bell's disk iterations*, § 5.2)

$$\hat{z}_i = z_i - \frac{\Delta_2(z_i)}{\Delta_3(z_i) - \frac{1}{3}S_{3,i}(Z,Z) - \frac{1}{2}S_{2,i}(Z,Z)\,S_{1,i}(Z,Z) - \frac{1}{6}S_{1,i}^3(Z,Z)}$$

$$(r(n) = 5), \tag{I7}$$

(*SS Bell's disk iterations*, § 5.2)

$$\hat{z}_i = z_i - \frac{\Delta_2(z_i)}{\Delta_3(z_i) - \frac{1}{3}S_{3,i}(\hat{Z},Z) - \frac{1}{2}S_{2,i}(\hat{Z},Z)\,S_{1,i}(\hat{Z},Z) - \frac{1}{6}S_{1,i}^3(\hat{Z},Z)}. \tag{I7a}$$

The lower bounds of the R-order of convergence for the single-step methods (I1a), (I3a), (I4a), (I5a), (I6a) and (I7a) are given in Table 6.1 (§ 6.2).

As it was emphasized at the beginning, the operations of circular arithmetic are executed by real arithmetic operations. The number of basic real arithmetic operations AS(n), M(n) and D(n) for the interval methods (I1) - (I7a) is displayed in Table 6.6. The symbol * appearing in the iterative formulas for (I4), (I4a), (I5) and (I5a) indicates that one "appropriate" disk (among k disks, k = 2 or 3) has to be chosen (see § 4.1). The operations necessary for this selecting root procedure are also included.

| Interval SIP | AS(n) | M(n) | D(n) |
|---|---|---|---|
| (I1),(I1a) | $22n^2 - 6n$ | $25n^2 - 6n$ | $8n^2 - n$ |
| (I2) | $23n^2 - 4n$ | $23n^2 + 2n$ | $7n^2 + 2n$ |
| (I3),(I3a) | $15n^2 - 4n$ | $11n^2 + 2n$ | $3n^2 + 2n$ |
| (I4),(I4a) | $27n^2 + 43n$ | $29n^2 + 50n$ | $7n^2 + 21n$ |
| (I5),(I5a) | $34n^2 + 94n$ | $39n^2 + 149n$ | $7n^2 + 32n$ |
| (I6),(I6a) | $30n^2 + 3n$ | $29n^2 + 20n$ | $7n^2 + 8n$ |
| (I7),(I7a) | $48n^2 + 5n$ | $53n^2 + 28n$ | $11n^2 + 23n$ |

Table 6.6 The number of basic arithmetic operations

The interval methods (I1) - (I7a) were compared by calculating the coefficients of efficiency (6.5) for the same computers as in the previous section. We note that the normalized cost of the evaluation $\Theta(n)$, given by (6.4), has the form

$$\Theta(n) = a + \frac{b}{n},$$

where a ( > 0 ) and b are real constants.

The values of the coefficient of efficiency

$$E(SIP,n) = r(n)^{1/\Theta(n)}$$

were computed for n = 3(1)10, 15 and 20 and shown in Table 6.7 for various computers.

The *rating vectors* $R_I = (i_1,\ldots,i_{14})$ $(i_j \in \{I1, I1a, \ldots, I7a\},\ j = 1,\ldots,13)$, formed according to the ordering

$$E((i_1),n) > E((i_2),n) > \cdots > E((i_{13}),n),$$

are given in Table 6.8 for each of the considered computers and n=3(1)10, 15 and 20. The prefix "I" in Table 6.8 as well as the prefix "P" in Table 6.4 are omitted for brevity.

| | Interval SIP | n = 3 | 4 | 5 | 6 | 7 | 8 | 9 | 10 | 15 | 20 |
|---|---|---|---|---|---|---|---|---|---|---|---|
| HONEYWELL DPS 6/92 | (I1) | 1.088 | 1.086 | 1.085 | 1.084 | 1.084 | 1.084 | 1.083 | 1.083 | 1.082 | 1.082 |
| | (I1a) | 1.108 | 1.100 | 1.095 | 1.093 | 1.091 | 1.089 | 1.088 | 1.088 | 1.085 | 1.084 |
| | (I2) | 1.138 | 1.139 | 1.140 | 1.141 | 1.141 | 1.141 | 1.142 | 1.142 | 1.142 | 1.142 |
| | (I3) | 1.289 | 1.294 | 1.297 | 1.300 | 1.301 | 1.302 | 1.303 | 1.304 | 1.307 | 1.308 |
| | (I3a) | 1.337 | 1.328 | 1.324 | 1.321 | 1.320 | 1.318 | 1.317 | 1.317 | 1.315 | 1.314 |
| | (I4) | 1.090 | 1.101 | 1.108 | 1.114 | 1.119 | 1.122 | 1.125 | 1.128 | 1.136 | 1.140 |
| | (I4a) | 1.101 | 1.109 | 1.115 | 1.120 | 1.124 | 1.127 | 1.129 | 1.131 | 1.138 | 1.142 |
| | (I5) | 1.062 | 1.072 | 1.080 | 1.086 | 1.092 | 1.096 | 1.100 | 1.103 | 1.114 | 1.120 |
| | (I5a) | 1.067 | 1.077 | 1.084 | 1.090 | 1.095 | 1.099 | 1.102 | 1.105 | 1.116 | 1.122 |
| | (I6) | 1.122 | 1.129 | 1.133 | 1.136 | 1.138 | 1.140 | 1.141 | 1.142 | 1.145 | 1.147 |
| | (I6a) | 1.137 | 1.139 | 1.141 | 1.143 | 1.144 | 1.145 | 1.145 | 1.146 | 1.148 | 1.149 |
| | (I7) | 1.078 | 1.083 | 1.086 | 1.088 | 1.090 | 1.091 | 1.092 | 1.093 | 1.095 | 1.096 |
| | (I7a) | 1.086 | 1.088 | 1.090 | 1.092 | 1.093 | 1.094 | 1.094 | 1.095 | 1.097 | 1.097 |
| VAX 11/780 | (I1) | 1.076 | 1.075 | 1.074 | 1.073 | 1.073 | 1.073 | 1.072 | 1.072 | 1.072 | 1.071 |
| | (I1a) | 1.093 | 1.086 | 1.083 | 1.080 | 1.079 | 1.078 | 1.077 | 1.076 | 1.074 | 1.073 |
| | (I2) | 1.119 | 1.120 | 1.121 | 1.121 | 1.122 | 1.122 | 1.122 | 1.122 | 1.123 | 1.123 |
| | (I3) | 1.243 | 1.248 | 1.250 | 1.252 | 1.254 | 1.255 | 1.255 | 1.256 | 1.258 | 1.259 |
| | (I3a) | 1.284 | 1.276 | 1.272 | 1.270 | 1.269 | 1.268 | 1.267 | 1.266 | 1.265 | 1.264 |
| | (I4) | 1.079 | 1.088 | 1.095 | 1.100 | 1.104 | 1.107 | 1.110 | 1.112 | 1.120 | 1.124 |
| | (I4a) | 1.088 | 1.095 | 1.101 | 1.105 | 1.109 | 1.111 | 1.114 | 1.116 | 1.122 | 1.126 |
| | (I5) | 1.057 | 1.066 | 1.074 | 1.080 | 1.084 | 1.088 | 1.092 | 1.095 | 1.105 | 1.111 |
| | (I5a) | 1.062 | 1.071 | 1.077 | 1.083 | 1.087 | 1.091 | 1.094 | 1.097 | 1.106 | 1.112 |
| | (I6) | 1.108 | 1.113 | 1.117 | 1.120 | 1.121 | 1.123 | 1.124 | 1.125 | 1.128 | 1.129 |
| | (I6a) | 1.121 | 1.123 | 1.124 | 1.126 | 1.127 | 1.127 | 1.128 | 1.129 | 1.130 | 1.131 |
| | (I7) | 1.069 | 1.073 | 1.076 | 1.078 | 1.080 | 1.081 | 1.082 | 1.083 | 1.085 | 1.087 |
| | (I7a) | 1.076 | 1.078 | 1.080 | 1.082 | 1.083 | 1.084 | 1.084 | 1.085 | 1.087 | 1.088 |
| IBM 4341 | (I1) | 1.048 | 1.047 | 1.046 | 1.046 | 1.046 | 1.046 | 1.046 | 1.046 | 1.045 | 1.045 |
| | (I1a) | 1.058 | 1.054 | 1.052 | 1.051 | 1.050 | 1.049 | 1.048 | 1.048 | 1.047 | 1.046 |
| | (I2) | 1.075 | 1.076 | 1.077 | 1.077 | 1.077 | 1.078 | 1.078 | 1.078 | 1.078 | 1.079 |
| | (I3) | 1.154 | 1.159 | 1.162 | 1.163 | 1.165 | 1.166 | 1.167 | 1.167 | 1.169 | 1.170 |
| | (I3a) | 1.179 | 1.176 | 1.175 | 1.175 | 1.174 | 1.174 | 1.174 | 1.174 | 1.174 | 1.174 |
| | (I4) | 1.050 | 1.057 | 1.061 | 1.065 | 1.068 | 1.070 | 1.072 | 1.074 | 1.079 | 1.082 |
| | (I4a) | 1.056 | 1.061 | 1.065 | 1.068 | 1.071 | 1.073 | 1.074 | 1.076 | 1.080 | 1.083 |
| | (I5) | 1.039 | 1.046 | 1.051 | 1.055 | 1.059 | 1.062 | 1.064 | 1.066 | 1.074 | 1.078 |
| | (I5a) | 1.043 | 1.049 | 1.054 | 1.058 | 1.061 | 1.064 | 1.066 | 1.068 | 1.075 | 1.079 |
| | (I6) | 1.070 | 1.074 | 1.077 | 1.079 | 1.081 | 1.082 | 1.083 | 1.083 | 1.086 | 1.087 |
| | (I6a) | 1.078 | 1.080 | 1.082 | 1.083 | 1.084 | 1.085 | 1.085 | 1.086 | 1.087 | 1.088 |
| | (I7) | 1.044 | 1.048 | 1.050 | 1.052 | 1.053 | 1.054 | 1.055 | 1.056 | 1.058 | 1.059 |
| | (I7a) | 1.048 | 1.051 | 1.052 | 1.054 | 1.055 | 1.056 | 1.056 | 1.057 | 1.059 | 1.060 |
| CRAY X-MP/2 | (I1) | 1.097 | 1.095 | 1.094 | 1.093 | 1.093 | 1.093 | 1.092 | 1.092 | 1.091 | 1.091 |
| | (I1a) | 1.120 | 1.111 | 1.106 | 1.103 | 1.101 | 1.099 | 1.098 | 1.097 | 1.095 | 1.093 |
| | (I2) | 1.152 | 1.153 | 1.153 | 1.154 | 1.154 | 1.154 | 1.154 | 1.154 | 1.154 | 1.155 |
| | (I3) | 1.305 | 1.307 | 1.309 | 1.310 | 1.310 | 1.311 | 1.311 | 1.312 | 1.313 | 1.313 |
| | (I3a) | 1.356 | 1.343 | 1.336 | 1.332 | 1.329 | 1.327 | 1.326 | 1.324 | 1.321 | 1.319 |
| | (I4) | 1.099 | 1.110 | 1.119 | 1.125 | 1.130 | 1.133 | 1.137 | 1.139 | 1.148 | 1.153 |
| | (I4a) | 1.111 | 1.120 | 1.126 | 1.131 | 1.135 | 1.138 | 1.141 | 1.143 | 1.151 | 1.155 |
| | (I5) | 1.069 | 1.080 | 1.089 | 1.096 | 1.101 | 1.106 | 1.110 | 1.113 | 1.125 | 1.131 |
| | (I5a) | 1.075 | 1.085 | 1.093 | 1.099 | 1.105 | 1.109 | 1.113 | 1.116 | 1.127 | 1.133 |
| | (I6) | 1.135 | 1.141 | 1.144 | 1.147 | 1.149 | 1.150 | 1.152 | 1.153 | 1.156 | 1.157 |
| | (I6a) | 1.151 | 1.152 | 1.153 | 1.155 | 1.155 | 1.156 | 1.157 | 1.157 | 1.158 | 1.159 |
| | (I7) | 1.088 | 1.092 | 1.095 | 1.097 | 1.099 | 1.100 | 1.101 | 1.101 | 1.104 | 1.105 |
| | (I7a) | 1.096 | 1.098 | 1.100 | 1.101 | 1.102 | 1.103 | 1.103 | 1.104 | 1.105 | 1.106 |

Table 6.7 The values of E(SIP,n) (rounded to the third decimal digit)

| | | n | | | | | | | | | |
|---|---|---|---|---|---|---|---|---|---|---|---|
| | Interval SIP | 3 | 4 | 5 | 6 | 7 | 8 | 9 | 10 | 15 | 20 |
| HONEYWELL DPS 11/780 | 1. | 3a | 3a | 3a | 3a | 3a | 3a | 3a | 3a | 3a | 3a |
| | 2. | 3 | 3 | 3 | 3 | 3 | 3 | 3 | 3 | 3 | 3 |
| | 3. | 2 | 2 | 6a | 6a | 6a | 6a | 6a | 6a | 6a | 6a |
| | 4. | 6a | 6a | 2 | 2 | 2 | 2 | 2 | 6 | 6 | 6 |
| | 5. | 6 | 6 | 6 | 6 | 6 | 6 | 6 | 2 | 2 | 2 |
| | 6. | 1a | 4a | 4a | 4a | 4a | 4a | 4a | 4a | 4a | 4a |
| | 7. | 4a | 4 | 4 | 4 | 4 | 4 | 4 | 4 | 4 | 4 |
| | 8. | 4 | 1a | 1a | 1a | 5a | 5a | 5a | 5a | 5a | 5a |
| | 9. | 1 | 7a | 7a | 7a | 7a | 5 | 5 | 5 | 5 | 5 |
| | 10. | 7a | 1 | 7 | 5a | 5 | 7a | 7a | 7a | 7a | 7a |
| | 11. | 7 | 7 | 1 | 7 | 1a | 7 | 7 | 7 | 7 | 7 |
| | 12. | 5a | 5a | 5a | 5 | 7 | 1a | 1a | 1a | 1a | 1a |
| | 13. | 5 | 5 | 5 | 1 | 1 | 1 | 1 | 1 | 1 | 1 |
| VAX 11/780 | 1. | 3a | 3a | 3a | 3a | 3a | 3a | 3a | 3a | 3a | 3a |
| | 2. | 3 | 3 | 3 | 3 | 3 | 3 | 3 | 3 | 3 | 3 |
| | 3. | 6a | 6a | 6a | 6a | 6a | 6a | 6a | 6a | 6a | 6a |
| | 4. | 2 | 2 | 2 | 2 | 2 | 6 | 6 | 6 | 6 | 6 |
| | 5. | 6 | 6 | 6 | 6 | 6 | 2 | 2 | 2 | 2 | 4a |
| | 6. | 1a | 4a | 4a | 4a | 4a | 4a | 4a | 4a | 4a | 4 |
| | 7. | 4a | 4 | 4 | 4 | 4 | 4 | 4 | 4 | 4 | 2 |
| | 8. | 4 | 1a | 1a | 5a | 5a | 5a | 5a | 5a | 5a | 5a |
| | 9. | 1 | 7a | 7a | 7a | 5 | 5 | 5 | 5 | 5 | 5 |
| | 10. | 7a | 1 | 5a | 1a | 7a | 7a | 7a | 7a | 7a | 7a |
| | 11. | 7 | 7 | 7 | 5 | 7 | 7 | 7 | 7 | 7 | 7 |
| | 12. | 5a | 5a | 1 | 7 | 1a | 1a | 1a | 1a | 1a | 1a |
| | 13. | 5 | 5 | 5 | 1 | 1 | 1 | 1 | 1 | 1 | 1 |
| IBM 4341 | 1. | 3a | 3a | 3a | 3a | 3a | 3a | 3a | 3a | 3a | 3a |
| | 2. | 3 | 3 | 3 | 3 | 3 | 3 | 3 | 3 | 3 | 3 |
| | 3. | 6a | 6a | 6a | 6a | 6a | 6a | 6a | 6a | 6a | 6a |
| | 4. | 2 | 2 | 6 | 6 | 6 | 6 | 6 | 6 | 6 | 6 |
| | 5. | 6 | 6 | 2 | 2 | 2 | 2 | 2 | 2 | 4a | 4a |
| | 6. | 1a | 4a | 4a | 4a | 4a | 4a | 4a | 4a | 4 | 4 |
| | 7. | 4a | 4 | 4 | 4 | 4 | 4 | 4 | 4 | 2 | 5a |
| | 8. | 4 | 1a | 5a | 5a | 5a | 5a | 5a | 5a | 5a | 2 |
| | 9. | 7a | 7a | 7a | 5 | 5 | 5 | 5 | 5 | 5 | 5 |
| | 10. | 1 | 5a | 1a | 7a | 7a | 7a | 7a | 7a | 7a | 7a |
| | 11. | 7 | 7 | 5 | 7 | 7 | 7 | 7 | 7 | 7 | 7 |
| | 12. | 5a | 1 | 7 | 1a | 1a | 1a | 1a | 1a | 1a | 1a |
| | 13. | 5 | 5 | 1 | 1 | 1 | 1 | 1 | 1 | 1 | 1 |
| CRAY X-MP/2 | 1. | 3a | 3a | 3a | 3a | 3a | 3a | 3a | 3a | 3a | 3a |
| | 2. | 3 | 3 | 3 | 3 | 3 | 3 | 3 | 3 | 3 | 3 |
| | 3. | 2 | 2 | 6a | 6a | 6a | 6a | 6a | 6a | 6a | 6a |
| | 4. | 6a | 6a | 2 | 2 | 2 | 2 | 2 | 2 | 6 | 6 |
| | 5. | 6 | 6 | 6 | 6 | 6 | 6 | 6 | 6 | 2 | 4a |
| | 6. | 1a | 4a | 4a | 4a | 4a | 4a | 4a | 4a | 4a | 2 |
| | 7. | 4a | 1a | 4 | 4 | 4 | 4 | 4 | 4 | 4 | 4 |
| | 8. | 4 | 4 | 1a | 1a | 5a | 5a | 5a | 5a | 5a | 5a |
| | 9. | 1 | 7a | 7a | 7a | 7a | 5 | 5 | 5 | 5 | 5 |
| | 10. | 7a | 1 | 7 | 5a | 5 | 7a | 7a | 7a | 7a | 7a |
| | 11. | 7 | 7 | 1 | 7 | 1a | 7 | 7 | 7 | 7 | 7 |
| | 12. | 5a | 5a | 5a | 5 | 7 | 1a | 1a | 1a | 1a | 1a |
| | 13. | 5 | 5 | 5 | 1 | 1 | 1 | 1 | 1 | 1 | 1 |

Table 6.8 The rating vectors $R_T$

From Table 6.8 the following conclusions may be deduced:

($C_1$) The behavior of the coefficients of efficiency for the interval methods (I1) - (I7a) is nearly the same for all considered computers so that the rating vector $R_I$ is almost unchangeable.

($C_2$) The interval methods (I3a) and (I3) of Maehly's type are extremely superior compared to the other methods for *all* n and *all* considered machines. E((I3a),n) and E((I3),n) are remarkably larger than the coefficients of efficiency of the remaining methods.

($C_3$) The Halley-like algorithms (I6a) and (I6) follow the superior SIP (I3a) and (I3), with a slightly larger coefficients of efficiency in relation to (I2), (I4a) and (I4). We note that the method (I2) is more efficient for small n than for large n. On the other hand, the efficiency of the third root iterations (I5a) and (I5) is significantly improved when n increases.

($C_4$) The interval methods (I1), (I1a) and (I7) are the least efficient. The inefficiency of Weierstrass' methods (I1) and (I1a) is the result of numerous multiplications of disks, where the absolute values of complex numbers are required (see (2.11)).

($C_5$) The algorithms (I3), (I4) and (I5) belong to the family of the k-th root iterations, given by (4.15) $(k = 1,2,3)$. Also, the SIP (I3), (I6) and (I7) are special cases of the family of Bell's disk iterations (5.12), which are obtained for $k = 1,2$ and 3. From Table 6.8 we have

$$E((I3),n) > E((I4),n) > E((I5),n)$$

and

$$E((I3),n) > E((I6),n) > E((I7),n).$$

Therefore, the most efficient method from the both families (in the class of total-step methods) is (I3), which has the lowest order of convergence among them, but it requires the smallest number of operations. From the above ordering of the coefficients of efficiency it could be concluded that the construction of higher-order methods (increasing k in (4.15) and (5.12)) is not necessary. Nevertheless, the situation is slightly different in practice; applying any SIP with a very high order of convergence, the approximations of zeros can be determined with the wanted accuracy after only three or even two iterations. Further, we observe from Table 6.8 that the Halley-like methods (I6) and (I6a) belong to the most efficient methods, while the square root method (I4) is efficient for polynomials of the high degree. On the other hand, taking into account the computational cost of the interval methods (I5) and (I7) (whose order of convergence is five), the cumbersome and robust software in their implementation on a computer as well as their positions on the rating list, we will apply some of the high rated methods rather than (I5) and (I7).

($C_6$) The interval method (I3a) is even more efficient than the point methods (P2), (P6) and (P7).

**REMARK 3.** Calculating the coefficient of efficiency of the interval methods (I1) - (I7a) we have presumed that circular complex operations are performed using the schemes with real arithmetic operations. Such realization makes circular arithmetic very costly and, consequently, the corresponding circular algorithms are insufficiently effective. The efficiency of interval methods would be considerably greater if interval arithmetic operations were executed as hardware machine arithmetic operations realized by microprogramming. This procedure is a major goal of current research in interval analysis and it has already become the reality.®

## 6.4. COMBINED METHODS

To increase the efficiency of iterative processes for the simultaneous inclusion of polynomial complex zeros, some combined methods have been established in the previous chapters. In this section we will compare seven combined methods, constructed by the point methods (P1), (P4), (P5) and the interval methods (I1), (I3), (I6). Since these methods have been listed in § 6.2 and § 6.3, we will present the combined methods (K1) - (K7) only by the ordered pairs of the corresponding methods as follows:

| (K1) | (K2) | (K3) | (K4) | (K5) | (K6) | (K7) |
|---|---|---|---|---|---|---|
| (P1,I1) | (P1,I3) | (P1,I6) | (P4,I3) | (P4,I6) | (P5,I3) | (P5,I6) |

Table 6.9 The combined methods

The first component of each pair denotes an iterative method in point complex arithmetic, while the second component is an iterative interval method which is applied in the final iterative step.

In the following, the relation $\alpha \sim \beta$ will denote that $\alpha$ is of the same order of magnitude as $\beta$, that is, $\alpha = O(\beta)$.

Let (P) be the iterative point method with the convergence order $k_P$ and let $z^{(0)}$ be the initial approximation to the zero $\xi$. For the improved approximation $z^{(m)}$, obtained after m iterations applying the iterative method (P), we have the estimation

$$|z^{(m)} - \xi| \sim |z^{(0)} - \xi|^{k_P^m} . \tag{6.7}$$

Assume now that the iterative interval method (I) of the convergence order $k_I$ produces the sequence $(Z^{(m)})$ of circular approximations, starting from the initial disk $Z^{(0)}$. After m iterations one obtains

$$\varepsilon^{(m)} \sim (\varepsilon^{(0)})^{k_I^m}, \tag{6.8}$$

where $\varepsilon^{(m)} = \operatorname{rad} Z^{(m)}$ $(m = 0,1,\dots)$.

Let $Z_1^{(0)},\dots,Z_n^{(0)}$ be the initial disks that contain the zeros $\xi_1$, $\dots,\xi_n$ of a polynomial of degree n, and let $z_i^{(0)} = \operatorname{mid} Z_i^{(0)}$ $(i=1,\dots,n)$. Assume that the centers $z_1^{(0)},\dots,z_n^{(0)}$ are sufficiently good initial approximations to the zeros $\xi_1,\dots,\xi_n$ so that the iterative point method (P) of the order $k_P$ is converging. Let us consider the combined method (P,I) where the point method (P) is applied M times successively, and the interval method (I) only once, in the final (M+1)-th iterative step. The resulting inclusion disks will be denoted by $Z_1^{(M,1)},\dots,Z_n^{(M,1)}$ and their radii by $\varepsilon_1^{(M,1)},\dots,\varepsilon_n^{(M,1)}$. Besides, we will use the notations

$$\varepsilon_i^{(0)} = \operatorname{rad} Z_i^{(0)} \quad (i=1,\dots,n), \qquad \varepsilon^{(M,1)} = \max_{1 \le i \le n} \varepsilon_i^{(M,1)} .$$

In our analysis we will assume that the initial approximations are correct to one decimal place, that is, $\varepsilon_i^{(0)} = \operatorname{rad} Z_i^{(0)} \sim 10^{-1}$. If (I) is one of the methods (I1), (I3) or (I6), according to the convergence analysis of these interval methods (see §3.1, §4.1 and §5.3) and (6.7), we find the following estimations concerning the combined method (P,I):

$$\varepsilon^{(M,1)} \sim (\varepsilon^{(0)})^{\alpha(k_P,M)}, \tag{6.9}$$

where *)

$$\alpha(k_P,M) = \begin{cases} k_P^M + 1 & \text{for } (P,I1), \\ 2k_P^M + 1 & \text{for } (P,I3), \\ 3k_P^M + 1 & \text{for } (P,I6). \end{cases}$$

Suppose that the combined method (P,I) has to produce the approximations to the zeros correct to at least q decimal places, assuming that $\varepsilon^{(0)} \sim 10^{-1}$. From the relation

$$\varepsilon^{(M,1)} \sim 10^{-q}$$

and the estimation (6.9), we find the necessary number of "point" iterations M:

$$M = \left[\frac{\log(q-1)}{\log k_P}\right] + 1 \qquad \text{for } (P,I1),$$

$$M = \left[\frac{\log((q-1)/2)}{\log k_P}\right] + 1 \qquad \text{for } (P,I3), \tag{6.10}$$

$$M = \left[\frac{\log((q-1)/3)}{\log k_P}\right] + 1 \qquad \text{for } (P,I6).$$

The notation [a] in (6.10) denotes the greatest integer not exceeding a.

To compare the combined method (P,I) with the interval method (I), in this section we have also considered the interval methods (I1), (I3) and (I6) which appear in the combined methods (K1) - (K7). The necessary number of interval iterations N for obtaining at least q correct decimal digits (that is, $\varepsilon^{(N)} \sim 10^{-q}$), applying the interval method (I) of the order $k_I$, is equal to

$$N = \left[\frac{\log q}{\log k_I}\right] + 1. \tag{6.11}$$

Using the estimations (6.8) and (6.9) it is possible to appraise the "expected" accuracy for the interval methods (I1), (I3) and (I6) and the combined methods (K1) - (K7), given by the radii $\varepsilon^{(N)}$ and $\varepsilon^{(M,1)}$ in the final iterative step (assuming, again, $\varepsilon^{(0)} \sim 10^{-1}$). The values of $\varepsilon^{(N)}$ and $\varepsilon^{(M,1)}$ are shown in Table 6.10.

---

*) In general, $\alpha(k_P,k_I,M) = (k_I-1)k_P^M+1$.

| N | I1 | I3 | I6 | M | K1 | K2 | K3 | K4 | K5 | K6 | K7 |
|---|---|---|---|---|---|---|---|---|---|---|---|
| 1 | $10^{-2}$ | $10^{-3}$ | $10^{-4}$ | 1 | $10^{-3}$ | $10^{-5}$ | $10^{-7}$ | $10^{-7}$ | $10^{-10}$ | $10^{-9}$ | $10^{-13}$ |
| 2 | $10^{-4}$ | $10^{-9}$ | $10^{-16}$ | 2 | $10^{-5}$ | $10^{-9}$ | $10^{-13}$ | $10^{-19}$ | $10^{-28}$ | $10^{-33}$ | |
| 3 | $10^{-8}$ | $10^{-27}$ | | 3 | $10^{-9}$ | $10^{-17}$ | $10^{-25}$ | | | | |
| 4 | $10^{-16}$ | | | 4 | $10^{-17}$ | | | | | | |

Table 6.10 The "expected" accuracy of the interval and combined methods

Clearly, the results displayed in Table 6.10 are rather theoretical ones. For comparison, we have solved the polynomial equation of degree 8 with simple complex roots using quad precision arithmetic (about 34 significant decimal digits). The values $\varepsilon^{(N)}$ and $\varepsilon^{(M,1)}$ are shown in Table 6.11. The symbol * points out that the combined methods (K6) and (K7) break down in the second iteration because of the limited precision of the used arithmetic and the presence of rounding errors. This is a typical example where the very fast convergence may even be an obstruction.

| N | I1 | I3 | I6 | M | K1 | K2 | K3 | K4 | K5 | K6 | K7 |
|---|---|---|---|---|---|---|---|---|---|---|---|
| 1 | 8.2(-2) | 8.1(-3) | 1.3(-3) | 1 | 1.3(-2) | 3.7(-4) | 4.1(-6) | 7.9(-6) | 1.2(-9) | 9.2(-8) | 1.7(-11) |
| 2 | 6.1(-4) | 8.5(-8) | 3.5(-13) | 2 | 2.2(-4) | 4.2(-8) | 6.3(-12) | 4.5(-17) | 6.4(-26) | * | * |
| 3 | 7.5(-7) | 7.3(-24) | | 3 | 1.5(-7) | 1.7(-16) | 2.4(-23) | | | | |
| 4 | 1.3(-14) | | | 4 | 5.9(-15) | | | | | | |

Table 6.11 The real accuracy - the example of a polynomial equation of degree 8 †)

Tables 6.10 and 6.11 merely demonstrate the ability of the considered methods in terms of the accuracy which can be expected in their application. But, a complete estimation of their efficiency requires also a computational cost of these methods. Definition of the coefficient of efficiency (6.5) cannot be applied to the methods (K1) - (K7) because they combine floating-point and interval arithmetic during the iterative procedure, which means that the iterative steps have different weights. For this reason, we introduce a *global* measure of efficiency for combined

†) A(-h) in Table 6.11 means $A\times 10^{-h}$.

methods (and for interval methods too, for comparison), which takes into consideration the complete iterative process (all iterations), where the required accuracy determines the stopping criterion (and, consequently, the total number of iterations). In a certain sense, this measure is proportional to the reciprocal value of total CPU time, necessary for finding all zeros with the desired accuracy.

Let $w_S \cong w_A = 1$ and let $w_M$ and $w_D$ be the operation weights for multiplication and division, normalized in relation to $w_A$. Denote the normalized costs of evaluation per iteration for the point and interval methods with $\theta^{(p)}(n)$ and $\theta^{(i)}(n)$ respectively, where

$$\theta^{(p)}(n) = \frac{AS^{(p)}(n) + w_M M^{(p)}(n) + w_D D^{(p)}(n)}{G(n)},$$

$$\theta^{(i)}(n) = \frac{AS^{(i)}(n) + w_M M^{(i)}(n) + w_D D^{(i)}(n)}{G(n)},$$

and $G(n) = 4n^2(1 + w_M)$ as in (6.4). The number of operations is given in Table 6.2 (for the point methods, the upper index (p)) and Table 6.6 (for the interval methods, the upper index (i)).

A measure of efficiency of any combined method (K) will be defined by

$$E_q((K),n) = \alpha(k_p,M)^{1/(M\theta^{(p)}(n) + \theta^{(i)}(n))}, \tag{6.12}$$

where M is the number of point iterations, given by (6.10), which is necessary (together with the final interval iteration) to provide the accuracy $10^{-q}$ (q correct decimal digits of each zero).

For comparison, we introduce the corresponding measure of efficiency for the interval method (I),

$$E_q((I),n) = (k_I^N)^{1/(N\theta^{(i)}(n))} = k_I^{1/\theta^{(i)}(n)}; \tag{6.13}$$

hence, we note that $E_q((I),n)$ reduces to (6.5) for interval methods.

We computed the measures of efficiency $E_q((K_i),n)$ $(i = 1,...,7)$ and $E_q((I_j),n)$ $(j = 1,3,6)$ for the computers HONEYWELL DPS 6/92, VAX 11/780 and IBM 4341, and for $n = 5(1)15$. Four values of the required accuracy were used, $10^{-15}$, $10^{-20}$, $10^{-25}$ and $10^{-30}$ (that is, $q = 15, 20, 25, 30$). Regarding any of the estimated methods in this section we observed that the variations of the values of the computational efficiency are insignificant changing n for each q = 15,20,25,30 and for all considered computers. This convenient fact enables us to make a simplified table of rating vectors (compared to Tables 6.5 and 6.8). As an example, we give Table 6.12 with the values $E_{15}$, calculated for the VAX computer.

| n | 5 | 6 | 7 | 8 | 9 | 10 | 11 | 12 | 13 | 14 | 15 |
|---|---|---|---|---|---|---|---|---|---|---|---|
| (I1) | 1.074 | 1.073 | 1.073 | 1.073 | 1.072 | 1.072 | 1.072 | 1.072 | 1.072 | 1.072 | 1.072 |
| (I3) | 1.250 | 1.252 | 1.254 | 1.255 | 1.255 | 1.256 | 1.257 | 1.257 | 1.257 | 1.258 | 1.258 |
| (I6) | 1.117 | 1.120 | 1.121 | 1.123 | 1.124 | 1.125 | 1.126 | 1.126 | 1.127 | 1.127 | 1.128 |
| (K1) | 1.162 | 1.163 | 1.164 | 1.164 | 1.165 | 1.165 | 1.166 | 1.166 | 1.166 | 1.166 | 1.166 |
| (K2) | 1.272 | 1.276 | 1.280 | 1.282 | 1.285 | 1.286 | 1.288 | 1.289 | 1.290 | 1.291 | 1.291 |
| (K3) | 1.181 | 1.185 | 1.188 | 1.190 | 1.192 | 1.193 | 1.195 | 1.196 | 1.197 | 1.197 | 1.198 |
| (K4) | 1.253 | 1.256 | 1.257 | 1.259 | 1.259 | 1.260 | 1.261 | 1.261 | 1.262 | 1.262 | 1.263 |
| (K5) | 1.175 | 1.178 | 1.180 | 1.182 | 1.183 | 1.184 | 1.185 | 1.186 | 1.187 | 1.187 | 1.188 |
| (K6) | 1.291 | 1.296 | 1.300 | 1.303 | 1.305 | 1.307 | 1.309 | 1.310 | 1.311 | 1.312 | 1.313 |
| (K7) | 1.200 | 1.205 | 1.208 | 1.211 | 1.213 | 1.215 | 1.216 | 1.217 | 1.218 | 1.219 | 1.220 |

Table 6.12 The values of $E_{15}$ for the interval and combined methods

HONEYWELL DPS 6/92

| q=15 | q=20 | q=25 | q=30 |
|---|---|---|---|
| K6 | K6 | K6 | K6 |
| K2 | K4 (<7); K2 (≥7) | K4 (<7); K2 (≥7) | K4 (<7); K2 (≥7) |
| I3 (<13); K4 (≥13) | K2 (<7); K4 (≥7) | K2 (<7); K4 (≥7) | K2 (<7); K4 (≥7) |
| K4 (<13); I3 (≥13) | I3 | I3 | I3 |
| K7 | K7 | K7 | K7 |
| K5 (<9); K3 (≥9) | K5 (<9); K3 (≥9) | K3 | K5 |
| K3 (<9); K5 (≥9) | K3 (<9); K5 (≥9) | K5 | K3 |
| K1 | K1 | K1 | K1 |
| I6 | I6 | I6 | I6 |
| I1 | I1 | I1 | I1 |

VAX 11/780

| q=15,30 | q=20,25 |
|---|---|
| K6 | K6 |
| K2 | K2 |
| K4 | K4 |
| I3 | I3 |
| K7 | K7 |
| K3 | K3 |
| K5 | K1 (<10); K5 (≥10) |
| K1 | K5 (<10); K1 (≥10) |
| I6 | I6 |
| I1 | I1 |

IBM 4341

| q=15 | q=20 | q=25 | q=30 |
|---|---|---|---|
| K2 | K2 | K2 | K2 |
| K6 | K6 | K6 | K6 |
| K4 | K4 | K4 | K4 |
| I3 | I3 | I3 | I3 |
| K7 | K7 | K3 | K3 |
| K3 | K3 | K7 | K7 |
| K5 | K1 | K1 | K1 (<9); K5 (≥9) |
| K1 | K5 | K5 | K5 (<9); K1 (≥9) |
| I6 | I6 | I6 | I6 |
| I1 | I1 | I1 | I1 |

Table 6.13 The rating vectors $R_K(q)$ and $R_I(q)$ [*]

*) Numbers in brackets denote the polynomial degrees n indicating dominant range.

In accordance with the evaluated values of $E_q$ $(q = 15,20,25,30)$, the *rating vectors* $R_K(q)$ and $R_I(q)$, concerning the combined and interval methods, have been formed for the three computers and shown in Table 6.13.

We present some conclusions which may be drawn from Table 6.13.

($C_1$) The combined methods are more efficient compared to the interval methods (in the sense of definitions (6.12) and (6.13)). Only Maehly's interval method (I3) can compete with the combined methods.

($C_2$) Maehly's interval method (I3) requires the smallest number of operations among all interval methods (see Table 6.6). For that reason, the combined methods (K6), (K2) and (K4), which use this interval method, are the most efficient for all three computers as well as for all q and n.

($C_3$) The combined method (K2) (that is, (P1,I3)) is the most efficient at the IBM computer due to the same reason which has already been discussed in § 6.2 (see ($C_1$)) concerning the Weierstrass point process (P1).

($C_4$) The inclusion of multiple zeros by the combined methods (K1), (K2) and (K3) is not possible because Weierstrass' method (P1) cannot be applied for multiple zeros. This fact is inconvenient because (K2) is one of the most efficient methods. However, using Newton's point method with the known multiplicity $\mu_i$

$$\hat{z}_i = z_i - \mu_i \frac{P(z_i)}{P'(z_i)},$$

which requires the same number of basic operations as Weierstrass' method (P1), this deficiency can be overcome in the case of the combined methods (K2) and (K3).

($C_5$) The combined methods (K6) and (K7) converge very fast because the point algorithm (P5) has the high order of convergence. But, this (usually convenient) property may become even uncomfortable leading to the break of the iterative process in the implementation of the methods (K6) and (K7) on computers with arithmetic of small precision (see Table 6.11 and the remark concerning the symbol *).

# BIBLIOGRAPHY


1. O. Aberth: *Iteration methods for finding all zeros of a polynomial simultaneously.* Math. Comp. 27 (1973), 339-344.

2. M. Abramowitz, I. A. Stegun: Handbook of mathematical functions. Dover-New York 1965.

3. G. Alefeld: *Über Eigenschaften und Anwendungsmöglichkeiten einer komplexen Intervallarithmetik.* Z. Angew. Math. Mech. 50 (1970), 455-465.

4. G. Alefeld, J. Herzberger: *On the convergence speed of some algorithms for the simultaneous approximation of polynomial zeros.* SIAM J. Numer. Anal. 11 (1974), 237-243.

5. G. Alefeld, J. Herzberger: *Über Simultanverfahren zur Bestimmung reeller Polynomwurzeln.* Z. Angew. Math. Mech. 54 (1974), 413-420.

6. G. Alefeld, J. Herzberger: Introduction to interval computations. Academic Press, New York 1983.

7. E. T. Bell: *Exponential polynomials.* Math. Ann. 35 (1934), 258-277.

8. I. S. Berezin, N. P. Zhidkov: Computing methods, Vol. II. Pergamon Press, Oxford 1965.

9. W. Börsch-Supan: *A posteriori error bounds for the zeros of polynomials.* Numer. Math. 5 (1963), 380-398.

10. W. Börsch-Supan: *Residuenabschätzung für Polynom-Nullstellen mittels Lagrange-Interpolation.* Numer. Math. 14 (1970), 287-296.

11. D. Braess, K. P. Hadeler: *Simultaneous inclusion of the zeros of a polynomial.* Numer. Math. 21 (1973), 161-165.

12. W. Burmeister, J. W. Schmidt: *Determination of the cone radius for positive concave operators.* Computing 33 (1984), 37-49.

13. P. Byrnev, K. Dochev: *Some modifications of Newton's method for the approximate solution of algebraic equations.* U.S.S.R. Comput. Math. and Math. Phys. 4 (1964), 915-920.

14. O. Caprani, K. Madsen: *Iterative methods for interval inclusion of fixed points.* BIT 18 (1978), 42-51.

15. T. J. Dekker: *Newton-Laguerre iteration.* Programmation en Mathématiques Numériques, Centre Nat. Recherche Sci., Paris 1968, pp. 189-200.

16. E. Deutsch: *Lower bounds for the Perron root of a non-negative irreducible matrix.* Math. Proc. Cambridge Philos. Soc. 92 (1982), 49-54.

17. K. Dočev: *Modified Newton method for the simultaneous approximate calculation of all roots of a given algebraic equation* (in Bulgarian). Mat. Spis. B"lgar. Akad. Nauk 5 (1962), 136-139.

18. E. Durand: Solutions numériques des équations algébraiques, Tome I: Équations du Type F(x) = 0; Racines d'un Polynôme. Masson, Paris 1960.

19. L. W. Ehrlich: *A modified Newton method for polynomials.* Comm. ACM 10 (1967), 107-108.

20. H. Ehrmann: *Konstruktion und Durchführung von Iterations-verfahren höherer Ordnung.* Arch. Rational Mech. Anal. 4 (1959), 65-88.

21. L. Elzner: *A remark on simultaneous inclusions of the zeros of a polynomial by Gershgorin's theorem.* Numer. Math. 21 (1973), 425-427.

22. G. H. Ellis, L. T. Watson: *A parallel algorithm for simple roots of polynomials.* Comput. Math. Appl. 10 (1984), 107-121.

23. M. R. Farmer, G. Loizou: *A class of iteration functions for improving, simultaneously, approximations to the zeros of a polynomial.* BIT 15 (1975), 250-258.

24. M. R. Farmer, G. Loizou: *An algorithm for the total, or partial, factorization of a polynomial.* Math. Proc. Cambridge Philos. Soc. 82 (1977), 427-437.

25. M. R. Farmer, G. Loizou: *Locating multiple zeros interactively.* Comput. Math. Appl. 11 (1985), 595-603.

26. H. Fischer: *Intervall-Arithmetiken für komplexe Zahlen.* Z. Angew. Math. Mech. 53 (1973), 190-191.

27. A. Friedli: *Optimal covering algorithms in methods of search for solving polynomial equations.* J. Assoc. Comput. Mach. 20 (1973), 290-300.

28. I. Gargantini: *Parallel algorithms for the determination of polynomial zeros.* Proc. III Manitoba Conf. on Numer. Math., Winnipeg 1973 (eds. R. Thomas and H. C. Williams), Utilitas Mathematica Publ. Inc., Winnipeg 1974, pp. 195-211.

29. I. Gargantini: *Recent results on the square-root iteration.* Proc. VIII Manitoba Conf. on Numer. Math. and Computing, Winnipeg 1978 (eds. D. McCartney and H. C. Williams), Utilitas Mathematica Publ. Inc., Winnipeg 1979, pp. 205-215.

30. I. Gargantini: *Parallel square-root iterations.* Proc. I Symp. on Interval Mathematics, Karlsruhe 1975 (ed. K. Nickel), Lecture Notes in Comput. Sci. 29, Springer-Verlag, Berlin 1975, pp. 196-204.

31. I. Gargantini: *Parallel Laguerre iterations: Complexe case.* Numer. Math. 26 (1976), 317-323.

32. I. Gargantini: *Comparing parallel Newton's method with parallel Laguerre's method.* Comput. Math. Appl. 2 (1976), 201-206.

33. I. Gargantini: *Further applications of circular arithmetic: Schroeder-like algorithms with error bounds for finding zeros of polynomials.* SIAM J. Numer. Anal. 15 (1978), 497-510.

34. I. Gargantini: *The numerical stability of simultaneous iteration via square-rooting.* Comput. Math. Appl. 5 (1979), 25-31.

35. I. Gargantini: *Parallel square-root iterations for multiple roots.* Comput. Math. Appl. 6 (1980), 279-288.

36. I. Gargantini: *An application of interval mathematics: A polynomial solver with degree four convergence.* Freiburger Intervall-Berichte 7 (1981), 15-25.

37. I. Gargantini, P. Henrici: *Circular arithmetic and the determination of polynomial zeros.* Numer. Math. 18 (1972), 305-320.

38. I. Gargantini, W. Münzner: *An experimental program for the simultaneous determination of all zeros of a polynomial.* IBM Research Report RZ-237 (1967).

39. S. Gerschgorin: *Über die Abgrenzung der Eigenwerte einer Matrix.* Izv. Akad. Nauk. SSSR Ser. Mat. 7 (1931), 749-754.

40. G. Glatz: *Newton-Algorithmen zur Bestimmung von Polynomwurzeln unter Verwendung komplexer Kreisarithmetik.* Proc. I Symp. on Interval Mathematics, Karlsruhe 1975 (ed. K. Nickel), Lecture Notes in Comput. Sci. 29, Springer-Verlag, Berlin 1975, pp. 205-214.

41. J. A. Grant, G. D. Hitchins: *The solution of polynomial equations in interval arithmetic.* Comput. J. 16 (1973), 69-72.

42. M. Gutknecht: *A posteriori error bounds for the zeros of a polynomial.* Numer. Math. 20 (1972), 139-148.

43. R. Güting: *Polynomials with multiple zeros.* Mathematika 14 (1967), 181-196.

44. E. Hansen: Topics in interval analysis. Oxford University Press (Clarendon), London-New York 1969.

45. E. Hansen, M. Patrick: *A family of root finding methods.* Numer. Math. 27 (1977), 257-269.

46. M. Hauenschild: *Arithmetiken für komplexe Kreise.* Computing 13 (1974), 299-312.

47. A. S. Householder: Principles of numerical analysis. Mc Graw-Hill, New York 1953.

48. P. Henrici: *Uniformly convergent algorithms for the simultaneous determination of all zeros of a polynomial.* Proc. Symp. on Numerical Solution of Nonlinear Problems (eds. J. M. Ortega and W. C. Rheinboldt), SIAM Studies in Numerical Analysis 2, Philadelphia 1970, pp. 1-8.

49. P. Henrici: *Methods of search for solving polynomial equations.* J. Assoc. Comput. Mach. 17 (1970), 273-283.

50. P. Henrici: *Circular arithmetic and the determination of polynomial zeros.* Proc. Conf. on Applications of Numerical Analysis, Lecture Notes in Math. 228, Springer-Verlag, Berlin 1971, pp. 86-92.

51. P. Henrici: Applied and computational complex analysis, Vol. I. John Wiley and Sons Inc., New York 1974.

52. P. Henrici, I. Gargantini: *Uniformly convergent algorithms for the simultaneous determination of all zeros of a polynomial.* Proc. Symp. on Constructive Aspects of the Fundamental Theorem of Algebra (eds. B. Dejon and P. Henrici), Wiley-Interscience, London 1969, pp. 77-114.

53. C. Hoffmann: *Weyl-Newton, ein globales Verfahren zur Bestimmung von Polynomnullstellen mit Fehlerschranken.* Ph. D. Theses, Eidgenössische Technische Hochschule Zürich, Zürich 1976.

54. M. Igarishi: *Zeros of a polynomial and an estimation of its accuracy.* J. Inform. Process 3 (1982), 172-175.

55. M. A. Jenkins, J. F. Traub: *A three-stage variable-shift iteration for polynomial zeros.* SIAM J. Numer. Anal. 7 (1970), 545-566.

56. I. O. Kerner: *Ein Gesamtschrittverfahren zur Berechnung der Nullstellen von Polynomen.* Numer. Math. 8 (1966), 290-294.

57. N. Krier. *Komplexe Kreisarithmetik.* Ph. D. Thesis, Universität Karlsruhe, Karlsruhe 1973.

58. E. V. Krishnamurthy, H. Venkateswaran: *A parallel Wilf algorithm for complex zeros of a polynomial.* BIT 21 (1981), 104-111.

59. H. W. Kuhn: *Finding roots of polynomial by pivoting.* Proc. I Internat. Conf. on Fixed Points: Algorithms and Applications, Clemson 1974 (ed. S. Karamadian), Academic Press, New York 1977, pp. 11-39.

60. J. L. Lagouanelle: *Sur une méthode de calcul de l'ordre de multiplicité des zéros d'un polynôme.* C. R. Acad. Sci. Paris Sér. A. 262 (1966), 626-627.

61. D. H. Lehmer: *A machine method for solving polynomial equations.* J. Assoc. Comput. Mach. 8 (1961), 151-162.

62. D. H. Lehmer: *Search procedures for polynomial equation solving.* Proc. Symp. on Constructive Aspects of the Fundamental Theorem of Algebra (eds. B. Dejon and P. Henrici), Wiley-Interscience, London 1969, pp. 193-208.

63. G. Loizou: *Higher-order iteration functions for simultaneously approximating polynomial zeros.* Internat. J. Comput. Math. 14 (1983), 45-58.

64. O. Lubeck, J. Moore, R. Mendez: *A benchmark comparison of three supercomputers: Fujitsu VP-200, Hitachi S810/20, and CRAY X-MP/2.* Computer 18 (1985), 10-24.

65. V. H. Maehly: *Zur iterativen Auflösung algebraischer Gleichungen.* Z. Angew. Math. Phys. 5 (1954), 260-263.

66. K. Mahler: *An inequality for the discriminant of a polynomial.* Michigan Math. J. 11 (1964), 257-262.

67. M. Marden: Geometry of polynomials. Amer. Math. Soc., Providence, Rhode Island 1966.

68. M. Mignote: *Some useful bounds.* Computer Algebra, Symbolic and Algebra Computation (eds. B. Buchberger, G. E. Collins and R. Loos). Computing Supplementum 4, Springer-Verlag, Wien-New York 1983, pp. 259-263.

69. G. V. Milovanović, M. S. Petković: *On the convergence order of a modified method for simultaneous finding polynomial zeros.* Computing 30 (1983), 171-178.

70. G. V. Milovanović, M. S. Petković: *On computational efficiency of the iterative methods for the simultaneous approximation of polynomial zeros.* ACM Trans. Math. Software 12 (1966), 295-306.

71. D. S. Mitrinović: Analytic inequalities. Springer-Verlag, Berlin - Heidelberg - New York 1970.

72. R. E. Moore: Interval analysis. Prentice-Hall, Englewood Cliffs, New Jersey 1966.

73. R. E. Moore: Methods and applications of interval analysis. SIAM Studies in Applied Mathematics, Philadelphia 1979.

74. A. Neumaier: *An existence test for root clusters and multiple roots.* Z. Angew. Math. Mech. 68 (1988), 256-257.

75. A. W. M. Nourein: *An iteration formula for the simultaneous determination of the zeroes of a polynomial.* J. Comput. Appl. Math. 4 (1975), 251-254.

76. A. W. M. Nourein: *An improvement on Nourein's method for the simultaneous determination of the zeros of a polynomial (an algorithm).* J. Comput. Appl. Math. 3 (1977), 109-110.

77. A. W. M. Nourein: *An improvement on two iteration methods for simultaneous determination of the zeros of a polynomial.* Internat. J. Comput. Math. 6 (1977), 241-252.

78. J. M. Ortega, W. C. Rheinboldt: Iterative solution of nonlinear equations in several variables. Academic Press, New York 1970.

79. A. M. Ostrowski: Solution of equations and systems of equations. Academic Press, New York 1966.

80. A. M. Ostrowski: *A theorem on clusters of roots of polynomial equations.* SIAM J. Numer. Anal. 7 (1970), 567-570.

81. C. T. Pan, K. S. Chao: *Multiple solutions of nonlinear equations: Roots of polynomials.* IEEE Trans. Circuits and Systems 27 (1980), 825-832.

82. L. Pasquini, D. Trigiante: *A globally convergent method for simultaneosly finding polynomial roots.* Math. Comp. 44 (1985), 135-149.

83. R. I. Peluso: *Una famiglia di metodi iterativi per ogni ordine superiore al seconde.* Calcolo 10 (1973), 145-149.

84. L. D. Petković: *A note on the evaluation in circular arithmetic.* Z. Angew. Math. Mech. 66 (1986), 371-373.

85. L. D. Petković, M. S. Petković: *On the k-th root in circular arithmetic.* Computing 33 (1984), 27-35.

86. M. S. Petković: *Some interval methods of the second order for the simultaneous approximation of polynomial roots.* Univ. Beograd. Publ. Elektrotehn. Fak. Ser. Mat. Fiz. No. 634-No. 677 (1979), 75-82.

87. M. S. Petković: *Some iterative interval methods for solving equations* (in Serbo-Croatian). Ph. D. Thesis, University of Niš, Niš 1980.

88. M. S. Petković: *On a generalisation of the root iterations for polynomial complex zeros in circular interval arithmetic.* Computing 27 (1981), 37-55.

89. M. S. Petković: *On an iterative method for simultaneous inclusion of polynomial complex zeros.* J. Comput. Appl. Math. 8 (1982), 51-56.

90. M. S. Petković: *A family of simultaneous methods for the determination of polynomial complex zeros.* Internat. J. Comput. Math. 11 (1982), 285-296.

91. M. S. Petković: *Generalised root iterations for the simultaneous determination of multiple complex zeros.* Z. Angew. Math. Mech. 62 (1982), 627-630.

92. M. S. Petković: *On some interval iterations for finding a zero of a polynomial with error bounds.* Comput. Math. Appl. 14 (1987), 479-495.

93. M. S. Petković: *On the Halley-like algorithms for the simultaneous approximation of polynomial complex zeros.* SIAM J. Numer. Anal. (to appear).

94. M. S. Petković, G. V. Milovanović: *A note on some improvements of the simultaneous methods for determination of polynomial zeros.* J. Comput. Appl. Math. 9 (1983), 65-69.

95. M. S. Petković, G. V. Milovanović, L. V. Stefanović: *On the convergence order of an accelerated simultaneous method for polynomial complex zeros.* Z. Angew. Math. Mech. 66 (1986), 428-429.

96. M. S. Petković, G. V. Milovanović, L. V. Stefanović: *On some higher-order methods for the simultaneous approximation of multiple polynomial zeros.* Comput. Math. Appl. 12 (1986), 951-962.

97. M. S. Petković, L. D. Petković: *On a representation of the k-th root in complex circular interval arithmetic.* Proc. II Symp. on Interval Mathematics, Freiburg 1980 (ed. K. Nickel), Academic Press, New York 1980, pp. 473-479.

98. M. S. Petković, L. D. Petković: *On a computational test for the existence of polynomial zero.* Comput. Math. Appl. 7 (1989), 1109-1114.

99. M. S. Petković, L. D. Petković, L. V. Stefanović: *On the R-order of a class of simultaneous iterative processes.* Z. Angew. Math. Mech. 69 (1989), 199-201.

100. M. S. Petković, L. V. Stefanović: *On the convergence order of accelerated root iterations.* Numer. Math. 44 (1984), 463-476.

101. M. S. Petković, L. V. Stefanović: *The numerical stability of the generalised root iterations for polynomial zeros.* Comput. Math. Appl. 10 (1984), 97-106.

102. M. S. Petković, L. V. Stefanović: *On some improvements of square root iteration for polynomial complex zeros.* J. Comput. Appl. Math. 15 (1986), 13-25.

103. M. S. Petković, L. V. Stefanović: *On the simultaneous method of the second order for finding polynomial complex zeros in circular arithmetic.* Freiburger Intervall-Berichte 3 (1985), 63-95.

104. M. S. Petković, L. V. Stefanović: *On a second order method for the simultaneous inclusion of a polynomial complex zeros in rectangular arithmetic.* Computing 36 (1986), 249-261.

105. M. S. Petković, L. V. Stefanović: *On some iteration functions for the simultaneous computation of multiple complex polynomial zeros.* BIT 27 (1987), 111-122.

106. J. R. Pinkert: *An exact method of finding the roots of a complex polynomial.* ACM Trans. Math. Software 2 (1976), 351-363.

107. S. B. Prešić: *Un procédé itératif pour la factorisation des polynômes.* C. R. Acad. Sci. Paris Sér. A 262 (1966), 862-863.

108. J. Riordan: Combinatorial identities. John Wiley and Sons, Inc., New York-London-Sydney 1968.

109. J. Rokne, P. Lancaster: *Complex interval arithmetic.* Comm. ACM 14 (1971), 111-112.

110. K. Samelson: *Factorisierung von Polynomen durch funktionale Iteration.* Bayer. Akad. Wiss. Math.-Natur. Kl. Abh. 95 (1958), 1-25.

111. J. W. Schmidt: *Eine Anwendung des Brouwer'schen Fixpunktsatzes zur Gewinnung von Fehlerschranken für Näherungen von Polynomnullstellen.* Beitr. zur Numer. Math. 6 (1977), 158-163.

112. J. W. Schmidt, H. Dressel: *Fehlerabschätzung bei Polynomgleichungen mit dem Fixpunktsatz von Brouwer.* Numer. Math. 10 (1967), 42-50.

113. E. Schröder: *Über unendlich viele Algorithmen zur Auflösung der Gleichungen.* Math. Ann. 2 (1870), 317-365.

114. A. van der Sluis: *Upperbounds for roots of polynomials.* Numer. Math. 17 (1970), 250-262.

115. B. T. Smith: *Error bounds for zeros of a polynomial based upon Gerschgorin's theorem.* J. Assoc. Comput. Mach. 17 (1970), 661-674.

116. L. V. Stefanović, M. S. Petković: *On the simultaneous improving k inclusive disks for polynomial complex zeros.* Freiburger Intervall-Berichte 7 (1982), 1-13.

117. L. V. Stefanović, M. S. Petković: *The R-order of convergence of a modified iterative method of Weierstrass' type for finding polynomial zeros.* Proc. II Conf. on Numerical Methods and Approximation Theory, Novi Sad 1985 (ed. D. Herceg), Faculty of Science, Institute of Mathematics, Novi Sad 1985.

118. G. W. Stewart: *On Lehmer's method for finding the zeros of a polynomial.* Math. Comp. 23 (1969), 829-835.

119. J. Stoer, R. Bulirsch: Einführung in die Numerische Mathematik II. Springer-Verlag, Berlin 1973.

120. E. C. Titchmarsh: The theory of functions. Oxford University Press 1939.

121. D. D. Tošić, G. V. Milovanović: *An application of Newton's method to simultaneous determination of zeros of a polynomial.* Univ. Beograd. Publ. Elektrotehn. Fak. Ser. Mat. Fiz. No. 412-No. 460 (1973), 175-177.

122. J. F. Traub: Iterative methods for the solution of equations. Prentice Hall, Englewood Cliffs, New Jersey 1964.

123. H. van de Vel: *A method for computing a root of a single nonlinear equation, including its multiplicity.* Computing 14 (1975), 167-171.

124. R. S. Varga: Matrix iterative analysis. Prentice Hall, Englewood Cliffs, New Jersey 1962.

125. D. Wang, Y. Wu: *A parallel circular algorithm for the simultaneous determination of all zeros of a complex polynomial* (in Chinese). J. Engrg. Math. 2 (1985), 22-31.

126. D. Wang, Y. Wu: *Some modifications of the parallel Halley iteration method and their convergence.* Computing 38 (1987), 75-87.

127. X. Wang, S. Zheng: *A family of parallel and interval iterations for finding all roots of a polynomial simultaneously with rapid convergence (I).* J. Comput. Math. 1 (1984), 70-76.

128. X. Wang, S. Zheng: *The quasi-Newton method in parallel circular iteration.* J. Comput. Math. 4 (1984), 305-309.

129. X. Wang, S. Zheng: *Parallel Halley iteration method with circular arithmetic for finding all zeros of a polynomial.* Numer. Math., J. Chinese Univ. 4 (1985), 308-314.

130. X. Wang, S. Zheng: *A family of parallel and interval iterations for finding all roots of a polynomial simultaneously with rapid convergence (II)* (in Chinese). J. Comput. Math. 4 (1985), 433-444.

131. X. Wang, S. Zheng, G. Shen: *Bell's disk polynomial and parallel disk iteration.* Numer. Math., J. Chinese Univ. 4 (1987), 328-345.

132. G. N. Watson: Bessel functions. Cambridge University Press 1952.

133. J. J. Wavrik: *Computers and the multiplicity of polynomial roots.* Amer. Math. Monthly 89 (1982), 34-56.

134. K. Weierstrass: *Neuer Beweis des Satzes, dass jede ganze rationale Funktion einer Veränderlichen dargestellt werden kann als ein Produkt aus linearen Funktionen derselben Veränderlichen.* Ges. Werke 3 (1903), 251-269 (Johnson Reprint Corp., New York 1967).

135. **F. Weisenhorn:** *Ein Beitrag zur Bestimmung der Nullstellen aus einem Polynom in Summenform und aus der Summe von Polynomen in Produktform.* **AËU - Arch. Elektron. Übertragungstech 24 (1970), 372-378.**

136. **H. Weyl:** *Randbemerkungen zu Hauptproblemen der Mathematik, II Fundamentalsatz der Algebra und Grundlagen der Mathematik.* **Math. Z. 20 (1924), 131-150.**

137. **H. S. Wilf:** *A global bisection algorithm for computing the zeros of polynomials in the complex plane.* **J. Assoc. Comput. Mach. 25 (1978), 415-420.**

138. **J. H. Wilkinson:** *Rigorous error bounds for computed eigensystems.* **Computer J. 4 (1961), 230-241.**

139. **J. H. Wilkinson:** Rounding errors in algebraic processes. **Prentice Hall, Englewood Cliffs, New Jersey 1963.**

140. **J. M. Yohe:** *The interval arithmetic package.* **MRC Technical Summary Report 1755, Mathematics Research Center, University of Wisconsin, Madison 1977.**

# INDEX OF NOTATION

## SUBJECT INDEX

A



B



C



D



E

F



G



H



I

R



S



T



W

Z